D1922409

Hongtian Xiao, Zhongqi Yue

Fracture Mechanics in Layered and Graded Solids

Preface

In general, all solid materials can be considered as non-homogeneous because their properties can vary with locations in a three-dimensional space. One special type of solid material is characterized by the variations of its physical and mechanical components, structures and properties along only one given coordinate; the material properties have very small or no variations in any other direction perpendicular to the given coordinate. These types of solid materials are called functionally graded materials (FGMs). For example, plant and tree stems, animal bones and other biological hard tissues have gradient variations in their microstructures and functions. Bamboo is a self-optimizing graded structure constructed with a cell-based system for sensing external mechanical stimuli. Learning from nature, material scientists have increasingly aimed to design and fabricate graded materials that are more damage-resistant than their conventional homogeneous counterparts. As a design concept, FGMs were originally proposed as an alternative to conventional ceramic thermal barrier coatings to overcome their well-documented shortcomings and to meet the demands of new technologies.

The mechanical responses of FGMs have an important significance in many engineering fields and are of great interest to material scientists, and design and manufacturing engineers. The problems of fracture and crack propagation in FGMs are particularly important and have been studied in depth. The boundary element method (BEM), also known as the boundary integral equation method, is now firmly established in many engineering disciplines and is increasingly used as an effective and accurate numerical tool. Fracture mechanics has been the most active, specialized area of research in BEM and is probably the one most exploited by industry. The traditional BEM is based on the Kelvin's fundamental solution and meets the difficulties encountered when analyzing the fracture mechanics of FGMs.

Since 1983, the second author of this book has devoted much of his research to understanding the elasticity of a multilayered medium and has achieved important results. One of these results is the analytical and closed-form formulation of fundamental solutions for a multilayered elastic medium and a transversely isotropic bi-material. These solutions can be applied to investigate and analyze many problems in multilayered media encountered in the science and engineering disciplines using the BEM. Since 2000, the authors have dedicated their efforts to the development of the new BEMs based on these fundamental solutions under the funding of The University Grants Committee of Hong Kong, The University of Hong Kong and the National Natural Science Foundation of China.

This book brings together the descriptions of the new boundary element formulation based on the two fundamental solutions and new analyses and results for the fracture mechanics of layered and functionally graded materials. This method overcomes the mathematical degeneration that is associated with the solitary use of the displacement boundary integral equation for cracked bodies by developing the multi-region and single-region methods of BEMs. Effective implementation of the methods is detailed, devoting special attention to the description of accurate algorithms for the evaluation of various singular integrals in the boundary element formulations. The layered discretization technique is used to simulate the variations of the material property of FGMs with depth. The proposed numerical methods, together with fracture mechanics theories, are used to calculate the stress intensity factors of three-dimensional cracks in FGMs and to analyze the crack growth. The influence of the material parameters and crack dimensions on the fracture properties has been analyzed and quantified. The new material presented in this book is supported by recent articles published in relevant peer-reviewed journals in English or Chinese.

Hongtian Xiao, Zhongqi Yue
September 1st, 2013

Contents

Chapter 4

Yue's Solution-based Boundary Element Method

4.1 Introduction

The boundary element method (BEM) is a numerical computational method for solving partial differential equations that are formulated as boundary integral equations. The boundary integral equations may be regarded as an exact solution of the governing partial differential equations. Based on the boundary integral equations, BEM attempts to use discretization techniques similar to those employed in the finite element method (FEM). In the general case, a formulation is obtained that does not contain integrals over the problem domain and consequently does not require domain discretization. Therefore, only the boundary of the problem has to be discretized into elements; this results in a substantial reduction in the model preparation time as well as the fact that a much smaller algebraic system of equations needs to be solved. Once this is done, in the post-processing stage, an integral equation can be used to directly calculate, numerically, the solution for any desired point in the interior of the solution domain.

Since its introduction in the late 1960s, BEM has matured into a powerful alternative to FEM in the fields of solid and structure mechanics, etc. The idea that the boundary value problems of partial differential equations are transformed into boundary integral equations is not new; scientists had already developed different types of boundary integral equations, for example, Kinoshita and Mura (1956) derived the singular boundary integral equations for elasticity. The boundary element method has received much attention since Rizzo (1967) proposed a numerical treatment of the boundary integral equation for plane strain of an elastic material. Undoubtedly inspired by the work of Rizzo, Cruse (1969) presented numerical solutions of boundary integral equations in three dimensional elastostatics. Lachat and Watson (1976) developed effective numerical methods for treating boundary integral equations including weakly and strongly singular integrals. Their works provided the foundations for the use of BEM numerical methods in elastostatics.

The early BEM was based on Kelvin's fundamental solution of a homogeneous isotropic medium of infinite extent. In dealing with a multilayered composite structure, all interfaces between the different layers need to be discretized and these layers are then joined together such that the equilibrium of tractions and the compatibility of displacements are enforced. The main drawback of this multi-region method is that a large system of algebraic equations is generated and an automatic procedure cannot be implemented easily.

To overcome these shortcomings, Green's functions for layered materials, which satisfy the interfacial conditions and hence eliminate the need for interface discretization, need to be developed instead of using a Kelvin-type Green's function. Green's functions for several types of layered materials have been proposed and implemented into BEM formulations. Lou and Zhang (1992) incorporated the fundamental solution for an isotropic bi-material of infinite extent into a BEM formulation for elasto-plastic analysis. Benitez et al. (1993) implemented the 3D fundamental solution for a point load acting in the interior of an infinite layer of uniform thickness into a BEM formulation and thus analyzed cavities in plate structures. Benitez and Wideberg (1996) developed a specialization of the BE technique by using a 3D fundamental solution for a point load acting in the interior of an infinite orthotropic multilayered space and on that basis treated composite laminated material problems. Pan et al. (2001) implemented a 3D layered Green's function, with the materials of each layer being generally anisotropic, into a 3D BEM formulation for the analysis of composite laminates with holes. Yang et al. (2003) developed the fundamental solution for anisotropic composite laminates and established the corresponding boundary element formulation to calculate the stresses in composite laminates with an elastically pinned hole.

As mentioned in Chapter 3, Yue (1995) obtained the fundamental singular solution for a multilayered medium, of any number of layers, subjected to a point load, achieving high accuracy and efficiency. In this chapter, we will establish the boundary integral equations by using the fundamental solution and develop the corresponding numerical methods. The computer codes in FORTRAN will be written to calculate the displacements and stresses in an n-layered elastic medium. Numerical examples will be presented to verify the accuracy and efficiency of this proposed method.

4.2 Betti's reciprocal work theorem

Betti's reciprocal theorem is often used to obtain specific results to problems in elasticity without obtaining a complete solution for the stress and displacement fields. Indeed, in many cases results can be obtained for problems in which a complete solution would otherwise be impossible.

To develop Yue's solution-based boundary integral equations, we first derive the classical integral identity, known as Betti's reciprocal work theorem, using the method of integration by parts. We begin by considering two equilibrium states in a region V bounded by a surface S. We now consider a given elastic body under one set of given surface forces $t_i^{(1)}$ and body forces $f_i^{(1)}$ and regard the displacements, strains, and stresses as known. These will be denoted by $u_i^{(1)}$, $\varepsilon_{ij}^{(1)}$, and $\sigma_{ij}^{(1)}$. Then, independently, we consider a second set of surface and body forces, $t_i^{(2)}$ and $f_i^{(2)}$, and indicate the results for this second problem as $u_i^{(2)}$, $\varepsilon_{ij}^{(2)}$, and $\sigma_{ij}^{(2)}$.

The work of the forces $t_i^{(2)}$ and $f_i^{(2)}$ of the second state on the displacements $u_i^{(1)}$ of the first state is defined by

$$I_1 = \int_S t_i^{(2)} u_i^{(1)} \mathrm{d}S + \int_V f_i^{(2)} u_i^{(1)} \mathrm{d}V, \quad (i = x, y, z) \tag{4.1a}$$

Using the differential equations of equilibrium (2.1a), boundary conditions (2.4b) and Gauss's theorem, we have

$$\begin{aligned} I_1 &= \int_S \sigma_{ij}^{(2)} n_j u_i^{(1)} \mathrm{d}S - \int_V \sigma_{ij,j}^{(2)} u_i^{(1)} \mathrm{d}V = \int_V (\sigma_{ij}^{(2)} u_i^{(1)})_{,j} \mathrm{d}V - \int_V \sigma_{ij,j}^{(2)} u_i^{(1)} \mathrm{d}V \\ &= \int_V \sigma_{ij}^{(2)} u_{i,j}^{(1)} \mathrm{d}V = \frac{1}{2}\int_V (\sigma_{ij}^{(2)} u_{i,j}^{(1)} + \sigma_{ij}^{(2)} u_{i,j}^{(1)}) \mathrm{d}V = \frac{1}{2}\int_V (\sigma_{ij}^{(2)} u_{i,j}^{(1)} + \sigma_{ji}^{(2)} u_{j,i}^{(1)}) \mathrm{d}V \\ &= \frac{1}{2}\int_V \sigma_{ij}^{(2)} (u_{i,j}^{(1)} + u_{j,i}^{(1)}) \mathrm{d}V = \int_V \sigma_{ij}^{(2)} \varepsilon_{ij}^{(1)} \mathrm{d}V, \quad (i,j = x,y,z) \end{aligned} \tag{4.1b}$$

The work of the forces of the first state on the displacements of the second state is defined by

$$I_2 = \int_V f_i^{(1)} u_i^{(2)} \mathrm{d}V + \int_S t_i^{(1)} u_i^{(2)} \mathrm{d}S, \quad (i = x,y,z) \tag{4.1c}$$

Similar to Eq. (4.1b), for Eq. (4.1c), we have

$$I_2 = \int_V \sigma_{ij}^{(1)} \varepsilon_{ij}^{(2)} \mathrm{d}V, \quad (i,j = x,y,z) \tag{4.1d}$$

Using Hooke's law of the first state and multiplying both sides by $\varepsilon_{ij}^{(2)}$, we have

$$\begin{aligned} \sigma_{ij}^{(1)} \varepsilon_{ij}^{(2)} &= 2\mu \varepsilon_{ij}^{(1)} \varepsilon_{ij}^{(2)} + \lambda \delta_{ij} \varepsilon_{kk}^{(1)} \varepsilon_{ij}^{(2)} \\ &= 2\mu \varepsilon_{ij}^{(1)} \varepsilon_{ij}^{(2)} + \lambda \varepsilon_{kk}^{(1)} \varepsilon_{mm}^{(2)} \\ &= 2\mu \varepsilon_{ij}^{(1)} \varepsilon_{ij}^{(2)} + \lambda \delta_{ij} \varepsilon_{ij}^{(1)} \varepsilon_{mm}^{(2)} \\ &= \sigma_{ij}^{(2)} \varepsilon_{ij}^{(1)}, \quad (i,j,k,m = x,y,z) \end{aligned} \tag{4.2}$$

where μ and λ are Lamé's constants.

Combining Eqs. (4.1) and (4.2), we obtain the following integral equation

$$\int_V f_i^{(1)} u_i^{(2)} \mathrm{d}V + \int_S t_i^{(1)} u_i^{(2)} \mathrm{d}S = \int_V f_i^{(2)} u_i^{(1)} \mathrm{d}V + \int_S t_i^{(2)} u_i^{(1)} \mathrm{d}S, \quad (i = x,y,z) \tag{4.3}$$

This is Betti's second reciprocal work theorem, which expresses the equality of the reciprocal work done by two equilibrium states through the domain V bounded by the surface S.

Although it is assumed that σ_{ij}, ε_{ij}, and u_i are continuous in the domain V when deriving Betti's reciprocal work theorem, Eq. (4.3) can be further used in the analysis of a multilayered elastic medium. In this case, the integral terms along the interfaces of different materials are not included in Eq. (4.3). When applying the reciprocal theorem, one state is the actual problem to be solved and another state is an auxiliary solution, which must in general represent a sufficiently simple state of the system for the stress and displacement fields to be obtainable in closed form. Thus, this theorem forms the basis of establishing the boundary element method.

4.3 Yue's solution-based integral equations

To develop Yue's solution-based integral equations for a layered elastic body shown in Fig. 4.1, we first rewrite the reciprocal theorem (4.3) in a more explicit form

$$\begin{aligned}&\int_S t_j^{(1)}(Q)u_j^{(2)}(Q)\mathrm{d}S(Q)+\int_V f_j^{(1)}(q)u_j^{(2)}(q)\mathrm{d}V(q)\\&=\int_S t_j^{(2)}(Q)u_j^{(1)}(Q)\mathrm{d}S(Q)+\int_V f_j^{(2)}(q)u_j^{(1)}(q)\mathrm{d}V(q),\quad (j=x,y,z)\end{aligned} \tag{4.4}$$

where q and Q are points in the domain V and on the boundary S, respectively. We now assume that one set (u_j, t_j, f_j) is the actual state whilst a second set (u_j^*, t_j^*, f_j^*) corresponds to that produced by a unit force system e_i^* $(i=x, y, z)$ in a multilayered elastic medium of infinite extent.

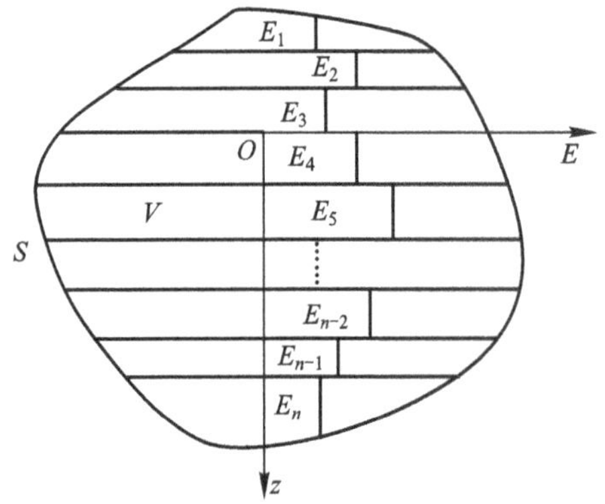

Fig. 4.1 A layered elastic body with the domain V and the boundary S. E_i is the elastic modulus for the i-th layer

According to the definitions of Yue's solution, we have

$$\begin{aligned}t_j^*(q)&=t_{ij}^Y(p,q)e_i^*(p)\\u_j^*(q)&=u_{ij}^Y(p,q)e_i^*(p),\quad (i,j=x,y,z)\end{aligned} \tag{4.5}$$

where p is the source point on which the unit load is located and q is the field point whose displacements and stresses need to be obtained; the superscript Y corresponds to Yue's solution. $t_{ij}^Y(p,q)$ and $u_{ij}^Y(p,q)$ are, respectively, tractions and displacements in the j direction of the field point q under the unit load in the i direction at the source point p.

Yue's solution should satisfy the following equations of equilibrium

$$(\lambda+\mu)u_{k,ki}^*(q)+\mu u_{i,kk}^*(q)+e_i^*(p)\Delta(p,q)=0,\quad (i,k=x,y,z) \tag{4.6}$$

where $\Delta(p,q)$ is the Dirac delta function, and λ and μ are piecewise continuous functions in the z direction.

Substituting Eq. (4.5) in Eq. (4.4), we have

$$\begin{aligned}&\int_S u_{ij}^Y(p,Q)e_i^*(p)t_j(Q)\mathrm{d}S(Q)+\int_V u_{ij}^Y(p,q)e_i^*(p)f_j(q)\mathrm{d}V(q)\\&=\int_S t_{ij}^Y(p,Q)e_i^*(p)u_j(Q)\mathrm{d}S(Q)+\int_V f_j^*(q)u_j(q)\mathrm{d}V(q),\quad (i,j=x,y,z)\end{aligned}\tag{4.7}$$

Considering Eq. (4.6), the second term on the right-hand side of Eq. (4.7) can be rewritten as

$$\begin{aligned}\int_V f_j^*(q)u_j(q)\mathrm{d}V(q)&=\int_V e_j^*(p)\Delta(p,q)u_j(q)\mathrm{d}V(q)\\&=\int_V \delta_{ij}e_i^*(p)\Delta(p,q)u_j(q)\mathrm{d}V(q),\quad (i,j=x,y,z)\end{aligned}\tag{4.8}$$

Substituting Eq. (4.8) into Eq. (4.7), we find that the integrand in each integral contains the vector of the unit concentrated load $e_i^*(p)$; we then take three sets of unit concentrated loads, i.e., $(e_x^*,e_y^*,e_z^*)=(1,0,0)$, $(0,1,0)$, and $(0,0,1)$. In order to illustrate how we obtain the integral equations, we consider Eq. (4.7) for each set of the unit concentrated loads. For each source point p, we have, from Eqs. (4.7) and (4.8),

$$\begin{aligned}\int_V \delta_{ij}\Delta(p,q)u_j(q)\mathrm{d}V(q)&=\int_V u_{ij}^Y(p,q)f_j(q)\mathrm{d}V(q)+\int_S u_{ij}^Y(p,Q)t_j(Q)\mathrm{d}S(Q)-\\&\int_S t_{ij}^Y(p,Q)u_j(Q)\mathrm{d}S(Q),\quad (i,j=x,y,z)\end{aligned}\tag{4.9}$$

Using the properties of $\Delta(p,q)$ and the Kronecker δ_{ij}, we obtain the following expression for the term on the left-hand side of Eq. (4.9)

$$\int_V \delta_{ij}\Delta(p,q)u_j(q)\mathrm{d}V(q)=\int_V \Delta(p,q)u_i(q)\mathrm{d}V(q)=u_i(p),\quad (i,j=x,y,z)\tag{4.10}$$

Substituting Eq. (4.10) into Eq. (4.9), we obtain Yue's solution-based integral equations

$$\begin{aligned}u_i(p)=&\int_V u_{ij}^Y(p,q)f_j(q)\mathrm{d}V(q)+\int_S u_{ij}^Y(p,Q)t_j(Q)\mathrm{d}S(Q)-\\&\int_S t_{ij}^Y(p,Q)u_j(Q)\mathrm{d}S(Q)\quad (i,j=x,y,z)\end{aligned}\tag{4.11a}$$

When the body forces are absent, the boundary integral equations in Eq. (4.11a) can be simplified as follows

$$u_i(p)=\int_S u_{ij}^Y(p,Q)t_j(Q)\mathrm{d}S(Q)-\int_S t_{ij}^Y(p,Q)u_j(Q)\mathrm{d}S(Q)\quad (i,j=x,y,z)\tag{4.11b}$$

It should be noted that Eq. (4.11) do not contain any integration on the layer interfaces because Yue's solution strictly satisfies the interface conditions. Thus, the analysis of a multilayered elastic body can be simplified by using Yue's solution-based integral equations. In Eq. (4.11), if a multilayered elastic medium degenerates into a homogeneous one, Yue's solution becomes the traditional Kelvin-type solution. In this case, Eq. (4.11) are known as Somigliana's identity.

By using Eq. (4.11), the displacement at any internal point can be obtained only when both tractions and displacements are known at every boundary point. In a well-posed boundary value problem, exactly one-half of these boundary conditions will be specified and, as a consequence, Eq. (4.11) are insufficient to solve such problems. Since Eq. (4.11) are valid for every point in the domain V bounded by the surface S, a boundary integral expression can be obtained by taking the source point in Eq. (4.11) to the boundary. This expression is then applied at different points on the boundary to produce a system of equations that can generate the necessary boundary values.

4.4 Yue's solution-based boundary integral equations

To derive the boundary integral equations, we begin with Eq. (4.11a) and determine its limiting form when the point load p approaches the boundary. If we let p approach a point P on the boundary, we arrive at the boundary integral equation

$$
\begin{aligned}
c_{ij}(P)u_j(P) = {} & \int_V u_{ij}^Y(P,q)f_j(q)\mathrm{d}V(q) + \int_S u_{ij}^Y(P,Q)t_j(Q)\mathrm{d}S(Q) - \\
& \int_S t_{ij}^Y(P,Q)u_j(Q)\mathrm{d}S(Q), \quad (i,j = x,y,z)
\end{aligned} \tag{4.12}
$$

where $c_{ij}(P)$ are coefficients that depend only upon the local geometry of the boundary at the source point P and can be evaluated using the following equations

$$
c_{ij}(P) = \lim_{\varepsilon \to 0} \int_{S^\varepsilon} t_{ij}^Y(P,Q)\mathrm{d}S(Q)
$$

in which S^ε is an infinitesimal spherical surface of center P and radius ε enclosed in the solid shown in Fig. 4.2. For a smooth boundary, $c_{ij}(P) = 0.5\delta_{ij}$, in which δ_{ij} is the Kronecker delta function.

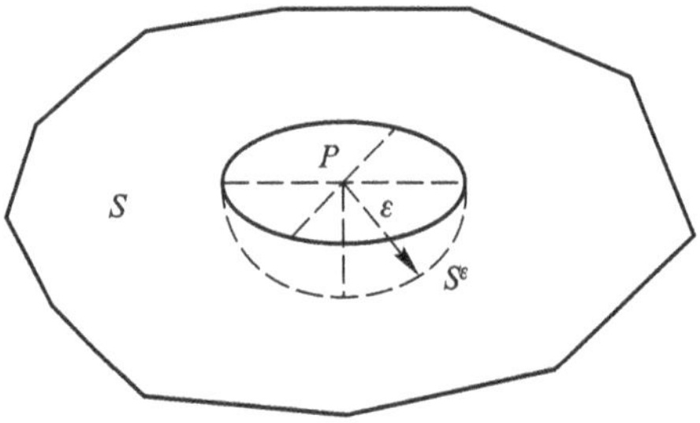

Fig. 4.2 A hemispherical surface S^ε of radius ε within the elastic body

When body forces are absent, Eq. (4.12) can be simplified as follows

$$c_{ij}(P)u_j(P) = \int_S u_{ij}^Y(P,Q)t_j(Q)\mathrm{d}S(Q) - \int_S t_{ij}^Y(P,Q)u_j(Q)\mathrm{d}S(Q), \quad (i,j=x,y,z) \qquad (4.13)$$

If the boundary S is divided into S_u on which the displacements are prescribed and S_σ on which the tractions are prescribed, Eq. (4.13) can be rewritten as

$$\begin{aligned} c_{ij}(P)u_j(P) = & \int_{S_u} [u_{ij}^Y(P,Q)t_j(Q) - t_{ij}^Y(P,Q)\bar{u}_j(Q)]\mathrm{d}S(Q) + \\ & \int_{S_\sigma} [u_{ij}^Y(P,Q)\bar{t}_j(Q) - t_{ij}^Y(P,Q)u_j(Q)]\mathrm{d}S(Q), \quad (i,j=x,y,z) \end{aligned} \qquad (4.14)$$

where u_j and t_j are unknown variables on the boundary, and $\bar{u}_j$ and $\bar{t}_j$ are known quantities.

4.5 Discretized boundary integral equations

It is as difficult to solve the boundary integral equation (4.14) as it is to solve the partial differential equations describing elastostatics. However, it is easy to discretize the boundary integral equations. For an elastic body without body forces, only the boundary needs to be discretized. Since the boundary integral equations only contain the components of displacements and tractions, they do not include differential terms. Thus, the accuracy of the numerical method is not lowered when calculating differentials in discretization.

The first step is to subdivide the boundary S into a sufficient number of elements. These elements form a piecewise continuous approximation to the boundary. Assuming that there are ne elements on the discretized boundary, we have

$$S = \sum_{e=1}^{ne} S_e \qquad (4.15)$$

Note that the term on the right-hand side of Eq. (4.15) indicates summation over all the ne elements on the surface and S_e is the surface of the e-th element. It is assumed that

$$S_1, S_2, \cdots, S_{n_1} \in S_\sigma, \quad S_{n_1+1}, S_{n_1+2}, \cdots, S_{ne} \in S_u \qquad (4.16)$$

In discretizing the boundary, different types of elements, such as quadrilateral or Serendipity quadrilateral elements, can be used and the boundary elements in a three-dimensional BEM are similar to those in a two-dimensional FEM. In this text we consider four- to eight-node isoparametric elements based on an isoparametric formulation; the elements included are illustrated in Fig. 4.3 and are

(1) the four-node isoparametric quadrilateral element with a linear variation of displacements and tractions,

(2) the eight-node Serendipity quadrilateral element with a quadratic variation of the displacement and traction fields within the element,

(3) the five- to seven-node element, which is obtained from the eight-node element in Fig. 4.3 by making one to three midside nodes disappear.

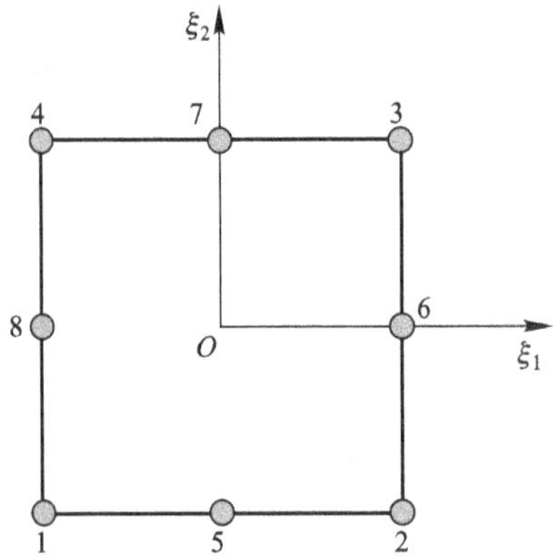

Fig. 4.3 A Serendipity quadrilateral element

An element is defined in the (ξ_1, ξ_2) coordinates (or local coordinates) and is square shaped. The local coordinate of an element normally takes a value in the range $[-1, +1]$. The shape functions can also be thought of as functions that map the global coordinates of the element into the local coordinate system. The eight shape functions can be written as

$$\begin{aligned}
N_\alpha(\xi_1, \xi_2) &= \frac{1}{2}(1-\xi_1^2)(1+\xi_2^\alpha \xi_2), \quad (\alpha = 5,7),\\
N_\alpha(\xi_1, \xi_2) &= \frac{1}{2}(1-\xi_2^2)(1+\xi_1^\alpha \xi_1), \quad (\alpha = 6,8),\\
N_1(\xi_1, \xi_2) &= N_1^* - \frac{1}{2}(N_8 + N_5),\\
N_2(\xi_1, \xi_2) &= N_2^* - \frac{1}{2}(N_5 + N_6),\\
N_3(\xi_1, \xi_2) &= N_3^* - \frac{1}{2}(N_6 + N_7),\\
N_4(\xi_1, \xi_2) &= N_4^* - \frac{1}{2}(N_7 + N_8)
\end{aligned} \tag{4.17}$$

in which $N_\alpha^*(\xi_1, \xi_2) = \frac{1}{4}(1+\xi_1^\alpha \xi_1)(1+\xi_2^\alpha \xi_2)$ $(\alpha = 1,2,3,4)$. ξ_β^α $(\beta = 1,2)$ denotes the local coordinates in the β direction at node α. The shape functions of a variable-node element are obtained by letting the functions of the disappearing midside nodes be zero.

Within each element, the global coordinates (x, y, z) of any point can be obtained by interpolation between the coordinates $(x^\alpha, y^\alpha, z^\alpha)$ of the nodes of that element through the shape functions (4.17). Thus

$$x=\sum_{\alpha=1}^{m} N_\alpha(\xi_1,\xi_2)x^\alpha,\quad y=\sum_{\alpha=1}^{m} N_\alpha(\xi_1,\xi_2)y^\alpha,\quad z=\sum_{\alpha=1}^{m} N_\alpha(\xi_1,\xi_2)z^\alpha \tag{4.18a}$$

We express the displacement and traction fields within the element in terms of the nodal values, in much the same way as the geometry is interpolated between element nodal values. This type of element is called an iso-parametric element. Thus

$$u_i=\sum_{\alpha=1}^{m} N_\alpha(\xi_1,\xi_2)u_i^\alpha,\quad (i=x,y,z) \tag{4.18b}$$

$$t_i=\sum_{\alpha=1}^{m} N_\alpha(\xi_1,\xi_2)t_i^\alpha,\quad (i=x,y,z) \tag{4.18c}$$

where α denotes the α-th node of the m-node element ($m=4\sim 8$).

Having discretized the boundary S into ne elements (and N nodes), we can rewrite the boundary integral equation (4.13). Substituting Eq. (4.18) for u_i and t_i and splitting the integrals into a sum of integrals over the element, we have, from Eqs. (4.13) and (4.18),

$$\begin{aligned} c_{ij}(P)u_j(P) = &\sum_{e=1}^{ne}\left\{\sum_{\alpha=1}^{m} t_j^\alpha(Q^\alpha)\int_{S_e} u_{ij}^Y(P,Q)N_\alpha(Q)\mathrm{d}S(Q)\right\} - \\ &\sum_{e=1}^{ne}\left\{\sum_{\alpha=1}^{m} u_j^\alpha(Q^\alpha)\int_{S_e} t_{ij}^Y(P,Q)N_\alpha(Q)\mathrm{d}S(Q)\right\},\quad (i,j=x,y,z) \end{aligned} \tag{4.19a}$$

where Q and Q^α are, respectively, the field points within an element and at node α, and P is the source point at the element node in the general case.

Considering the boundary conditions (4.16), Eq. (4.19a) can be rewritten as

$$\begin{aligned} c_{ij}(P)u_j(P) = &\sum_{e=1}^{n_1}\left\{\sum_{\alpha=1}^{m} \bar{t}_j^\alpha(Q^\alpha)\int_{S_e} u_{ij}^Y(P,Q)N_\alpha(Q)\mathrm{d}S(Q)\right\} + \\ &\sum_{e=n_1+1}^{ne}\left\{\sum_{\alpha=1}^{m} t_j^\alpha(Q^\alpha)\int_{S_e} u_{ij}^Y(P,Q)N_\alpha(Q)\mathrm{d}S(Q)\right\} - \\ &\sum_{e=1}^{n_1}\left\{\sum_{\alpha=1}^{m} u_j^\alpha(Q^\alpha)\int_{S_e} t_{ij}^Y(P,Q)N_\alpha(Q)\mathrm{d}S(Q)\right\} - \\ &\sum_{e=n_1+1}^{ne}\left\{\sum_{\alpha=1}^{m} \bar{u}_j^\alpha(Q^\alpha)\int_{S_e} t_{ij}^Y(P,Q)N_\alpha(Q)\mathrm{d}S(Q)\right\},\quad (i,j=x,y,z) \end{aligned} \tag{4.19b}$$

To produce a set of linear equations, we write this equation for each node in turn. It can be observed that the integrands in Eq. (4.19) do not contain the unknowns and can be carried out over each of the elements. The techniques employed to carry out these integrations are critically important and are discussed in detail in the following sections of this chapter. Let us assume that the integrals to be computed are denoted by the symbols

$$\int_{S_e} u_{ij}^Y(P,Q)N_\alpha(Q)\mathrm{d}S(Q) = G_{ij}^{e\alpha}(P,Q) \tag{4.20a}$$

$$\int_{S_e} t_{ij}^Y(P,Q)N_\alpha(Q)\mathrm{d}S(Q) = H_{ij}^{e\alpha}(P,Q) \tag{4.20b}$$

Substituting Eq. (4.20) in Eq. (4.19a), we have

$$c_{ij}(P)u_j(P) = \sum_{e=1}^{ne}\sum_{\alpha=1}^{m} t_j^\alpha(Q^\alpha)G_{ij}^{e\alpha}(P,Q) - \sum_{e=1}^{ne}\sum_{\alpha=1}^{m} u_j^\alpha(Q^\alpha)H_{ij}^{e\alpha}(P,Q), \quad (i,j=x,y,z) \tag{4.21}$$

Now, we need to transform the integrals Eq. (4.20) in the global coordinate system into ones in the local coordinate system. From Eq. (4.20), we have

$$G_{ij}^{e\alpha}(P,Q) = \int_{-1}^{1}\int_{-1}^{1} u_{ij}^Y(P,Q(\xi_1,\xi_2))N_\alpha(\xi_1,\xi_2)J(\xi_1,\xi_2)\mathrm{d}\xi_1\mathrm{d}\xi_2 \tag{4.22a}$$

$$H_{ij}^{e\alpha}(P,Q) = \int_{-1}^{1}\int_{-1}^{1} t_{ij}^Y(P,Q(\xi_1,\xi_2))N_\alpha(\xi_1,\xi_2)J(\xi_1,\xi_2)\mathrm{d}\xi_1\mathrm{d}\xi_2 \tag{4.22b}$$

where $J(\xi_1,\xi_2)$ is the Jacobian determinant.

We then have to determine $J(\xi_1,\xi_2)$. In Fig. 4.4, a differential of area S will be given by

$$\mathrm{d}S = \left|\frac{\partial \boldsymbol{r}}{\partial \xi_1} \times \frac{\partial \boldsymbol{r}}{\partial \xi_2}\right| \mathrm{d}\xi_1\mathrm{d}\xi_2 = J(\xi_1,\xi_2)\mathrm{d}\xi_1\mathrm{d}\xi_2 \tag{4.23}$$

where $\boldsymbol{r}$ is a radius vector.

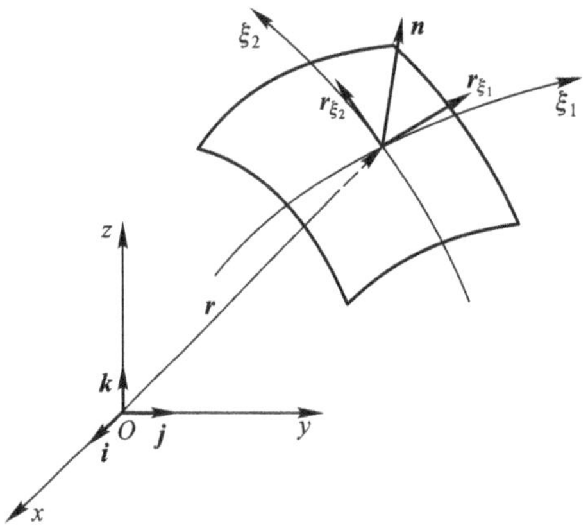

Fig. 4.4 Coordinate systems for an element on a curved surface

Taking $\dfrac{\partial \boldsymbol{r}}{\partial \xi_1} = \boldsymbol{r}_{\xi_1}$ and $\dfrac{\partial \boldsymbol{r}}{\partial \xi_2} = \boldsymbol{r}_{\xi_2}$, we have

$$J(\xi_1,\xi_2) = |\boldsymbol{r}_{\xi_1} \times \boldsymbol{r}_{\xi_2}| \tag{4.24}$$

where

$$\begin{aligned}
\boldsymbol{r}_{\xi_1} &= \frac{\partial x(\xi_1,\xi_2)}{\partial \xi_1}\boldsymbol{i} + \frac{\partial y(\xi_1,\xi_2)}{\partial \xi_1}\boldsymbol{j} + \frac{\partial z(\xi_1,\xi_2)}{\partial \xi_1}\boldsymbol{k}, \\
\boldsymbol{r}_{\xi_2} &= \frac{\partial x(\xi_1,\xi_2)}{\partial \xi_2}\boldsymbol{i} + \frac{\partial y(\xi_1,\xi_2)}{\partial \xi_2}\boldsymbol{j} + \frac{\partial z(\xi_1,\xi_2)}{\partial \xi_2}\boldsymbol{k}
\end{aligned} \tag{4.25}$$

where $\boldsymbol{i}$, $\boldsymbol{j}$, and $\boldsymbol{k}$ are the orthogonal unit basis vectors of the global coordinate system.

By using Eq. (4.25), the Jacobian determinant is found from Eq. (4.24)

$$J(\xi_1,\xi_2) = \begin{vmatrix} \boldsymbol{i} & \boldsymbol{j} & \boldsymbol{k} \\ \dfrac{\partial x(\xi_1,\xi_2)}{\partial \xi_1} & \dfrac{\partial y(\xi_1,\xi_2)}{\partial \xi_1} & \dfrac{\partial z(\xi_1,\xi_2)}{\partial \xi_1} \\ \dfrac{\partial x(\xi_1,\xi_2)}{\partial \xi_2} & \dfrac{\partial y(\xi_1,\xi_2)}{\partial \xi_2} & \dfrac{\partial z(\xi_1,\xi_2)}{\partial \xi_2} \end{vmatrix} = (J_1^2 + J_2^2 + J_3^2)^{1/2} \tag{4.26}$$

Notice that the values of J_1, J_2, and J_3 are given by

$$\begin{aligned}
J_1 &= \frac{\partial y}{\partial \xi_1}\frac{\partial z}{\partial \xi_2} - \frac{\partial z}{\partial \xi_1}\frac{\partial y}{\partial \xi_2}, \\
J_2 &= \frac{\partial z}{\partial \xi_1}\frac{\partial x}{\partial \xi_2} - \frac{\partial x}{\partial \xi_1}\frac{\partial z}{\partial \xi_2}, \\
J_3 &= \frac{\partial x}{\partial \xi_1}\frac{\partial y}{\partial \xi_2} - \frac{\partial y}{\partial \xi_1}\frac{\partial x}{\partial \xi_2}
\end{aligned} \tag{4.27}$$

Considering Eq. (4.18a), Eq. (4.27) can be rewritten as

$$\begin{aligned}
J_1 &= \sum_{\alpha=1}^{m} \frac{\partial N_\alpha}{\partial \xi_1} y^\alpha \sum_{\alpha=1}^{m} \frac{\partial N_\alpha}{\partial \xi_2} z^\alpha - \sum_{\alpha=1}^{m} \frac{\partial N_\alpha}{\partial \xi_1} z^\alpha \sum_{\alpha=1}^{m} \frac{\partial N_\alpha}{\partial \xi_2} y^\alpha, \\
J_2 &= \sum_{\alpha=1}^{m} \frac{\partial N_\alpha}{\partial \xi_1} z^\alpha \sum_{\alpha=1}^{m} \frac{\partial N_\alpha}{\partial \xi_2} x^\alpha - \sum_{\alpha=1}^{m} \frac{\partial N_\alpha}{\partial \xi_1} x^\alpha \sum_{\alpha=1}^{m} \frac{\partial N_\alpha}{\partial \xi_2} z^\alpha, \\
J_3 &= \sum_{\alpha=1}^{m} \frac{\partial N_\alpha}{\partial \xi_1} x^\alpha \sum_{\alpha=1}^{m} \frac{\partial N_\alpha}{\partial \xi_2} y^\alpha - \sum_{\alpha=1}^{m} \frac{\partial N_\alpha}{\partial \xi_1} y^\alpha \sum_{\alpha=1}^{m} \frac{\partial N_\alpha}{\partial \xi_2} x^\alpha
\end{aligned} \tag{4.28}$$

As shown in Fig. 4.4, the vectors $\boldsymbol{r}_{\xi_1}$ and $\boldsymbol{r}_{\xi_2}$ are directed in the local tangent plane to the surface. Denoting the vector cross-product of the two vectors as a new vector $\boldsymbol{n}^*$, which is normal to the surface, we have the components of the unit vector $\boldsymbol{n} = \boldsymbol{n}^*/|\boldsymbol{n}^*|$

$$n_1 = J_1/J, \quad n_2 = J_2/J, \quad n_3 = J_3/J \tag{4.29}$$

4.6 Assembly of the equation system

For Eq. (4.21), we recall the notation used

$$\underset{\substack{\uparrow \\ \text{The } i \text{ direction of point } P}}{\overset{\substack{\text{Element number} \\ \downarrow}}{G_{ij}^{e\alpha}}} \begin{array}{l} \leftarrow \text{Collocation point} \\ \leftarrow \text{The } j \text{ direction of point } Q \end{array} \tag{4.30}$$

The notations of $H_{ij}^{e\alpha}$ in Eq. (4.21) have the same meanings as the ones of $G_{ij}^{e\alpha}$ in (4.30). Eq. (4.21) can be rewritten in a matrix form

$$\begin{bmatrix} c_{xx}^P & c_{xy}^P & c_{xz}^P \\ c_{yx}^P & c_{yy}^P & c_{yz}^P \\ c_{zx}^P & c_{zy}^P & c_{zz}^P \end{bmatrix} \begin{Bmatrix} u_x^P \\ u_y^P \\ u_z^P \end{Bmatrix} + \sum_{e=1}^{ne} \begin{bmatrix} H_{xx}^{e1} & H_{xy}^{e1} & H_{xz}^{e1} & \cdots & H_{xx}^{em} & H_{xy}^{em} & H_{xz}^{em} \\ H_{yx}^{e1} & H_{yy}^{e1} & H_{yz}^{e1} & \cdots & H_{yx}^{em} & H_{yy}^{em} & H_{yz}^{em} \\ H_{zx}^{e1} & H_{zy}^{e1} & H_{zz}^{e1} & \cdots & H_{zx}^{em} & H_{zy}^{em} & H_{zz}^{em} \end{bmatrix} \begin{Bmatrix} u_x^1 \\ u_y^1 \\ u_z^1 \\ \vdots \\ u_x^m \\ u_y^m \\ u_z^m \end{Bmatrix}$$

$$= \sum_{e=1}^{ne} \begin{bmatrix} G_{xx}^{e1} & G_{xy}^{e1} & G_{xz}^{e1} & \cdots & G_{xx}^{em} & G_{xy}^{em} & G_{xz}^{em} \\ G_{yx}^{e1} & G_{yy}^{e1} & G_{yz}^{e1} & \cdots & G_{yx}^{em} & G_{yy}^{em} & G_{yz}^{em} \\ G_{zx}^{e1} & G_{zy}^{e1} & G_{zz}^{e1} & \cdots & G_{zx}^{em} & G_{zy}^{em} & G_{zz}^{em} \end{bmatrix} \begin{Bmatrix} t_x^1 \\ t_y^1 \\ t_z^1 \\ \vdots \\ t_x^m \\ t_y^m \\ t_z^m \end{Bmatrix}, \quad (P = 1, 2, \cdots, N) \tag{4.31}$$

where N is the node number on the boundary surface S and the superscript P corresponds to point P.

Thus, for any node P on the discretized boundary, three of the linear equations (4.31) can be established. In the general case, a node is shared by several elements. In Eq. (4.31), the coefficients of the displacements and tractions of the same node can be added together. Equation (4.31) can thus be rewritten as

$$\begin{bmatrix} c_{xx}^P & c_{xy}^P & c_{xz}^P \\ c_{yx}^P & c_{yy}^P & c_{yz}^P \\ c_{zx}^P & c_{zy}^P & c_{zz}^P \end{bmatrix} \begin{Bmatrix} u_x^P \\ u_y^P \\ u_z^P \end{Bmatrix} + \begin{bmatrix} \bar{h}_{xx}^{P1} & \bar{h}_{xy}^{P1} & \bar{h}_{xz}^{P1} & \cdots & \bar{h}_{xx}^{PN} & \bar{h}_{xy}^{PN} & \bar{h}_{xz}^{PN} \\ h_{yx}^{P1} & \bar{h}_{yy}^{P1} & \bar{h}_{yz}^{P1} & \cdots & \bar{h}_{yx}^{PN} & \bar{h}_{yy}^{PN} & \bar{h}_{yz}^{PN} \\ h_{zx}^{P1} & \bar{h}_{zy}^{P1} & \bar{h}_{zz}^{P1} & \cdots & \bar{h}_{zx}^{PN} & \bar{h}_{zy}^{PN} & \bar{h}_{zz}^{PN} \end{bmatrix} \begin{Bmatrix} u_x^1 \\ u_y^1 \\ u_z^1 \\ \vdots \\ u_x^N \\ u_y^N \\ u_z^N \end{Bmatrix}$$

$$
= \begin{bmatrix} g_{xx}^{P1} & g_{xy}^{P1} & g_{xz}^{P1} & \cdots & g_{xx}^{PN} & g_{xy}^{PN} & g_{xz}^{PN} \\ g_{yx}^{P1} & g_{yy}^{P1} & g_{yz}^{P1} & \cdots & g_{yx}^{PN} & g_{yy}^{PN} & g_{yz}^{PN} \\ g_{zx}^{P1} & g_{zy}^{P1} & g_{zz}^{P1} & \cdots & g_{zx}^{PN} & g_{zy}^{PN} & g_{zz}^{PN} \end{bmatrix} \begin{Bmatrix} t_x^1 \\ t_y^1 \\ t_z^1 \\ \vdots \\ t_x^N \\ t_y^N \\ t_z^N \end{Bmatrix}, \quad (P = 1,2,\cdots,N) \tag{4.32}
$$

where $\bar{h}_{ij}^{Pn}$ and g_{ij}^{Pn} $(n = 1,2,\cdots,N)$ are, respectively, the summations of $H_{ij}^{e\alpha}$ and $G_{ij}^{e\alpha}$ of all the elements having the node n.

Let point P collocate at the element node. For the two terms on the left-hand side of Eq. (4.32), the coefficients of displacements at the same node are combined together, i.e.,

$$
h_{ij}^{Pn} = \begin{cases} c_{ij}^{P} + \bar{h}_{ij}^{Pn}, & (P = n) \\ \bar{h}_{ij}^{Pn}, & (P \neq n) \end{cases} \tag{4.33}
$$

As a result, we have

$$
\begin{bmatrix} h_{xx}^{P1} & h_{xy}^{P1} & h_{xz}^{P1} & \cdots & h_{xx}^{PN} & h_{xy}^{PN} & h_{xz}^{PN} \\ h_{yx}^{P1} & h_{yy}^{P1} & h_{yz}^{P1} & \cdots & h_{yx}^{PN} & h_{yy}^{PN} & h_{yz}^{PN} \\ h_{zx}^{P1} & h_{zy}^{P1} & h_{zz}^{P1} & \cdots & h_{zx}^{PN} & h_{zy}^{PN} & h_{zz}^{PN} \end{bmatrix} \begin{Bmatrix} u_x^1 \\ u_y^1 \\ u_z^1 \\ \vdots \\ u_x^N \\ u_y^N \\ u_z^N \end{Bmatrix}
$$

$$
= \begin{bmatrix} g_{xx}^{P1} & g_{xy}^{P1} & g_{xz}^{P1} & \cdots & g_{xx}^{PN} & g_{xy}^{PN} & g_{xz}^{PN} \\ g_{yx}^{P1} & g_{yy}^{P1} & g_{yz}^{P1} & \cdots & g_{yx}^{PN} & g_{yy}^{PN} & g_{yz}^{PN} \\ g_{zx}^{P1} & g_{zy}^{P1} & g_{zz}^{P1} & \cdots & g_{zx}^{PN} & g_{zy}^{PN} & g_{zz}^{PN} \end{bmatrix} \begin{Bmatrix} t_x^1 \\ t_y^1 \\ t_z^1 \\ \vdots \\ t_x^N \\ t_y^N \\ t_z^N \end{Bmatrix}, \quad (P = 1,2,3\cdots,N) \tag{4.34}
$$

Putting Eq. (4.34) of all the nodes together, we have

$$
\begin{bmatrix}
h_{xx}^{11} & h_{xy}^{11} & h_{xz}^{11} & \cdots & h_{xx}^{1N} & h_{xy}^{1N} & h_{xz}^{1N} \\
h_{yx}^{11} & h_{yy}^{11} & h_{yz}^{11} & \cdots & h_{yx}^{1N} & h_{yy}^{1N} & h_{yz}^{1N} \\
h_{zx}^{11} & h_{zy}^{11} & h_{zz}^{11} & \cdots & h_{zx}^{1N} & h_{zy}^{1N} & h_{zz}^{1N} \\
\vdots & \vdots & \vdots & & \vdots & \vdots & \vdots \\
h_{xx}^{N1} & h_{xy}^{N1} & h_{xz}^{N1} & \cdots & h_{xx}^{NN} & h_{xy}^{NN} & h_{xz}^{NN} \\
h_{yx}^{N1} & h_{yy}^{N1} & h_{yz}^{N1} & \cdots & h_{yx}^{NN} & h_{yy}^{NN} & h_{yz}^{NN} \\
h_{zx}^{N1} & h_{zy}^{N1} & h_{zz}^{N1} & \cdots & h_{zx}^{NN} & h_{zy}^{NN} & h_{zz}^{NN}
\end{bmatrix}
\begin{Bmatrix} u_x^1 \\ u_y^1 \\ u_z^1 \\ \vdots \\ u_x^N \\ u_y^N \\ u_z^N \end{Bmatrix}
=
\begin{bmatrix}
g_{xx}^{11} & g_{xy}^{11} & g_{xz}^{11} & \cdots & g_{xx}^{1N} & g_{xy}^{1N} & g_{xz}^{1N} \\
g_{yx}^{11} & g_{yy}^{11} & g_{yz}^{11} & \cdots & g_{yx}^{1N} & g_{yy}^{1N} & g_{yz}^{1N} \\
g_{zx}^{11} & g_{zy}^{11} & g_{zz}^{11} & \cdots & g_{zx}^{1N} & g_{zy}^{1N} & g_{zz}^{1N} \\
\vdots & \vdots & \vdots & & \vdots & \vdots & \vdots \\
g_{xx}^{N1} & g_{xy}^{N1} & g_{xz}^{N1} & \cdots & g_{xx}^{NN} & g_{xy}^{NN} & g_{xz}^{NN} \\
g_{yx}^{N1} & g_{yy}^{N1} & g_{yz}^{N1} & \cdots & g_{yx}^{NN} & g_{yy}^{NN} & g_{yz}^{NN} \\
g_{zx}^{N1} & g_{zy}^{N1} & g_{zz}^{N1} & \cdots & g_{zx}^{NN} & g_{zy}^{NN} & g_{zz}^{NN}
\end{bmatrix}
\begin{Bmatrix} t_x^1 \\ t_y^1 \\ t_z^1 \\ \vdots \\ t_x^N \\ t_y^N \\ t_z^N \end{Bmatrix}
\tag{4.35}
$$

In Eq. (4.35), the displacements u_j^n and tractions t_j^n are prescribed for some nodes and are unknown for other nodes. Thus, Eq. (4.35) needs to be rearranged; that is, the unknown quantities t_j^n and the corresponding coefficients are moved to the left-hand side of the equations whilst the given quantities $\bar{u}_j^n$ and the corresponding coefficients are moved to the right-hand side of the equations. Assume that the quantities $u_x^1 = \bar{u}_x^1$, $u_y^1 = \bar{u}_y^1$, $t_z^1 = \bar{t}_z^1$, $t_x^N = \bar{t}_x^N$, $t_y^N = \bar{t}_y^N$, $u_z^N = \bar{u}_z^N$ and the corresponding quantities t_x^1, t_y^1, u_z^1, u_x^N, u_y^N, t_z^N are unknown. The equations can be rewritten as

$$
\begin{bmatrix}
-g_{xx}^{11} & -g_{xy}^{11} & h_{xz}^{11} & \cdots & h_{xx}^{1N} & h_{xy}^{1N} & -g_{xz}^{1N} \\
-g_{yx}^{11} & -g_{yy}^{11} & h_{yz}^{11} & \cdots & h_{yx}^{1N} & h_{yy}^{1N} & -g_{yz}^{1N} \\
-g_{zx}^{11} & -g_{zy}^{11} & h_{zz}^{11} & \cdots & h_{zx}^{1N} & h_{zy}^{1N} & -g_{zz}^{1N} \\
\vdots & \vdots & \vdots & & \vdots & \vdots & \vdots \\
-g_{xx}^{N1} & -g_{xy}^{N1} & h_{xz}^{N1} & \cdots & h_{xx}^{NN} & h_{xy}^{NN} & -g_{xz}^{NN} \\
-g_{yx}^{N1} & -g_{yy}^{N1} & h_{yz}^{N1} & \cdots & h_{yx}^{NN} & h_{yy}^{NN} & -g_{yz}^{NN} \\
-g_{zx}^{N1} & -g_{zy}^{N1} & h_{zz}^{N1} & \cdots & h_{zx}^{NN} & h_{zy}^{NN} & -g_{zz}^{NN}
\end{bmatrix}
\begin{Bmatrix} t_x^1 \\ t_y^1 \\ u_z^1 \\ \vdots \\ u_x^N \\ u_y^N \\ t_z^N \end{Bmatrix}
=
\begin{bmatrix}
-h_{xx}^{11} & -h_{xy}^{11} & g_{xz}^{11} & \cdots & g_{xx}^{1N} & g_{xy}^{1N} & -h_{xz}^{1N} \\
-h_{yx}^{11} & -h_{yy}^{11} & g_{yz}^{11} & \cdots & g_{yx}^{1N} & g_{yy}^{1N} & -h_{yz}^{1N} \\
-h_{zx}^{11} & -h_{zy}^{11} & g_{zz}^{11} & \cdots & g_{zx}^{1N} & g_{zy}^{1N} & -h_{zz}^{1N} \\
\vdots & \vdots & \vdots & & \vdots & \vdots & \vdots \\
-h_{xx}^{N1} & -h_{xy}^{N1} & g_{xz}^{N1} & \cdots & g_{xx}^{NN} & g_{xy}^{NN} & -h_{xz}^{NN} \\
-h_{yx}^{N1} & -h_{yy}^{N1} & g_{yz}^{N1} & \cdots & g_{yx}^{NN} & g_{yy}^{NN} & -h_{yz}^{NN} \\
-h_{zx}^{N1} & -h_{zy}^{N1} & g_{zz}^{N1} & \cdots & g_{zx}^{NN} & g_{zy}^{NN} & -h_{zz}^{NN}
\end{bmatrix}
\begin{Bmatrix} \bar{u}_x^1 \\ \bar{u}_y^1 \\ \bar{t}_z^1 \\ \vdots \\ \bar{t}_x^N \\ \bar{t}_y^N \\ \bar{u}_z^N \end{Bmatrix}
\tag{4.36}
$$

From Eq. (4.36) we have, in the form of partitioned matrix,

$$\begin{bmatrix} \boldsymbol{A}_{11} & \boldsymbol{A}_{12} \\ \boldsymbol{A}_{21} & \boldsymbol{A}_{22} \end{bmatrix} \begin{Bmatrix} \boldsymbol{U} \\ \boldsymbol{T} \end{Bmatrix} = \begin{bmatrix} \boldsymbol{B}_{11} & \boldsymbol{B}_{12} \\ \boldsymbol{B}_{21} & \boldsymbol{B}_{22} \end{bmatrix} \begin{Bmatrix} \overline{\boldsymbol{T}} \\ \overline{\boldsymbol{U}} \end{Bmatrix} \tag{4.37}$$

where $\boldsymbol{U}$ and $\boldsymbol{T}$ are, respectively, the arrays of unknown displacements and tractions, and $\overline{\boldsymbol{U}}$ and $\overline{\boldsymbol{T}}$ are the arrays of given displacements and tractions, respectively. Notice that the tractions $\boldsymbol{T}$ on the boundary having $\overline{\boldsymbol{U}}$ are unknown whilst the tractions $\boldsymbol{U}$ on the boundary having $\overline{\boldsymbol{T}}$ are unknown.

By using a matrix multiplication, the matrices on the right-hand side of Eq. (4.37) are transformed into an array $\boldsymbol{F}$. On the left-hand side of Eq. (4.37), an array $\boldsymbol{X}$ is used to denote the unknown quantities and a square matrix $\boldsymbol{A}$ is used to denote the coefficients. Thus, Eq. (4.37) can be converted into the convenient form

$$\boldsymbol{A}\boldsymbol{X} = \boldsymbol{F} \tag{4.38}$$

Notice that the matrix $\boldsymbol{A}$ is fully populated since, in a single region, there is an interaction between every pair (P, Q) of nodal points. Solving the above linear equations, all boundary values are fully determined. The Gaussian elimination method can be used for the system of equations. However, as the size of the problem increases, the time required for the Gaussian elimination method to solve the final system of equations increase considerably. Recently, there has been a resurgence of iterative solvers, such as the Generalized Minimum Residual Algorithm (GMRES). The advantage of these solvers is that the number of operations decreases.

4.7 Numerical integration of non-singular integrals

4.7.1 Gaussian quadrature formulas

The distance between the source point P and the field point Q is denoted by r. In Eq. (4.22), the displacement kernel function $u_{ij}^{Y}(P,Q)$ is of order r^{-1} and the traction kernel function $t_{ij}^{Y}(P,Q)$ is of order r^{-2}. When P is not located in the same element as Q, the integrals (4.22) are not singular and can be calculated directly by Gaussian quadrature. Thus, the integrals (4.22) are given by the expressions

$$G_{ij}^{e\alpha}(P,Q) = \sum_{k=1}^{m_1}\sum_{l=1}^{m_2} u_{ij}^{Y}[P, Q(\xi_1^k, \xi_2^l)] N_\alpha(\xi_1^k, \xi_2^l) J(\xi_1^k, \xi_2^l) w_1^k w_2^l \tag{4.39a}$$

$$H_{ij}^{e\alpha}(P,Q) = \sum_{k=1}^{m_1}\sum_{l=1}^{m_2} t_{ij}^{Y}[P, Q(\xi_1^k, \xi_2^l)] N_\alpha(\xi_1^k, \xi_2^l) J(\xi_1^k, \xi_2^l) w_1^k w_2^l \tag{4.39b}$$

where (ξ_1^k, ξ_2^l) are the coordinates of the Gaussian point, w_1^k and w_2^l are the corresponding weights, m_1 and m_2 are the numbers of the Gaussian points employed in the directions ξ_1 and ξ_2, respectively.

4.7.2 Adaptive integration

In establishing the final system of linear equations, a large amount of numerical integrations Eq. (4.39) need to be calculated. Therefore, the accurate and efficient integration of the integrals in Eq. (4.39) is vitally important. The best method is that, for each integral, a numerical integration with different numbers of Gaussian points is employed to obtain the same accuracy. This method is referred to as the Gaussian quadrature with the same accuracy.

In estimating the accuracy of the numerical integration, the differentials of order $2n$ of the integrand are needed. In Eq. (4.39) we find that the integrand is very complicated. However, in Eq. (4.39) the shape functions are polynomials of a degree equal to or less than 2 and a weak variation of the Jacobian determinant exists. Only the kernel functions, i.e., the quantities related to the fundamental solutions, vary strongly. It can be found that there are obvious variations of the kernel functions when the element approaches the source point. For convenience, the same method can be used to calculate the number of the Gaussian points on any element for one given source point. The highest order of singularity of the fundamental solutions is $1/r^2$. To determine the number of the Gaussian points, the following integral is used

$$\int_{-1}^{1}\int_{-1}^{1}\frac{1}{r^2(\xi_1,\xi_2)}\mathrm{d}\xi_1\mathrm{d}\xi_2 \tag{4.40}$$

The upper bounds of the relative error of Eq. (4.40) are

$$\Delta_1 = 4(m_1+1)\frac{r^2(0,0)}{r^2(\xi_1^*,\xi_2^*)}\left(\frac{1}{2r}\frac{\partial r}{\partial \xi_1}\right)^{2m_1}\Bigg|_{\xi_1=\xi_1^*,\xi_2=\xi_2^*} \tag{4.41a}$$

$$\Delta_2 = 4(m_2+1)\frac{r^2(0,0)}{r^2(\xi_1^*,\xi_2^*)}\left(\frac{1}{2r}\frac{\partial r}{\partial \xi_2}\right)^{2m_2}\Bigg|_{\xi_1=\xi_1^*,\xi_2=\xi_2^*} \tag{4.41b}$$

where Δ_1 and Δ_2 are the accuracies of the Gaussian quadrature in the directions ξ_1 and ξ_2 respectively, the point (ξ_1^*,ξ_2^*) is the corner node, which has the minimum r value among the four corners, and the point $(0,0)$ is the origin of the local coordinate system attached at the element.

In writing the computer code, $\dfrac{\partial r}{\partial \xi_1}$ and $\dfrac{\partial r}{\partial \xi_2}$ are replaced by $\dfrac{\Delta r}{\Delta \xi_1}$ and $\dfrac{\Delta r}{\Delta \xi_2}$ respectively. Considering $\Delta_1 = \Delta_2$ in Eq. (4.41), we can determine m_1 and m_2 in order to obtain the same accuracy in the directions ξ_1 and ξ_2. In writing the code, let $\Delta \leqslant 10^{-4}$ and take m_1 and m_2 to be in the range of 3 to 6. When m_1 and m_2 exceed the given value, a subdivision scheme for integration is implemented.

4.7.3 Nearly singular integrals

When P lies close to but not in the element of Q, the displacements and tractions vary strongly as the distance r from P to Q becomes small. In this case the integrals are referred to as nearly singular integrals. To obtain the desired accuracy of the integrals of this type, more Gaussian points should be collocated. Another effective method is to implement a subdivision scheme. The number of subregions is determined by the distance from point P to the element and the magnitude of the element. Lachat and Watson (1976), Gao et al. (2002), and Beer et al. (2008) proposed their own subdivision schemes. In the following, we incorporate the subdivision scheme given in Beer et al. (2008) into the Yue's solution-based boundary element method.

We subdivide the elements into subregions as shown in Fig. 4.5. The numbers of subregions N_{ξ_1} in the direction ξ_1 and N_{ξ_2} in the direction ξ_2 are determined by

$$N_{\xi_1} = INT[(R/L)_{\min}/(R/L_{\xi_1})], \quad N_{\xi_2} = INT[(R/L)_{\min}/(R/L_{\xi_2})] \tag{4.42}$$

where INT denotes rounding off the result, $(R/L)_{\min}$ is the minimum value when the number of Gaussian points is 6. Eberwien et al. (2005) discussed the values of $(R/L)_{\min}$.

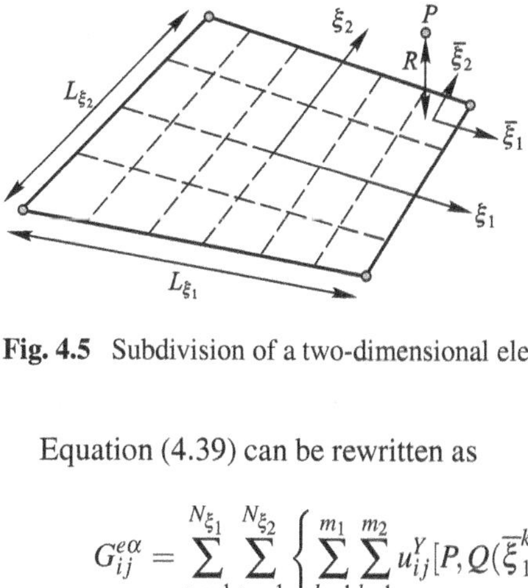

Fig. 4.5 Subdivision of a two-dimensional element

Equation (4.39) can be rewritten as

$$G_{ij}^{e\alpha} = \sum_{s_1=1}^{N_{\xi_1}} \sum_{s_2=1}^{N_{\xi_2}} \left\{ \sum_{k=1}^{m_1} \sum_{l=1}^{m_2} u_{ij}^Y[P, Q(\bar{\xi}_1^k, \bar{\xi}_2^l)] N_\alpha(\bar{\xi}_1^k, \bar{\xi}_2^l) J \cdot \bar{J} w_1^k w_2^l \right\}_{s_1, s_2} \tag{4.43a}$$

$$H_{ij}^{e\alpha} = \sum_{s_1=1}^{N_{\xi_1}} \sum_{s_2=1}^{N_{\xi_2}} \left\{ \sum_{k=1}^{m_1} \sum_{l=1}^{m_2} t_{ij}^Y[P, Q(\bar{\xi}_1^k, \bar{\xi}_2^l)] N_\alpha(\bar{\xi}_1^k, \bar{\xi}_2^l) J \cdot \bar{J} w_1^k w_2^l \right\}_{s_1, s_2} \tag{4.43b}$$

The relationships between the local coordinates (ξ_1, ξ_2) and $(\bar{\xi}_1, \bar{\xi}_2)$ are

$$\xi_1 = \frac{1}{2}(\xi_1^s + \xi_1^e) + \frac{\bar{\xi}_1}{N_{\xi_1}} \tag{4.44a}$$

$$\xi_2 = \frac{1}{2}(\xi_2^s + \xi_2^e) + \frac{\overline{\xi}_2}{N_{\xi_2}} \tag{4.44b}$$

where (ξ_1^s, ξ_2^s) and (ξ_1^e, ξ_2^e) define the subregion shown in Fig. 4.5. The Jacobian determinant is given by

$$\overline{J} = \frac{\partial \xi_1}{\partial \overline{\xi}_1} \frac{\partial \xi_2}{\partial \overline{\xi}_2} = \frac{1}{N_{\xi_1} N_{\xi_2}} \tag{4.45}$$

4.8 Numerical integration of singular integrals

4.8.1 General

When the source point P is located in the same element as Q, the integral in Eq. (4.39a) containing the displacement kernel function $u_{ij}^Y(P,Q)$, is weakly singular and the integral in Eq. (4.39b) containing the traction kernel function $t_{ij}^Y(P,Q)$, is strongly singular. In this case the direct application of the Gaussian quadrature is inadequate and special techniques must be employed to obtain these integrals with a very high degree of accuracy and efficiency.

4.8.2 Weakly singular integrals

Lachat and Watson (1976) developed an element subdivision technique to calculate the weakly singular integral, Eq. (4.39a). In this technique, the elements containing the source point P are divided into two or three triangular sub-elements where P is located at the common vertex of these sub-elements. Then, the triangular sub-elements are mapped into the square local elements. For this purpose, a new local coordinate system (η_1, η_2) is attached to each of the triangular sub-elements. After the transformation from element local coordinates to sub-element local coordinates, the integral in Eq. (4.22a) can be written, for simplification, as

$$\begin{aligned}
&\int_{S_e} F(\xi_1, \xi_2) \mathrm{d}\xi_1 \mathrm{d}\xi_2 \\
&= \sum_{k=1}^{2 \text{ or } 3} \int_{\Delta_k} F[\xi_1^k(\eta_1, \eta_2), \xi_2^k(\eta_1, \eta_2)] J_1^k(\eta_1, \eta_2) \mathrm{d}\eta_1 \mathrm{d}\eta_2 \\
&= \sum_{k=1}^{2 \text{ or } 3} \int_{-1}^{1} \int_{-1}^{1} F[\xi_1^k(\eta_1, \eta_2), \xi_2^k(\eta_1, \eta_2)] J_1^k(\eta_1, \eta_2) \mathrm{d}\eta_1 \mathrm{d}\eta_2
\end{aligned} \tag{4.46}$$

The integrand $F(\xi_1, \xi_2)$ of Eq. (4.46) is of order r^{-1}. After the coordinate transformation, $J_1^k(\eta_1, \eta_2)$ appears in the integrands of each sub-element and is of the order r. Thus, the singularities of the integrands are eliminated and the traditional Gaussian quadrature can be employed to calculate these integrals.

Because the transformation is linear from the element local coordinates to the sub-element local coordinates, linear shape functions can be used to determine the original local coordinates for a point in the new local coordinate system. The linear shape functions of the four-node element shown in Fig. 4.6 are

$$\begin{aligned}
L_1(\eta_1,\eta_2) &= \frac{1}{4}(1-\eta_1)(1-\eta_2),\\
L_2(\eta_1,\eta_2) &= \frac{1}{4}(1+\eta_1)(1-\eta_2),\\
L_3(\eta_1,\eta_2) &= \frac{1}{4}(1+\eta_1)(1+\eta_2),\\
L_4(\eta_1,\eta_2) &= \frac{1}{4}(1-\eta_1)(1+\eta_2)
\end{aligned} \tag{4.47}$$

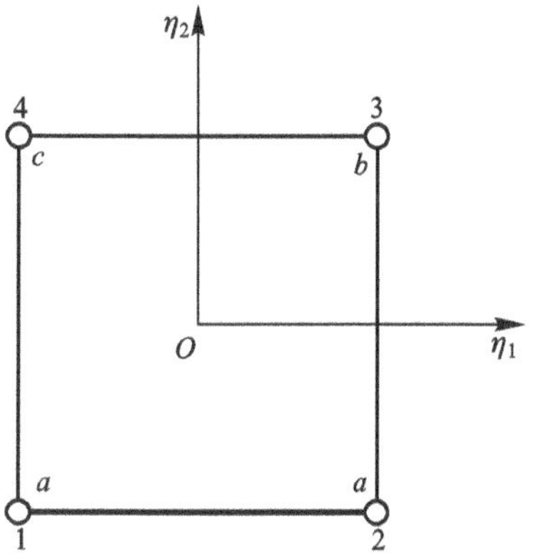

Fig. 4.6 A four-node linear element

Thus, there are the following interpolations

$$\begin{aligned}
\xi_1 &= \sum_{\alpha=1}^{4} L_\alpha(\eta_1,\eta_2)\xi_1^\alpha,\\
\xi_2 &= \sum_{\alpha=1}^{4} L_\alpha(\eta_1,\eta_2)\xi_2^\alpha
\end{aligned} \tag{4.48}$$

where ξ_1^α and ξ_2^α are the coordinates of node α in the original local coordinates (ξ_1,ξ_2).

Without loss of generality, we consider that P is located at corner point 1 and midside point 5, respectively. To establish the relationship between (ξ_1,ξ_2) and (η_1,η_2), the triangular sub-element is regarded as a degeneration from the square element. As shown in Fig. 4.6, the source point P corresponds to the side aa of the square element.

When P is located at node 1, shown in Fig. 4.7, the element is divided into two triangular sub-elements. For sub-element 134, we have

$$\begin{aligned}\xi_1 &= (L_1+L_2)\xi_1^1 + L_3\xi_1^3 + L_4\xi_1^4 \\ &= \frac{1}{2}(1+\eta_1)(1+\eta_2) - 1\end{aligned} \tag{4.49a}$$

$$\begin{aligned}\xi_2 &= (L_1+L_2)\xi_2^1 + L_3\xi_2^3 + L_4\xi_2^4 \\ &= \eta_2\end{aligned} \tag{4.49b}$$

$$J_1 = \frac{\partial(\xi_1,\xi_2)}{\partial(\eta_1,\eta_2)} = \begin{vmatrix} \frac{\partial\xi_1}{\partial\eta_1} & \frac{\partial\xi_2}{\partial\eta_1} \\ \frac{\partial\xi_1}{\partial\eta_2} & \frac{\partial\xi_2}{\partial\eta_2} \end{vmatrix} = \begin{vmatrix} \frac{1}{2}(1+\eta_2) & 0 \\ \frac{1}{2}(1+\eta_1) & 1 \end{vmatrix} = \frac{1}{2}(1+\eta_2) = o(r) \tag{4.49c}$$

For sub-element 123, we have

$$\begin{aligned}\xi_1 &= (L_1+L_2)\xi_1^1 + L_3\xi_1^2 + L_4\xi_1^3 \\ &= \eta_2\end{aligned} \tag{4.50a}$$

$$\begin{aligned}\xi_2 &= (L_1+L_2)\xi_2^1 + L_3\xi_2^2 + L_4\xi_2^3 \\ &= \frac{1}{2}(1-\eta_1)(1+\eta_2) - 1\end{aligned} \tag{4.50b}$$

$$J_1 = \frac{1}{2}(1+\eta_2) = o(r) \tag{4.50c}$$

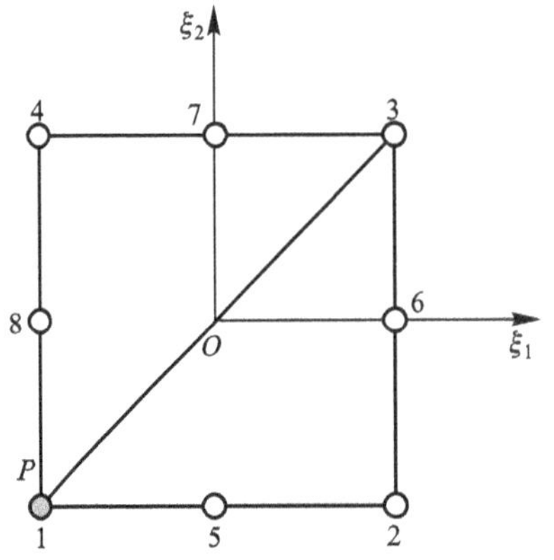

Fig. 4.7 Sub-elements for numerical integration when P is located at corner node 1

When P is located at node 5, shown in Fig. 4.8, the element is divided into three triangular sub-elements. For sub-element 541, we have

$$\begin{aligned}\xi_1 &= (L_1+L_2)\xi_1^5 + L_3\xi_1^4 + L_4\xi_1^1 \\ &= -\frac{1}{2}(1+\eta_2) \end{aligned} \tag{4.51a}$$

$$\begin{aligned}\xi_2 &= (L_1+L_2)\xi_2^5 + L_3\xi_2^4 + L_4\xi_2^1 \\ &= \frac{1}{2}(1+\eta_1)(1+\eta_2) - 1 \end{aligned} \tag{4.51b}$$

$$J_1 = \frac{1}{4}(1+\eta_2) = o(r) \tag{4.51c}$$

For sub-element 523, we have

$$\begin{aligned}\xi_1 &= (L_1+L_2)\xi_1^5 + L_3\xi_1^2 + L_4\xi_1^3 \\ &= \frac{1}{2}(1+\eta_2) \end{aligned} \tag{4.52a}$$

$$\begin{aligned}\xi_2 &= (L_1+L_2)\xi_2^5 + L_3\xi_2^2 + L_4\xi_2^3 \\ &= \frac{1}{2}(1-\eta_1)(1+\eta_2) - 1 \end{aligned} \tag{4.52b}$$

$$J_1 = \frac{1}{4}(1+\eta_2) = o(r) \tag{4.52c}$$

For sub-element 534, we have

$$\begin{aligned}\xi_1 &= (L_1+L_2)\xi_1^5 + L_3\xi_1^3 + L_4\xi_1^4 \\ &= \frac{1}{2}\eta_1(1+\eta_2) \end{aligned} \tag{4.53a}$$

$$\begin{aligned}\xi_2 &= (L_1+L_2)\xi_2^5 + L_3\xi_2^3 + L_4\xi_2^4 \\ &= \eta_2 \end{aligned} \tag{4.53b}$$

$$J_1 = \frac{1}{2}(1+\eta_2) = o(r) \tag{4.53c}$$

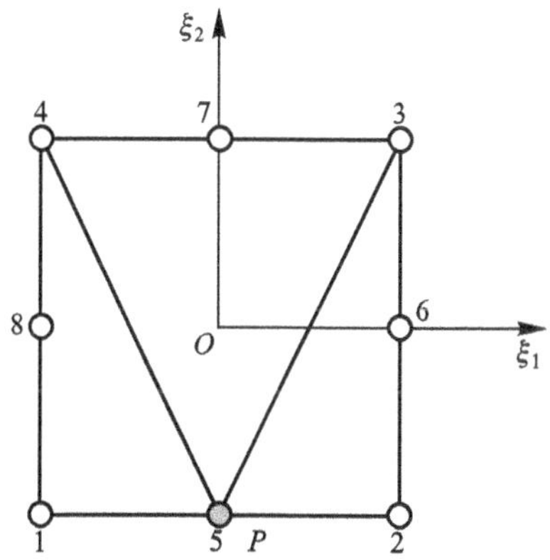

Fig. 4.8 Sub-elements for numerical integration when P is located at midside node 5

When the source point P is located at other nodes, similar transformations, Eqs. (4.49)~(4.53), can be employed.

4.8.3 Strongly singular integrals

For the integral having the kernel function $t_{ij}^{Y}(P,Q)$, a strongly singular integral arises only if the collocation point P coincides with one of the element nodes. Let us rewrite Eq. (4.21) as

$$c_{ij}(P)u_j(P) + \sum_{\substack{e=1\\ P\in S_e \text{ and } P=Q^\alpha}}^{ne} u_j^\alpha(Q^\alpha)H_{ij}^{e\alpha}(P,Q) + \sum_{e=1}^{ne}\sum_{\substack{\alpha=1\\ P\neq Q^\alpha}}^{m} u_j^\alpha(Q^\alpha)H_{ij}^{e\alpha}(P,Q)$$
$$= \sum_{e=1}^{ne}\sum_{\alpha=1}^{m} t_j^\alpha(Q^\alpha)G_{ij}^{e\alpha}(P,Q) \tag{4.54}$$

where Q^α stands for node α on the integrating element. Thus, in the second term on the left-hand side of Eq. (4.54), all the terms involving a strongly singular integration are obtained.

To generate a rigid body translation for a three-dimensional finite domain, we consider three sets of unit rigid-body displacements for all the nodes, i.e., $(u_x,u_y,u_z)=(1,0,0)$ in the x direction, $(u_x,u_y,u_z)=(0,1,0)$ in the y direction, and $(u_x,u_y,u_z)=(0,0,1)$ in the z direction. All tractions are set to zero, i.e., $(t_x,t_y,t_z)=(0,0,0)$. Considering the rigid body movement in the x direction, we have from Eq. (4.54)

$$\begin{bmatrix} c_{xx} & c_{xy} & c_{xz} \\ c_{yx} & c_{yy} & c_{yz} \\ c_{zx} & c_{zy} & c_{zz} \end{bmatrix}\begin{Bmatrix}1\\0\\0\end{Bmatrix} + \sum_{\substack{e=1\\ P\in S_e \text{ and } P=Q^\alpha}}^{ne} \begin{bmatrix} H_{xx}^{e\alpha} & H_{xy}^{e\alpha} & H_{xz}^{e\alpha} \\ H_{yx}^{e\alpha} & H_{yy}^{e\alpha} & H_{yz}^{e\alpha} \\ H_{zx}^{e\alpha} & H_{zy}^{e\alpha} & H_{zz}^{e\alpha} \end{bmatrix}\begin{Bmatrix}1\\0\\0\end{Bmatrix} +$$
$$\sum_{e=1}^{ne}\sum_{\substack{\alpha=1\\ P\neq Q^\alpha}}^{m} \begin{bmatrix} H_{xx}^{e\alpha} & H_{xy}^{e\alpha} & H_{xz}^{e\alpha} \\ H_{yx}^{e\alpha} & H_{yy}^{e\alpha} & H_{yz}^{e\alpha} \\ H_{zx}^{e\alpha} & H_{zy}^{e\alpha} & H_{zz}^{e\alpha} \end{bmatrix}\begin{Bmatrix}1\\0\\0\end{Bmatrix}$$
$$= \sum_{e=1}^{ne}\sum_{\alpha=1}^{m} \begin{bmatrix} G_{xx}^{e\alpha} & G_{xy}^{e\alpha} & G_{xz}^{e\alpha} \\ G_{yx}^{e\alpha} & G_{yy}^{e\alpha} & G_{yz}^{e\alpha} \\ G_{zx}^{e\alpha} & G_{zy}^{e\alpha} & G_{zz}^{e\alpha} \end{bmatrix}\begin{Bmatrix}0\\0\\0\end{Bmatrix} \tag{4.55}$$

Rearranging Eq. (4.55), we have

$$\begin{Bmatrix}c_{xx}\\c_{yx}\\c_{zx}\end{Bmatrix} + \sum_{\substack{e=1\\ P\in S_e \text{ and } P=Q^\alpha}}^{ne} \begin{Bmatrix}H_{xx}^{e\alpha}\\H_{yx}^{e\alpha}\\H_{zx}^{e\alpha}\end{Bmatrix} + \sum_{e=1}^{ne}\sum_{\substack{\alpha=1\\ P\neq Q^\alpha}}^{m} \begin{Bmatrix}H_{xx}^{e\alpha}\\H_{yx}^{e\alpha}\\H_{zx}^{e\alpha}\end{Bmatrix} = \begin{Bmatrix}0\\0\\0\end{Bmatrix} \tag{4.55a}$$

Considering the rigid body movements in the y and z directions, from Eq. (4.54) we have

$$\begin{Bmatrix} c_{xy} \\ c_{yy} \\ c_{zy} \end{Bmatrix} + \sum_{\substack{e=1 \\ P\in S_e \text{ and } P=Q^\alpha}}^{ne} \begin{Bmatrix} H_{xy}^{e\alpha} \\ H_{yy}^{e\alpha} \\ H_{zy}^{e\alpha} \end{Bmatrix} + \sum_{e=1}^{ne} \sum_{\substack{\alpha=1 \\ P\neq Q^\alpha}}^{m} \begin{Bmatrix} H_{xy}^{e\alpha} \\ H_{yy}^{e\alpha} \\ H_{zy}^{e\alpha} \end{Bmatrix} = \begin{Bmatrix} 0 \\ 0 \\ 0 \end{Bmatrix} \tag{4.55b}$$

$$\begin{Bmatrix} c_{xz} \\ c_{yz} \\ c_{zz} \end{Bmatrix} + \sum_{\substack{e=1 \\ P\in S_e \text{ and } P=Q^\alpha}}^{ne} \begin{Bmatrix} H_{xz}^{e\alpha} \\ H_{yz}^{e\alpha} \\ H_{zz}^{e\alpha} \end{Bmatrix} + \sum_{e=1}^{ne} \sum_{\substack{\alpha=1 \\ P\neq Q^\alpha}}^{m} \begin{Bmatrix} H_{xz}^{e\alpha} \\ H_{yz}^{e\alpha} \\ H_{zz}^{e\alpha} \end{Bmatrix} = \begin{Bmatrix} 0 \\ 0 \\ 0 \end{Bmatrix} \tag{4.55c}$$

Putting all the equations of (4.55) together gives

$$\begin{bmatrix} c_{xx} & c_{xy} & c_{xz} \\ c_{yx} & c_{yy} & c_{yz} \\ c_{zx} & c_{zy} & c_{zz} \end{bmatrix} + \sum_{\substack{e=1 \\ P\in S_e \text{ and } P=Q^\alpha}}^{ne} \begin{bmatrix} H_{xx}^{e\alpha} & H_{xy}^{e\alpha} & H_{xz}^{e\alpha} \\ H_{yx}^{e\alpha} & H_{yy}^{e\alpha} & H_{yz}^{e\alpha} \\ H_{zx}^{e\alpha} & H_{zy}^{e\alpha} & H_{zz}^{e\alpha} \end{bmatrix} = -\sum_{e=1}^{ne} \sum_{\substack{\alpha=1 \\ P\neq Q^\alpha}}^{m} \begin{bmatrix} H_{xx}^{e\alpha} & H_{xy}^{e\alpha} & H_{xz}^{e\alpha} \\ H_{yx}^{e\alpha} & H_{yy}^{e\alpha} & H_{yz}^{e\alpha} \\ H_{zx}^{e\alpha} & H_{zy}^{e\alpha} & H_{zz}^{e\alpha} \end{bmatrix} \tag{4.56}$$

From Eq. (4.56), we have

$$c_{ij}(P) + \sum_{\substack{e=1 \\ P\in S_e \text{ and } P=Q^\alpha}}^{ne} H_{ij}^{e\alpha} = -\sum_{e=1}^{ne} \sum_{\substack{\alpha=1 \\ P\neq Q^\alpha}}^{m} H_{ij}^{e\alpha} \tag{4.57}$$

Equations (4.56) and (4.57) indicate that the strongly singular terms and the free term $c_{ij}(P)$ can be determined by simply summing up all coefficients of one equation except the terms to be evaluated and changing the sign of the summation. The advantages of the indirect method are that we avoid a direct calculation of the strongly singular integration and get the free term at no additional expense.

For an infinite domain, we cannot apply a rigid body translation, so we need to consider a three-dimensional domain bounded by a complementary surface S_∞. The boundary S and the auxiliary surface S_∞ form the new finite domain. Employing the rigid body translation to the finite domain, we obtain

$$c_{ij}(P) + \int_S t_{ij}^Y(P,Q)\mathrm{d}S(Q) + \int_{S_\infty} t_{ij}^Y(P,Q)\mathrm{d}S(Q) = 0 \tag{4.58}$$

where S_∞ is a sphere of radius R approaching infinity. The second integral on the left-hand side of Eq. (4.58) can be integrated analytically, and has

$$\int_{S_\infty} t_{ij}^Y(P,Q)\mathrm{d}S(Q) = -\delta_{ij} \tag{4.59}$$

That is, for an infinite domain, δ_{ij} should be added to the right-hand side of Eq. (4.57).

4.9 Evaluation of displacements and stresses at an internal point

As shown in Eq. (4.11), after obtaining the displacements and tractions on the boundary, the displacements at any internal point can be determined by integration. In this case, the singular integrals of the boundary do not exist and the traditional Gaussian quadrature can be used for calculating the displacements at any internal point.

From the integral equations (4.11), the strain-displacement equations (2.2) and Hooke's law (2.3), we can get the formulae for calculating the stresses at any internal point. For an elastic body without body forces, substituting Eq. (4.11b) into Eq. (2.2), we obtain the strains at a source point

$$\varepsilon_{ij}(p) = \int_S U^{\varepsilon}_{ijk}(p,Q)t_k(Q)\mathrm{d}S(Q) - \int_S T^{\varepsilon}_{ijk}(p,Q)u_k(Q)\mathrm{d}S(Q), \quad (i,j,k=x,y,z) \tag{4.60}$$

where

$$U^{\varepsilon}_{ijk} = (u^Y_{ik,j} + u^Y_{jk,i})/2 \tag{4.61a}$$

$$T^{\varepsilon}_{ijk} = (t^Y_{ik,j} + t^Y_{jk,i})/2 \tag{4.61b}$$

Notice that the derivatives are given with respect to the coordinates at the source point p and not at the field point Q. For calculating the stresses at any internal point, substituting Eq. (4.60) in Eq. (2.3) yields the following equation

$$\sigma_{ij}(p) = \int_S U^Y_{ijk}(p,Q)t_k(Q)\mathrm{d}S(Q) - \int_S T^Y_{ijk}(p,Q)u_k(Q)\mathrm{d}S(Q), \quad (i,j,k=x,y,z) \tag{4.62}$$

Following through these steps, we have the new kernel functions U^Y_{ijk} and T^Y_{ijk} as follows

$$U^Y_{ijk} = \lambda\delta_{ij}U^{\varepsilon}_{mmk} + 2\mu U^{\varepsilon}_{ijk} \tag{4.63a}$$

$$T^Y_{ijk} = \lambda\delta_{ij}T^{\varepsilon}_{mmk} + 2\mu T^{\varepsilon}_{ijk} \tag{4.63b}$$

The derivatives of the displacements in Eq. (4.61a) are approximated as

$$\frac{\partial u^Y_{ik}}{\partial x} \approx \frac{1}{2D}[u^Y_{ik}(x+D,y,z) - u^Y_{ik}(x-D,y,z)] \tag{4.64a}$$

$$\frac{\partial u^Y_{ik}}{\partial y} \approx \frac{1}{2D}[u^Y_{ik}(x,y+D,z) - u^Y_{ik}(x,y-D,z)] \tag{4.64b}$$

$$\frac{\partial u^Y_{ik}}{\partial z} \approx \frac{1}{2D}[u^Y_{ik}(x,y,z+D) - u^Y_{ik}(x,y,z-D)] \tag{4.64c}$$

where x,y,z are the coordinates of the source point p in the global coordinate system and D is the distance between the two source points. The derivatives of the tractions in Eq. (4.61b) can also be calculated using the above methods.

In order to calculate the derivatives $u^Y_{ik,j}$ and $t^Y_{ik,j}$, it is necessary to compute the displacements and stresses of Yue's solution at six points in the neighborhood of p. The choice of the interval D is a crucial decision and an extensive numerical investigation is executed in Chapter 8.

4.10 Evaluation of boundary stresses

The kernel function T^Y_{ijk} in Eq. (4.62) is of order r^{-3}. Thus, the stresses on the boundary can not be calculated directly using the boundary integral equation (4.62). The most popular technique for overcoming this difficulty is the traction recovery method (Brebbia, 1978; Gao et al., 2002) and we employ this to evaluate the boundary stresses. This scheme uses the displacements and tractions obtained by solving Eq. (4.38) at the boundary nodes.

The stresses at any point (ξ_1, ξ_2) of element e can be determined by

$$\sigma_{ij} = \lambda \delta_{ij} u_{k,k} + \mu(u_{i,j} + u_{j,i}), \quad (i, j, k = x, y, z) \tag{4.65}$$

Using the boundary conditions, we have

$$\lambda n_i u_{k,k} + \mu(u_{i,j} + u_{j,i}) n_j = t_i \tag{4.66}$$

The derivatives of the displacements $u_i(\xi_1, \xi_2)$ with respect to the local coordinates are obtained from the equations

$$\frac{\partial u_i}{\partial \xi_\beta} = \frac{\partial u_i}{\partial x_j}\frac{\partial x_j}{\partial \xi_\beta} = u_{i,j}\frac{\partial x_j}{\partial \xi_\beta}, \quad (\beta = 1, 2) \tag{4.67}$$

In Eqs. (4.66) and (4.67), t_i, $\dfrac{\partial u_i}{\partial \xi_\beta}$, $\dfrac{\partial x_j}{\partial \xi_\beta}$, and n_i can be obtained directly, leaving only the nine unknown quantities $u_{i,j}(i, j = x, y, z)$. Combining Eqs. (4.66) and (4.67), we obtain a set of nine linear equations containing the unknowns. The Gaussian elimination method is used to solve the linear equations and substitution of $u_{i,j}$ in Eq. (4.65) yields six components of stress at any point on the boundary. If stress fields are continuous and a node is shared by several elements, the results from each of the elements should be averaged.

4.11 Multi-region method

The multi-region method plays an important role in traditional boundary element analysis of problems in which the material properties vary in some piecewise fashion. Yue's solution-based BEM can easily be used to analyze multilayered elastic media. However, a straightforward application of the BEM to crack problems in layered materials still results in a mathematical degeneration if the two crack surfaces are considered co-planar. The first widely applicable method for dealing with co-planar crack surfaces was devised by Blandford et al. (1981). This approach, which is based on a multi-region formulation, is general and can be applied to both symmetrical and anti-symmetrical crack problems in both 2D and 3D configurations. The multi-region method has to introduce artificial boundaries into the body; these connect the cracks to the boundary in such a way that each region contains a crack surface. The two regions are then joined together such that equilibrium of tractions and compatibility of displacements are enforced.

As shown in Fig. 4.9, the elastic body is divided into two subregions. For each subregion, the linear equation (4.37) can be given in a portioned form. In Eq. (4.37), we introduce the superscripts I and II for subregions I and II, respectively, and the subscript i for the interface between the two subregions.

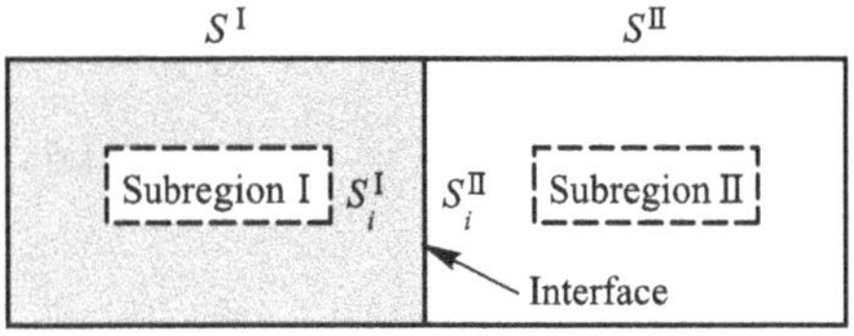

Fig. 4.9 Two subregions of a solid for the BEM

For subregion I, we have

$$\begin{bmatrix} \boldsymbol{A}_{11}^{\mathrm{I}} & \boldsymbol{A}_{12}^{\mathrm{I}} & \boldsymbol{A}_{13}^{\mathrm{I}} \\ \boldsymbol{A}_{21}^{\mathrm{I}} & \boldsymbol{A}_{22}^{\mathrm{I}} & \boldsymbol{A}_{23}^{\mathrm{I}} \\ \boldsymbol{A}_{31}^{\mathrm{I}} & \boldsymbol{A}_{32}^{\mathrm{I}} & \boldsymbol{A}_{33}^{\mathrm{I}} \end{bmatrix} \begin{Bmatrix} \boldsymbol{U}^{\mathrm{I}} \\ \boldsymbol{T}^{\mathrm{I}} \\ \boldsymbol{T}_i^{\mathrm{I}} \end{Bmatrix} = \begin{bmatrix} \boldsymbol{B}_{11}^{\mathrm{I}} & \boldsymbol{B}_{12}^{\mathrm{I}} & \boldsymbol{B}_{13}^{\mathrm{I}} \\ \boldsymbol{B}_{21}^{\mathrm{I}} & \boldsymbol{B}_{22}^{\mathrm{I}} & \boldsymbol{B}_{23}^{\mathrm{I}} \\ \boldsymbol{B}_{31}^{\mathrm{I}} & \boldsymbol{B}_{32}^{\mathrm{I}} & \boldsymbol{B}_{33}^{\mathrm{I}} \end{bmatrix} \begin{Bmatrix} \overline{\boldsymbol{T}}^{\mathrm{I}} \\ \overline{\boldsymbol{U}}^{\mathrm{I}} \\ \boldsymbol{U}_i^{\mathrm{I}} \end{Bmatrix} \tag{4.68}$$

For subregion II, we have

$$\begin{bmatrix} \boldsymbol{A}_{11}^{\mathrm{II}} & \boldsymbol{A}_{12}^{\mathrm{II}} & \boldsymbol{A}_{13}^{\mathrm{II}} \\ \boldsymbol{A}_{21}^{\mathrm{II}} & \boldsymbol{A}_{22}^{\mathrm{II}} & \boldsymbol{A}_{23}^{\mathrm{II}} \\ \boldsymbol{A}_{31}^{\mathrm{II}} & \boldsymbol{A}_{32}^{\mathrm{II}} & \boldsymbol{A}_{33}^{\mathrm{II}} \end{bmatrix} \begin{Bmatrix} \boldsymbol{U}_i^{\mathrm{II}} \\ \boldsymbol{U}^{\mathrm{II}} \\ \boldsymbol{T}^{\mathrm{II}} \end{Bmatrix} = \begin{bmatrix} \boldsymbol{B}_{11}^{\mathrm{II}} & \boldsymbol{B}_{12}^{\mathrm{II}} & \boldsymbol{B}_{13}^{\mathrm{II}} \\ \boldsymbol{B}_{21}^{\mathrm{II}} & \boldsymbol{B}_{22}^{\mathrm{II}} & \boldsymbol{B}_{23}^{\mathrm{II}} \\ \boldsymbol{B}_{31}^{\mathrm{II}} & \boldsymbol{B}_{32}^{\mathrm{II}} & \boldsymbol{B}_{33}^{\mathrm{II}} \end{bmatrix} \begin{Bmatrix} \boldsymbol{T}_i^{\mathrm{II}} \\ \overline{\boldsymbol{T}}^{\mathrm{II}} \\ \overline{\boldsymbol{U}}^{\mathrm{II}} \end{Bmatrix} \tag{4.69}$$

If the node number on the interface of subregion I corresponds to the one on the interface of subregion II, the field quantities on the interface of the two subregions have the following relationships

$$\boldsymbol{T}_i^{\mathrm{I}} = -\boldsymbol{T}_i^{\mathrm{II}}, \quad \boldsymbol{U}_i^{\mathrm{I}} = \boldsymbol{U}_i^{\mathrm{II}} \tag{4.70}$$

By considering the relationships (4.70), Eqs. (4.68) and (4.69) can be combined into the following linear equation

$$\begin{bmatrix} A^{\mathrm{I}}_{11} & A^{\mathrm{I}}_{12} & A^{\mathrm{I}}_{13} & -B^{\mathrm{I}}_{13} & 0 & 0 \\ A^{\mathrm{I}}_{21} & A^{\mathrm{I}}_{22} & A^{\mathrm{I}}_{23} & -B^{\mathrm{I}}_{23} & 0 & 0 \\ A^{\mathrm{I}}_{31} & A^{\mathrm{I}}_{32} & A^{\mathrm{I}}_{33} & -B^{\mathrm{I}}_{33} & 0 & 0 \\ 0 & 0 & B^{\mathrm{II}}_{11} & A^{\mathrm{II}}_{11} & A^{\mathrm{II}}_{12} & A^{\mathrm{II}}_{13} \\ 0 & 0 & B^{\mathrm{II}}_{21} & A^{\mathrm{II}}_{21} & A^{\mathrm{II}}_{22} & A^{\mathrm{II}}_{23} \\ 0 & 0 & B^{\mathrm{II}}_{31} & A^{\mathrm{II}}_{31} & A^{\mathrm{II}}_{32} & A^{\mathrm{II}}_{33} \end{bmatrix} \begin{Bmatrix} U^{\mathrm{I}} \\ T^{\mathrm{I}} \\ T^{\mathrm{I}}_i \\ U^{\mathrm{I}}_i \\ U^{\mathrm{II}} \\ T^{\mathrm{II}} \end{Bmatrix} = \begin{Bmatrix} B^{\mathrm{I}}_{11}\overline{T}^{\mathrm{I}} + B^{\mathrm{I}}_{12}\overline{U}^{\mathrm{I}} \\ B^{\mathrm{I}}_{21}\overline{T}^{\mathrm{I}} + B^{\mathrm{I}}_{22}\overline{U}^{\mathrm{I}} \\ B^{\mathrm{I}}_{31}\overline{T}^{\mathrm{I}} + B^{\mathrm{I}}_{32}\overline{U}^{\mathrm{I}} \\ B^{\mathrm{II}}_{12}\overline{T}^{\mathrm{II}} + B^{\mathrm{II}}_{13}\overline{U}^{\mathrm{II}} \\ B^{\mathrm{II}}_{22}\overline{T}^{\mathrm{II}} + B^{\mathrm{II}}_{23}\overline{U}^{\mathrm{II}} \\ B^{\mathrm{II}}_{32}\overline{T}^{\mathrm{II}} + B^{\mathrm{II}}_{33}\overline{U}^{\mathrm{II}} \end{Bmatrix} \tag{4.71}$$

After substituting the known boundary conditions in Eq. (4.71), the resulting linear equation, after some rearrangements, can be written in the form

$$\boldsymbol{AX} = \boldsymbol{F} \tag{4.72}$$

where $\boldsymbol{X}$ is an unknown array, $\boldsymbol{F}$ is a known array, and $\boldsymbol{A}$ is a square matrix of coefficients.

4.12 Symmetry

In many cases, the symmetry of a problem can be taken into account to considerably reduce the amount of the analysis effort required. We assume that both the geometry and the boundary conditions for the elastic body shown in Fig. 4.10, are symmetric about the two coordinate planes *Oyz* and *Oxz*.

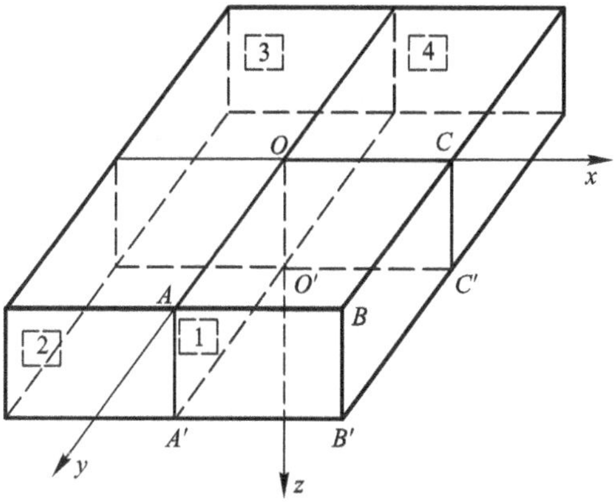

Fig. 4.10 A structure with the symmetric geometry and the symmetric boundary conditions

In the FEM, the conditions for the elastic body shown in Fig. 4.10 are simply implemented by generating only part of the mesh and providing the appropriate boundary conditions at the planes of symmetry. In the BEM, we can apply a completely different approach for the elastic body shown in Fig. 4.10. The elastic body is divided into the four regions, $R = 1,2,3,4$ and only a quarter of the elastic body, e.g., region $R = 1$, is considered. For region $R = 1$, only the boundaries exclusive of the symmetry planes need to be discretized and the influence of the other three regions ($R = 2$, 3 and 4) on region $R = 1$ is considered by using the mirror image method.

In the following, we will illustrate this approach. As shown in Fig. 4.11, we denote a four-node element on region $R = 1$ and automatically generate three mirrored elements on the boundaries of other three regions. The symbols 1^1, 1^2 and 1^3 denote the mirrored elements on which the coordinates, displacements and tractions of nodes have the relationships listed in Tables 4.1 and 4.2. Assume that region 1 has *ne* elements; all the elements on region 1 are projected onto the other three regions. Thus, there are *ne* mirrored elements on each of the other three regions. Relationships similar to the ones shown in Tables 4.1 and 4.2 can be established between all the mirrored and original elements.

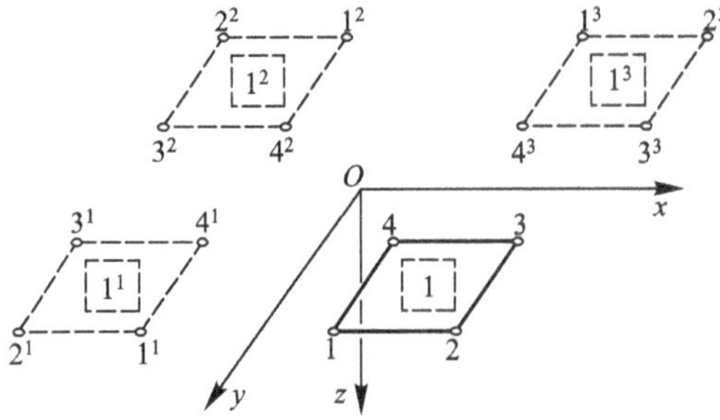

Fig. 4.11 Mirrored elements of an original element in a symmetric structure

Table 4.1 Relationships of coordinates at nodes for original and mirrored elements

Original element 1		Mirrored element 1^1		Mirrored element 1^2		Mirrored element 1^3	
Node 1	(x^1, y^1, z^1)	Node 1^1	$(-x^1, y^1, z^1)$	Node 1^2	$(-x^1, -y^1, z^1)$	Node 1^3	$(x^1, -y^1, z^1)$
Node 2	(x^2, y^2, z^2)	Node 2^1	$(-x^2, y^2, z^2)$	Node 2^2	$(-x^2, -y^2, z^2)$	Node 2^3	$(x^2, -y^2, z^2)$
Node 3	(x^3, y^3, z^3)	Node 3^1	$(-x^3, y^3, z^3)$	Node 3^2	$(-x^3, -y^3, z^3)$	Node 3^3	$(x^3, -y^3, z^3)$
Node 4	(x^4, y^4, z^4)	Node 4^1	$(-x^4, y^4, z^4)$	Node 4^2	$(-x^4, -y^4, z^4)$	Node 4^3	$(x^4, -y^4, z^4)$

For node P on region 1, we have, from Eq. (4.21),

$$c_{ij}(P)u_j(P) + \sum_{R=1}^{4}\sum_{e=1}^{ne}\sum_{\alpha=1}^{m}(u_j^\alpha H_{ij}^{e\alpha})_R = \sum_{R=1}^{4}\sum_{e=1}^{ne}\sum_{\alpha=1}^{m}(t_j^\alpha G_{ij}^{e\alpha})_R \tag{4.73}$$

where R denotes the four regions.

According to the relationships, shown in Tables 4.1 and 4.2, between the quantities of symmetric elements, the quantities of nodes on regions $R = 2,3,4$ are replaced by the ones on region $R = 1$. From Eq. (4.73) we obtain

Table 4.2 Relationships of displacements and tractions at nodes for original and mirrored elements

Original element 1		Mirrored element 1^1		Mirrored element 1^2		Mirrored element 1^3	
Node 1	(u_x^1, u_y^1, u_z^1) (t_x^1, t_y^1, t_z^1)	Node 1^1	$(-u_x^1, u_y^1, u_z^1)$ $(-t_x^1, t_y^1, t_z^1)$	Node 1^2	$(-u_x^1, -u_y^1, u_z^1)$ $(-t_x^1, -t_y^1, t_z^1)$	Node 1^3	$(u_x^1, -u_y^1, u_z^1)$ $(t_x^1, -t_y^1, t_z^1)$
Node 2	(u_x^2, u_y^2, u_z^2) (t_x^2, t_y^2, t_z^2)	Node 2^1	$(-u_x^2, u_y^2, u_z^2)$ $(-t_x^2, t_y^2, t_z^2)$	Node 2^2	$(-u_x^2, -u_y^2, u_z^2)$ $(-t_x^2, -t_y^2, t_z^2)$	Node 2^3	$(u_x^2, -u_y^2, u_z^2)$ $(t_x^2, -t_y^2, t_z^2)$
Node 3	(u_x^3, u_y^3, u_z^3) (t_x^3, t_y^3, t_z^3)	Node 3^1	$(-u_x^3, u_y^3, u_z^3)$ $(-t_x^3, t_y^3, t_z^3)$	Node 3^2	$(-u_x^3, -u_y^3, u_z^3)$ $(-t_x^3, -t_y^3, t_z^3)$	Node 3^3	$(u_x^3, -u_y^3, u_z^3)$ $(t_x^3, -t_y^3, t_z^3)$
Node 4	(u_x^4, u_y^4, u_z^4) (t_x^4, t_y^4, t_z^4)	Node 4^1	$(-u_x^4, u_y^4, u_z^4)$ $(-t_x^4, t_y^4, t_z^4)$	Node 4^2	$(-u_x^4, -u_y^4, u_z^4)$ $(-t_x^4, -t_y^4, t_z^4)$	Node 4^3	$(u_x^4, -u_y^4, u_z^4)$ $(t_x^4, -t_y^4, t_z^4)$

$$c_{ij}(P)u_j(P) + \sum_{R=1}^{4}\sum_{e=1}^{ne}\sum_{\alpha=1}^{m}[f_R^x(H_{ix}^{e\alpha})_R(u_x^\alpha)_{R=1} + f_R^y(H_{iy}^{e\alpha})_R(u_y^\alpha)_{R=1} + f_R^z(H_{iz}^{e\alpha})_R(u_z^\alpha)_{R=1}]$$
$$= \sum_{R=1}^{4}\sum_{e=1}^{ne}\sum_{\alpha=1}^{m}[f_R^x(G_{ix}^{e\alpha})_R(t_x^\alpha)_{R=1} + f_R^y(G_{iy}^{e\alpha})_R(t_y^\alpha)_{R=1} + f_R^z(G_{iz}^{e\alpha})_R(t_z^\alpha)_{R=1}] \quad (4.74)$$

where $f_R^d\,(d = x, y, z)$ has a positive or negative sign when the displacements and tractions on regions $R = 2, 3, 4$ are replaced by the ones of region $R = 1$. $(u_i^\alpha)_{R=1}$ and $(t_i^\alpha)_{R=1}$ are the displacements and tractions of nodes on region $R = 1$. From Table 4.3, we can extract the f_R^d values.

Table 4.3 Values of f_R^d

R	$d = x$	$d = y$	$d = z$
1	1	1	1
2	–1	1	1
3	–1	–1	1
4	1	–1	1

After executing the above steps, Eq. (4.74) contains only the displacements and tractions of the nodes on region $R = 1$. Thus, we can establish the boundary integral equation (4.74) for each node on region $R = 1$. By using the symmetric conditions and rearranging the coefficients of these equations, we obtain the linear equations that can be solved for the displacements and tractions on region $R = 1$. The displacements and tractions on regions $R = 2, 3, 4$ can be obtained analogically by using the symmetric conditions and the results from region $R = 1$.

4.13 Numerical evaluation and results

Based on the BEM expressions presented in this chapter, computer programs in FORTRAN have been developed to calculate the displacements and stresses in homogeneous and layered plates. Numerical examples are presented in the ensuing to validate the solutions obtained by the BEM and to illustrate the influence of the depth non-homogeneity on the elastic field.

4.13.1 A homogeneous rectangular plate

Consider a homogeneous rectangular plate shown in Fig. 4.12. The edges $x = \pm a/2$ are simply supported while the other two edges $y = \pm b/2$ are free. A load q is uniformly distributed on the top surface $(z/b = 0)$ and the bottom surface is free $(z/b = h)$. The deflection surface of the plate will be symmetrical with respect to the x and y axes. Because of symmetry, only a quarter of the plate needs to be analyzed such that only this quarter is discretized except the plane of symmetry. Herein, $b/a = 0.5$, $t/a = 0.1$ and $a = 2$. Figs. 4.13 and 4.14 show the surface discretization with 741 four-noded and 700 boundary elements (case 1) and with 2271 four-noded and 2220 boundary elements (case 2).

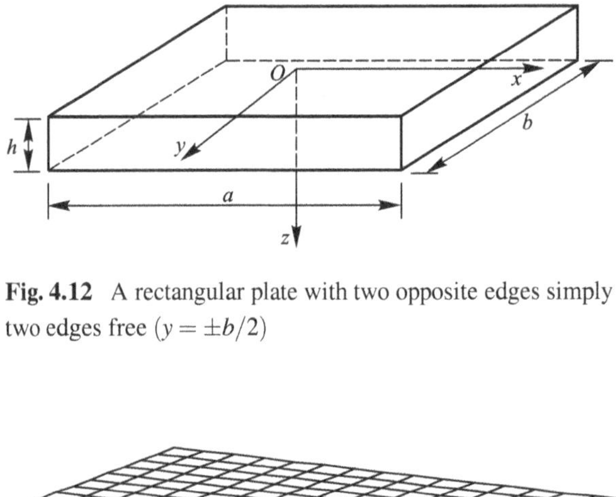

Fig. 4.12 A rectangular plate with two opposite edges simply supported $(x = \pm a/2)$ and the other two edges free $(y = \pm b/2)$

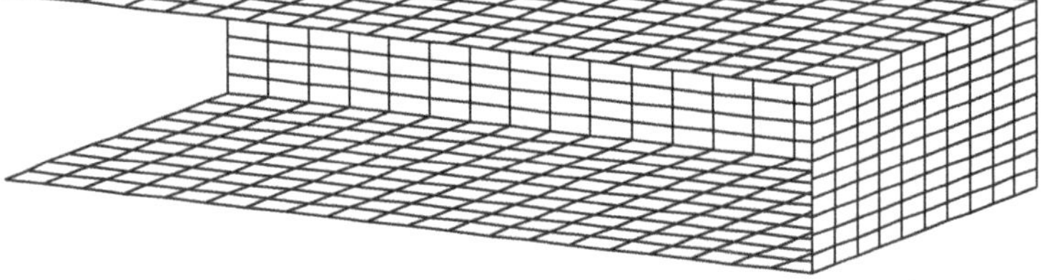

Fig. 4.13 Boundary mesh of the plate (case 1: 741 elements and 700 nodes)

The deflection of the elastic plate is given in the following expression (Timoshenko and Goodier, 1979)

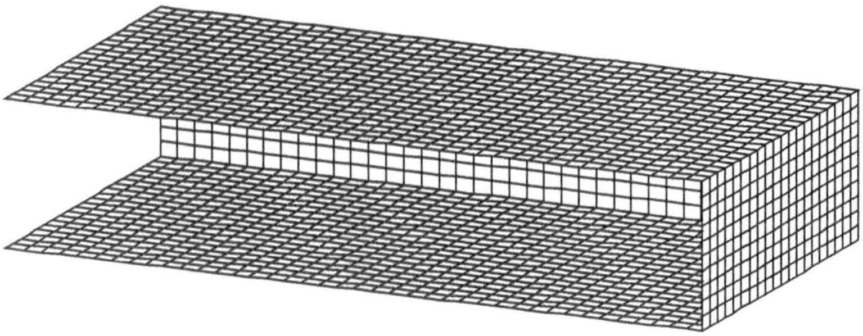

Fig. 4.14 Boundary mesh of the plate (case 2: 2271 elements and 2200 nodes)

$$w = \alpha_1 \frac{qa^4}{D}, \quad D = \frac{Et^3}{12(1-\nu^2)} \tag{4.75}$$

Assume that the Poisson's ratio $\nu = 0.3$. Figure 4.15 presents the coefficient α_1 along the lines $x/b = 0$ and $y/b = 0$ on the top and bottom surfaces $(z/b = 0, 0.2)$ of the plate. It can be seen that there are very small differences in the deflections between the top and bottom surfaces for cases 1 and 2. When compared with the exact solution given in Eq. (4.75), the maximum relative errors of α_1 along the lines $x/b = 0$ and $y/b = 0$ on the top surfaces are 9.57% and 0.46% for cases 1 and 2, respectively and both appear on the surface $z = 0$. At this point, the exact solution of α_1 is 0.013713 whilst the numerical solutions are 0.012401 and 0.013649. Obviously, as the numbers of nodes and elements increase, the numerical solutions tend towards a better agreement with the exact one.

4.13.2 A layered rectangular plate

Consider a layered rectangular plate with the two opposite edges simply supported and the other two edges free, as shown in Fig. 4.16. The plate consists of two dissimilar materials, i.e., material 1: $E_1 = 3.5$ GPa, $\nu_1 = 0.35$ and material 2: $E_2 = 70.0$ GPa, $\nu_2 = 0.22$. The plate is subjected to a uniform load q on the top surface.

The layered rectangular plate has the same dimensions as the ones depicted for the above homogeneous rectangular plate. Because the proposed BEM is used in this analysis, the interface of two dissimilar materials is not discretized. Thus, the boundary mesh shown in Fig. 4.14 is used to analyze the layered rectangular plate. The normalized deflection $\overline{w} = wE_0/(qb)$ $(E_0 = 0.1$ GPa) is defined. Fig. 4.17 presents the normalized deflection $\overline{w}$ along the lines $x = 0$ and $y = 0$ on the top and bottom surfaces $(z = 0)$ of the plate using the proposed BEM and a FEM. In using the FEM, the domain $(0 \leqslant x/b \leqslant 1$ and $0 \leqslant y/b \leqslant 0.5)$ is discretized with 8000 20-noded solid elements. The relative errors of the normalized deflections between the two methods are less than 5%. From the above analyses, it can be concluded that the results obtained with the present formulation are in very good agreement with the FEM results.

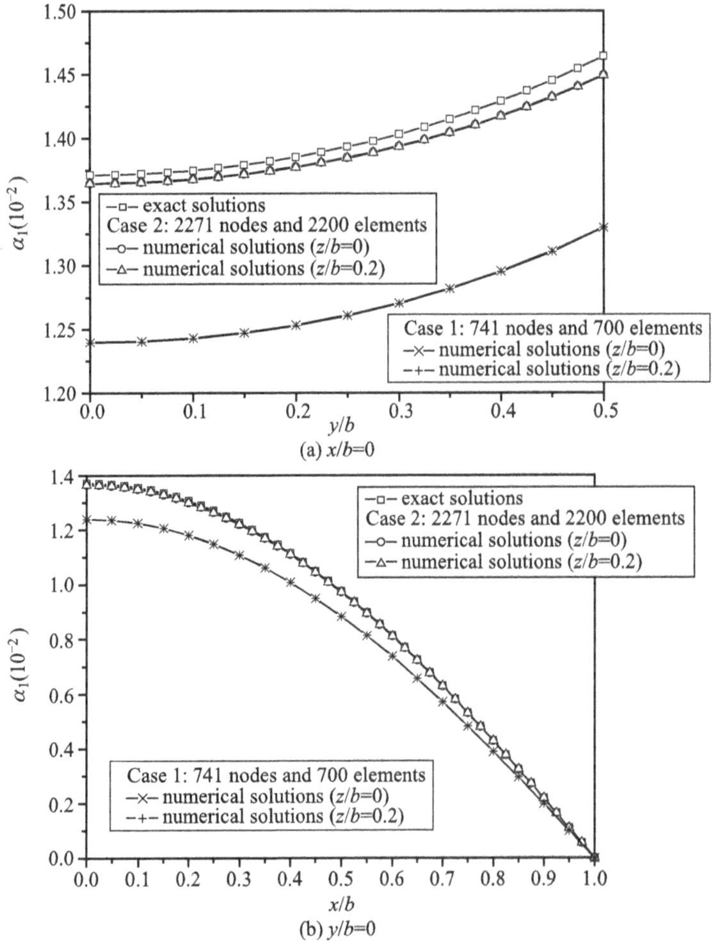

Fig. 4.15 Comparison of exact and numerical solutions for a homogeneous plate

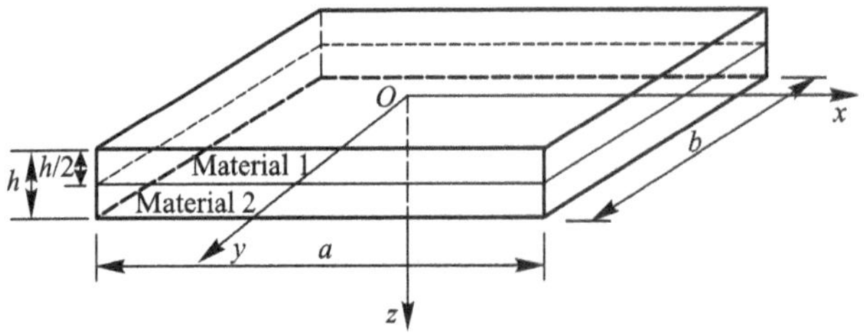

Fig. 4.16 A layered rectangular plate with two isotropic layers

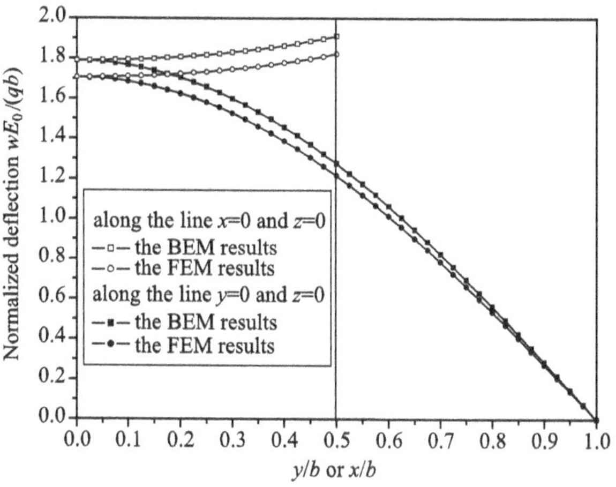

Fig. 4.17 Comparison of different numerical solutions for a two-layered plate

4.14 Summary

In this chapter, we have discussed Yue's solution-based BEM in some detail. Because Yue's solution for a layered material has the same singularities of displacements and stresses as the ones for Kelvin's solution in a homogeneous material, the traditional numerical method can be used to perform the integration of kernel-shape function products over the boundary elements. According to the ideas mentioned above, we have now developed a computer program in FORTRAN. The numerical examples of homogeneous and layered rectangular plates have been used to verify the proposed method. Numerical results presented in this chapter illustrate that a numerical evaluation of very high accuracy and efficiency can be easily achieved for layered elastic bodies. In the next chapter, we will introduce numerical methods to model cracks and incorporate them into the scheme developed in this chapter.

References

Blandford GE, Ingraffea AR, Liggett JA. Two-dimensional stress intensity factor computations using the boundary element method. *International Journal for Numerical Methods in Engineering*, 1981, 17(3): 387-404.

Beer G, Smith I, Duenser C. The Boundary Element Method with Programming: for Engineering and Scientists. New York: Springer, 2008.

Benitez FG, Lu L, Rosakis AJ. A boundary element formulation based on the three-dimensional elastostatic fundamental solution for the infinite layer: part 1 - theoretical and numerical development. *International Journal for Numerical Methods in Engineering*, 1993, 36: 3097-3130.

Benitez FG, Wideberg J. The boundary element method based on the three-dimensional elastostatic fundamental solution for the orthotropic multilayered space: application to composite materials. *Computational Mechanics*, 1996, 18: 24-45.

Brebbia CA. The Boundary Element Method for Engineers. London: Pentech Press, 1978.

Cruse TA. Numerical solutions in three dimensional elastostatics. *International Journal of Solids and Structures*, 1969, 5: 1259-1274.

Eberwien U, Duenser C, Moser W. Efficient calculation of internal results in 2D elasticity BEM. *Engineering Analysis with Boundary Elements*, 2005, 29: 447-453.

Gao XW, Davies TG. Boundary Element Programming in Mechanics. Cambridge: Cambridge University Press, 2002.

Kinoshita N, Mura T. On boundary value problem of elasticity. Research Report of the Faculty of Engineering, Meiji University, No.8, 56-82, 1956.

Lachat JC, Watson JO. Effective treatment of boundary integral equations: A formulation for three-dimensional elastostatics. *International Journal for Numerical Methods in Engineering*, 1976, 10: 991-1005.

Lou ZW, Zhang M. Elasto-plastic boundary element analysis with Hetenyi's fundamental solution. *Engineering Analysis with Boundary Elements*, 1992, 10: 231-239.

Rizzo FJ. An integral equation approach to boundary value problems of classical elastostatics. *The Quarterly Journal of Mathematics*, 1967, 25: 83-95.

Timoshenko SP, Goodier JN. Theory of Elasticity. 3rd ed. New York: McGraw-Hill, 1979.

Pan E, Yang B, Cai G, Yuan FG. Stress analyses around holes in composite laminates using boundary element method. *Engineering Analysis with Boundary Elements*, 2001, 25: 31-40.

Yang B, Pan E, Yuan FG. Three-dimensional stress analyses in composite laminates with an elastically pinned hole. *International Journal of Solids and Structures*, 2003, 40: 2017-2035.

Yue ZQ. On generalized Kelvin solutions in a multilayered elastic medium. *Journal of Elasticity*, 1995, 40: 1-43.

Chapter 5

Application of the Yue's Solution-based BEM to Crack Problems

5.1 Introduction

Aliabadi (1997) presented a complete review of the BEM applied to the solution of crack problems and concluded that fracture mechanics has been the most active specialized area of research in the BEM. This is because BEM can reduce the dimensionality of the problem. For 2D analyses, only the line-boundary of the domain needs to be discretized into elements and for 3D problems only the surface of the domain needs to be discretized. This means that high stress gradients can be modeled accurately and efficiently. The traditional Kelvin solution-based BEM has been widely employed to analyze the fracture mechanics of isotropic and homogeneous materials; many researchers incorporated other types of fundamental solutions into boundary element formulations and analyzed crack problems in different materials.

Yuuki and Cho (1989) employed Hetenyi's fundamental solution to develop the BEM and determined the stress intensity factors of an interface crack in isotropic bi-materials. Using the fundamental solutions of transversely isotropic and anisotropic solids, Pan and Yuan (2000) developed the dual BEM for analysis of the three-dimensional cracks. Ariza and Dominguez (2004) incorporated the fundamental solution for a transversely isotropic solid into the dual boundary element formulation for 3D transversely isotropic cracked bodies. Yue et al. (2005) and Xiao et al. (2005) developed the BEM by using the fundamental solution for transversely isotropic bi-materials and analyzed penny-shaped and elliptical-shaped cracks. Yue et al. (2007) further employed the same fundamental solution as that used in Yue et al. (2005) to develop the dual BEM and analyzed the rectangular crack in a transversely isotropic bi-material of infinite and semi-infinite extents.

In this chapter, we further develop the Yue's solution-based BEM to analyze crack problems in layered and graded materials. We will first introduce the traction-singular element and present the corresponding numerical methods. Two examples will be employed to verify the accuracy of the proposed BEM in analyzing crack problems in layered materials.

5.2 Traction-singular element and its numerical method

5.2.1 General

When considering a crack in an infinite linear elastic domain, Irwin (1957) and Williams (1957) showed that the displacement and stress fields in the crack region can be described by an infinite series expansion according to

$$u_i = \sqrt{r}\sum_{n=1}^{\infty} a_n r^{\frac{n-1}{2}} \tag{5.1}$$

and

$$\sigma_{ij} = \frac{1}{\sqrt{r}}\sum_{n=1}^{\infty} b_n r^{\frac{n-1}{2}} \tag{5.2}$$

repsectively. The terms a_n and b_n are constants and r represents the distance of the field point from the crack front. Equations (5.1) and (5.2) are strictly valid for very small values of r. Thus, in order to accurately model the stresses and displacements, any numerical approximation must take into account the $r^{1/2}$ displacement and $r^{-1/2}$ stress variations.

Referring back to Fig. 4.3, it can be seen that a standard quadratic element has the same distance between the nodes of the elements. Thus the shape functions employed to interpolate the nodal values of the displacements and tractions can only approximate variations of the type (Bains, 1992)

$$u_i = \sum_{n=1}^{3} a_{in} r^{n-1} \tag{5.3}$$

and

$$t_i = \sum_{n=1}^{3} b_{in} r^{n-1} \tag{5.4}$$

respectively. However, the above expressions do not contain the $r^{1/2}$ displacement and $r^{-1/2}$ traction variations as are given by the series expansion of the elastic fields near the crack front. Therefore, the standard polynomial shape functions cannot adequately approximate the known behavoiur of the displacement and traction fields near the crack front. If these types of elements are employed near the crack front, the mesh needs to be refined to account for the high stress gradients associated with the singularity. To accurately model the elastic fields of crack problems, some special boundary elements, that can cater to the elastic behaviour near the crack front, are introduced and incorporated into the boundary element formulation. Gao et al. (1992) reviewed methods to directly evaluate the stress intensity factors (SIFs) using quadratic quarter-point crack-tip elements and analyzed three general crack problems.

By shifting the midside nodes of an eight-noded element to the quarter-position, we obtain an eight-node quarter-point element, as shown in Fig. 5.1, which incorporates the $r^{1/2}$ variation of all displacement and traction fields in the region of crack front according to

$$u_i = \sqrt{r}\sum_{n=1}^{3} a_{in} r^{\frac{n-2}{2}} \tag{5.5}$$

and

$$t_i = \sqrt{r}\sum_{n=1}^{3} b_{in} r^{\frac{n-2}{2}} \tag{5.6}$$

respectively. In Eqs. (5.5) and (5.6), the proposed element is only capable of correctly modelling the displacement behaviour. Cruse and Wilson (1977) modified the shape functions of an eight-node quarter-point element by multiplying $\sqrt{l/r}$ where l is the element length and is proportional to $(1+\xi_2)^2$. Thus, this type of element can adequately approximate the traction fields near the crack front.

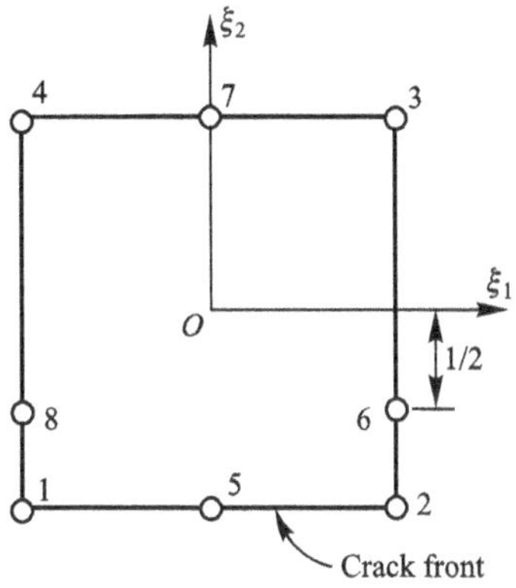

Fig. 5.1 Geometry of a quarter-point element with eight nodes

5.2.2 Traction-singular element

Through the use of proper shape functions, boundary elements have been formulated to correctly model displacement and traction variations near the crack front in three-dimensional geometries (Luchi and Rizzuti, 1987; Jia et al., 1989). Corresponding procedures, suggested for integrating the products of the kernel and shape functions over a special element, were also presented.

Figure 5.2 shows the special element where the side corresponding to $\xi_2 = -1$ lies on the crack front. The coordinates of any point within the element are interpolated by using the shape functions in Eq. (4.17); the displacements and tractions can be obtained by using different interpolation functions for modelling the displacement and traction variations accurately. Thus, the geometry of each element is represented in terms of the nodal values $(x^\alpha, y^\alpha, z^\alpha)$ as

$$x = \sum_{\alpha=1}^{8} N_\alpha(\xi_1, \xi_2) x^\alpha \tag{5.7a}$$

$$y = \sum_{\alpha=1}^{8} N_\alpha(\xi_1, \xi_2) y^\alpha \tag{5.7b}$$

$$z = \sum_{\alpha=1}^{8} N_\alpha(\xi_1, \xi_2) z^\alpha \tag{5.7c}$$

where $N_\alpha(\xi_1, \xi_2)$ is shape functions. The displacements and tractions of any point within an element are represented in terms of the nodal values $(u_x^\alpha, u_y^\alpha, u_z^\alpha)$ and $(t_x^\alpha, t_y^\alpha, t_z^\alpha)$, respectively, as

$$u_i = \sum_{\alpha=1}^{8} N_\alpha^d(\xi_1, \xi_2) u_i^\alpha \tag{5.8}$$

$$t_i = \sum_{\alpha=1}^{8} N_\alpha^t(\xi_1, \xi_2) t_i^\alpha \tag{5.9}$$

where $N_\alpha^d(\xi_1, \xi_2)$ and $N_\alpha^t(\xi_1, \xi_2)$ are defined to represent the displacement and traction variations, respectively.

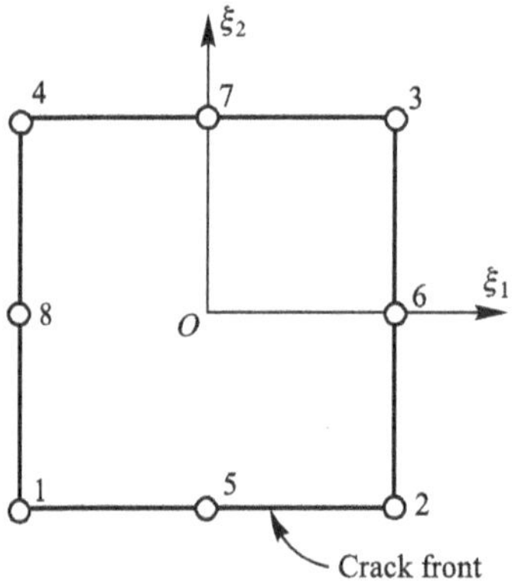

Fig. 5.2 Geometry of a traction-singular element with eight nodes

For displacements, we have

$$N_\alpha^d = \frac{1}{4}(1+\xi_1\xi_1^\alpha)[1-\xi_2^\alpha+\sqrt{2}\sqrt{1+\xi_2}\xi_2^\alpha][\xi_1\xi_1^\alpha + \sqrt{1+\xi_2}(\sqrt{1+\xi_2^\alpha}+\xi_2^\alpha)-\xi_1^\alpha\sqrt{1+\xi_2^\alpha}-\xi_1^\alpha(1+\xi_2^\alpha)], \quad (\alpha=1,2,3,4) \tag{5.10a}$$

$$N_\alpha^d = \frac{1}{2}\xi_1^\alpha\xi_1^\alpha(1+\xi_1\xi_1^\alpha)[(\sqrt{2}+2)\sqrt{1+\xi_2}-(1+\sqrt{2})(1+\xi_2)] + \frac{1}{2}\xi_2^\alpha\xi_2^\alpha[1-\xi_2^\alpha+\sqrt{2}\sqrt{1+\xi_2}\xi_2^\alpha](1-\xi_1\xi_1), \quad (\alpha=5,6,7,8) \tag{5.10b}$$

where $(\xi_1^\alpha, \xi_2^\alpha)$ is the local coordinate of node α.

For tractions, we have

$$N_\alpha^t = \frac{1}{\sqrt{1+\xi_2}} N_\alpha^d, \quad (\alpha = 1,2,5) \tag{5.11a}$$

$$N_\alpha^t = \frac{\sqrt{1+\xi_2^\alpha}}{\sqrt{1+\xi_2}} N_\alpha^d, \quad (\alpha = 3,4,6,7,8) \tag{5.11b}$$

where ξ_2^α is the ξ_2 value of node α.

By using the above interpolation functions, the $r^{-1/2}$ traction variation near the crack front can be described. This type of element is referred to as a traction-singular element. Of course, the $r^{1/2}$ displacement variation near the crack front can also be described. Jia et al. (1989) stated that it is probably best to use the shape functions defined above in a direct evaluation of the stress intensity factors.

The interpolation functions Eq. (5.10) of the displacements can be used on the elements near the crack front on the crack surfaces or the auxiliary surface. The interpolation functions Eq. (5.11) of tractions can be used on the elements near the crack front of the auxiliary surface, whereas the shape functions Eq. (4.17) should be used on the elements near the crack front on the crack surfaces in order to describe the traction variations. This is because traction is singular near the crack front on the auxiliary surface and is prescribed on the crack surfaces.

Luchi and Rizzuti (1987) and Jia et al. (1989) developed numerical procedures to evaluate the singular integrals extending over these special elements, respectively. In the following, we will introduce the numerical procedure suggested by Luchi and Rizzuti (1987) and then incorporate it into the Yue's solution-based boundary element formulation.

5.2.3 The numerical method of traction-singular elements

When the integral is executed on a traction-singular element, the integral (4.20a) can be written as

$$G_{ij}^{e\alpha} = \int_{-1}^{1}\int_{-1}^{1} u_{ij}^{Y}(P,Q) N_\alpha^t(\xi_1,\xi_2) J(\xi_1,\xi_2) \mathrm{d}\xi_1 \mathrm{d}\xi_2 \tag{5.12}$$

where $Q = Q(\xi_1,\xi_2)$ and the displacement kernel function $u_{ij}^Y(P,Q)$ is of the order r^{-1}. Because the new interpolation functions $N_\alpha^t(\xi_1,\xi_2)$ are introduced, the higher-order singular integral appears in the integrand. Hence, we need to introduce the corresponding numerical methods for dealing with the integral Eq. (5.12). In order to accurately evaluate the integral using standard Gaussian quadrature formulae, the integration domain $(-1 \leqslant \xi_1 \leqslant 1, -1 \leqslant \xi_2 \leqslant 1)$ must be mapped onto a different domain over which the integrand is bounded. Two cases of singular integrals will be discussed in the following case studies.

Case 1: The source point P is not located at the integrating traction-singular element

In this particular case, the source point P does not belong to the traction-singular element e. Integration can be carried out in the (η_1, η_2) plane where

$$\begin{aligned}\xi_1 &= \eta_1, \\ \xi_2 &= \frac{1}{2}(1+\eta_2)^2 - 1\end{aligned} \tag{5.13}$$

within the limits $-1 \leqslant \eta_1 \leqslant 1$ and $-1 \leqslant \eta_2 \leqslant 1$. In Eq. (5.11), the interpolation functions contain a term $(1+\xi_2)^{-1/2}$ or $(1+\eta_2)^{-1}$. In the (η_1, η_2) plane, the new Jacobian appears as follows

$$J^* = \frac{\partial(\xi_1, \xi_2)}{\partial(\eta_1, \eta_2)} = 1 + \eta_2 \tag{5.14}$$

In the new local coordinate system, the integral Eq. (5.12) can be transformed into

$$G_{ij}^{e\alpha} = \int_{-1}^{1}\int_{-1}^{1} u_{ij}^{Y}(P,Q) N_{\alpha}^{t}(Q) J(Q) J^*(\eta_1, \eta_2) \mathrm{d}\eta_1 \mathrm{d}\eta_2 \tag{5.15}$$

where $Q = Q(\xi_1, \xi_2) = Q[\xi_1(\eta_1, \eta_2), \xi_2(\eta_1, \eta_2)] = Q(\eta_1, \eta_2)$. The product of $N_{\alpha}^{t}(\alpha = 1,2,5)$ and J^* is, in fact, bounded. Thus, the integrand in Eq. (5.15) is non-singular and the standard Gaussian quadrature formulae can be used to evaluate the integral Eq. (5.15).

Case 2: The source point P is located at the integrating traction-singular element

In this case, P belongs to the traction-singular element e. In evaluating the singular integral, the element needs to be divided into two or three triangles and then new local coordinate systems are attached on each triangle. The singularity of the integrand can be removed by introducing two coordinate mappings. The first coordinate mapping is from (ξ_1, ξ_2) to (η_1, η_2) and the second coordinate mapping is from (η_1, η_2) to (η_1^*, η_2^*). The first coordinate mapping can be carried out using the same method as the one used to evaluate a weakly singular integral, which was outlined in Chapter 4. The second coordinate mapping is from one square ($-1 \leqslant \eta_1 \leqslant 1$ and $-1 \leqslant \eta_2 \leqslant 1$) to another square ($-1 \leqslant \eta_1^* \leqslant 1$ and $-1 \leqslant \eta_2^* \leqslant 1$). Thus the integral in Eq. (5.12) can be written as

$$G_{ij}^{e\alpha} = \sum_{k=1}^{2 \text{ or } 3} \int_{-1}^{1}\int_{-1}^{1} u_{ij}^{Y}[P, Q^{(k)}] N_{\alpha}^{t}(Q^{(k)}) J(Q^{(k)}) J^*(Q^{(k)}) \mathrm{d}\eta_1^* \mathrm{d}\eta_2^* \tag{5.16}$$

where $Q^{(k)}(\xi_1, \xi_2) = Q^{(k)}[\xi_1(\eta_1, \eta_2), \xi_2(\eta_1, \eta_2)] = Q^{(k)}[\eta_1(\eta_1^*, \eta_2^*), \eta_2(\eta_1^*, \eta_2^*)] = Q^{(k)}(\eta_1^*, \eta_2^*)$, the superscript k is the triangle number. J is the Jacobian determinant from the global coordinate system to the local coordinate system (ξ_1, ξ_2). In Eq. (5.16), we have

$$J^* = J_1 \cdot J_2 \tag{5.17}$$

where J_1 is the Jacobian from (ξ_1, ξ_2) to (η_1, η_2) and J_2 is the Jacobian from (η_1, η_2) to (η_1^*, η_2^*).

Without loss of generality, the two cases, i.e., where P is either a corner or midside node, will be discussed as follows.

(1) P is located at node 1 of the element e

In this case the element is divided into two triangles, as shown in Fig. 5.3. For each triangle, two coordinate mappings can be used to remove the singularity of the integrand.

For the triangle 134, $\eta_2 = -1$ is a degenerating side. The first coordinate mapping is given by

$$\xi_1 = \frac{1}{2}(1+\eta_1)(1+\eta_2) - 1,$$
$$\xi_2 = \eta_2 \tag{5.18}$$

and the second coordinate mapping is performed according to

$$\eta_1 = \eta_1^*,$$
$$\eta_2 = \frac{1}{2}(1+\eta_2^*)^2 - 1 \tag{5.19}$$

After the two coordinate mappings are finished, we have the following Jacobian

$$J^* = \frac{1}{4}(1+\eta_2^*)^3 \tag{5.20}$$

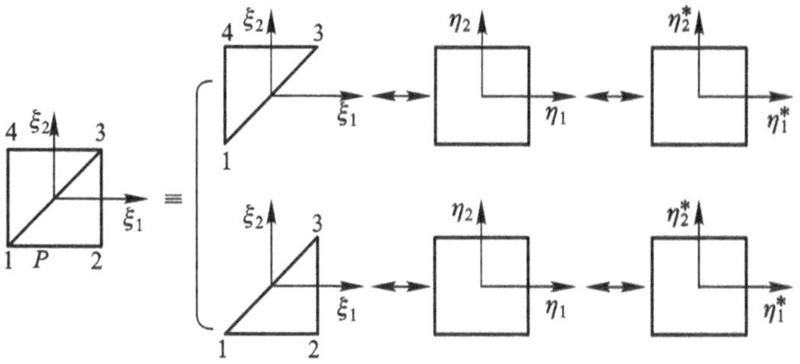

Fig. 5.3 Subdivision of the element when P is located at corner node 1 and subsequent coordinate mappings

For the triangle 123, the first coordinate mapping is given by

$$\xi_1 = \eta_2,$$
$$\xi_2 = \frac{1}{2}(1-\eta_1)(1+\eta_2) - 1 \tag{5.21}$$

and the second coordinate mapping is performed according to

$$\eta_1 = 1 - \frac{1}{2}(1-\eta_1^*)^2,$$
$$\eta_2 = \frac{1}{2}(1+\eta_2^*)^2 - 1 \qquad (5.22)$$

Completion of the two coordinate mappings gives the following Jacobian

$$J^* = \frac{1}{4}(1-\eta_1^*)(1+\eta_2^*)^3 \qquad (5.23)$$

(2) P is located at node 5 of the element e

In this case, the element is divided into three triangles, as shown in Fig. 5.4. For each triangle, two coordinate mappings are used to remove the singularity of the integrand.

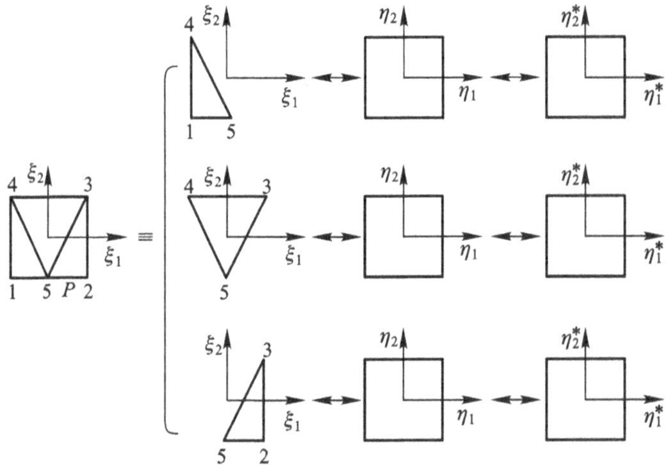

Fig. 5.4 Subdivision of the element when P is located at midside node 5 and subsequent coordinate mappings

For triangle 541, the first coordinate mapping is given by

$$\xi_1 = -\frac{1}{2}(1+\eta_2),$$
$$\xi_2 = \frac{1}{2}(1+\eta_1)(1+\eta_2) - 1 \qquad (5.24)$$

and the second coordinate mapping is performed according to

$$\eta_1 = \frac{1}{2}(1+\eta_1^*)^2 - 1,$$
$$\eta_2 = \frac{1}{2}(1+\eta_2^*)^2 - 1 \qquad (5.25)$$

After the two coordinate mappings we have the following Jacobian

$$J^* = \frac{1}{8}(1+\eta_1^*)(1+\eta_2^*)^3 \tag{5.26}$$

For triangle 523, the first coordinate mapping is given by

$$\begin{aligned} \xi_1 &= \frac{1}{2}(1+\eta_2), \\ \xi_2 &= \frac{1}{2}(1-\eta_1)(1+\eta_2) - 1 \end{aligned} \tag{5.27}$$

and the second coordinate mapping is performed according to

$$\begin{aligned} \eta_1 &= 1 - \frac{1}{2}(1-\eta_1^*)^2, \\ \eta_2 &= \frac{1}{2}(1+\eta_2^*)^2 - 1 \end{aligned} \tag{5.28}$$

Completion of the two coordinate mappings gives the following Jacobian

$$J^* = \frac{1}{8}(1-\eta_1^*)(1+\eta_2^*)^3 \tag{5.29}$$

For triangle 543, the first coordinate mapping is given by

$$\begin{aligned} \xi_1 &= \frac{1}{2}\eta_1(1+\eta_2), \\ \xi_2 &= \eta_2 \end{aligned} \tag{5.30}$$

and the second coordinate mapping is performed according to

$$\begin{aligned} \eta_1 &= \eta_1^*, \\ \eta_2 &= \frac{1}{2}(1+\eta_2^*)^2 - 1 \end{aligned} \tag{5.31}$$

After the two coordinate mappings are completed, we have the following Jacobian

$$J^* = \frac{1}{4}(1+\eta_2^*)^3 \tag{5.32}$$

In the above coordinate mappings, it can be seen that the triangle in the (ξ_1, ξ_2) plane is first mapped onto a square in the (η_1, η_2) plane and then this square is further mapped onto another square in the (η_1^*, η_2^*) plane.

5.3 Computation of stress intensity factors

After obtaining a numerical solution for the stresses and displacements in a cracked body by using the proposed BEM, we can determine the stress intensity factors (SIFs). The stress intensity factors are related to the asymptotic behavior of stresses and displacements near the crack tip, shown in Fig. 5.5. According to the displacement fields shown in Eq. (2.9), the SIFs can be described by

$$K_{\mathrm{I}} = \frac{E}{4(1-\nu^2)} \lim_{r\to 0} \sqrt{\frac{\pi}{2r}} \left(u_2\big|_{\theta=\pi} - u_2\big|_{\theta=-\pi} \right) \tag{5.33a}$$

$$K_{\mathrm{II}} = \frac{E}{4(1-\nu^2)} \lim_{r\to 0} \sqrt{\frac{\pi}{2r}} \left(u_1\big|_{\theta=\pi} - u_1\big|_{\theta=-\pi} \right) \tag{5.33b}$$

$$K_{\mathrm{III}} = \frac{E}{4(1+\nu)} \lim_{r\to 0} \sqrt{\frac{\pi}{2r}} \left(u_3\big|_{\theta=\pi} - u_3\big|_{\theta=-\pi} \right) \tag{5.33c}$$

where θ is the angle in the (x_1, x_2) plane and rotates from the x_1 axis.

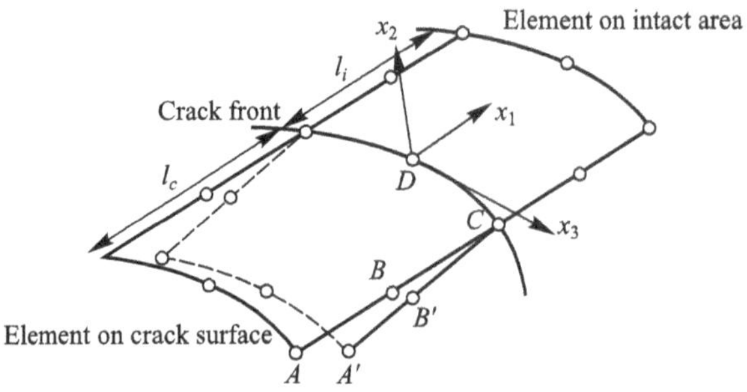

Fig. 5.5 Traction-singular elements at crack front

According to the stress fields shown in Eq. (2.8), the SIFs are given as

$$K_{\mathrm{I}} = \lim_{r\to 0} \sqrt{2\pi r}\sigma_{22}\big|_{\theta=0} \tag{5.34a}$$

$$K_{\mathrm{II}} = \lim_{r\to 0} \sqrt{2\pi r}\sigma_{12}\big|_{\theta=0} \tag{5.34b}$$

$$K_{\mathrm{III}} = \lim_{r\to 0} \sqrt{2\pi r}\sigma_{23}\big|_{\theta=0} \tag{5.34c}$$

Eqs. (5.33) and (5.34) can be used to calculate the SIFs. In order to obtain a more accurate stress intensity factor, extrapolation from two nodal values can be employed for the traction-singular elements. The extrapolation procedure involves the correlation of the computed displacements and tractions with the theoretical values and extrapolates them to the crack front. To demonstrate this procedure, we consider a crack front with the elements shown in Fig. 5.5.

For a mixed mode crack with mode I, II and III SIFs, K_{I}, K_{II} and K_{III} at the corner point C can be linearly extrapolated from the computed crack opening displacements at the nodes $A-A'$ and $B-B'$ as follows

$$K_{\mathrm{I}} = \frac{E}{4(1-\nu^2)}\sqrt{\frac{\pi}{2}}\frac{2\sqrt{2}(u_2^B - u_2^{B'}) - (u_2^A - u_2^{A'})}{\sqrt{l_c}} \tag{5.35a}$$

$$K_{\mathrm{II}} = \frac{E}{4(1-\nu^2)}\sqrt{\frac{\pi}{2}}\frac{2\sqrt{2}(u_1^B - u_1^{B'}) - (u_1^A - u_1^{A'})}{\sqrt{l_c}} \tag{5.35b}$$

$$K_{\mathrm{III}} = \frac{E}{4(1+\nu)}\sqrt{\frac{\pi}{2}}\frac{2\sqrt{2}(u_3^B - u_3^{B'}) - (u_3^A - u_3^{A'})}{\sqrt{l_c}} \tag{5.35c}$$

where l_c denotes the length of the traction-singular element on the crack surface and is measured perpendicular to the crack front. This linear extrapolation procedure decouples the mode I, II and III fracture when the intact plane coincides with the crack surface. In this case, the angle θ is zero.

K_{I}, K_{II} and K_{III} can also be evaluated from extrapolation of the tractions on the intact area to the crack front. For the points C and D in Fig. 5.5, we have

$$K_{\mathrm{I}}^{C,D} = \sqrt{\pi l_i} t_2^{C,D} \tag{5.36a}$$

$$K_{\mathrm{II}}^{C,D} = \sqrt{\pi l_i} t_1^{C,D} \tag{5.36b}$$

$$K_{\mathrm{III}}^{C,D} = \sqrt{\pi l_i} t_3^{C,D} \tag{5.36c}$$

where l_i denotes the length of the traction-singular element on the intact area and is measured perpendicular to the crack front.

5.4 Numerical examples and results

In this chapter, we have incorporated the traction-singular element into the proposed Yue's solution-based BEM outlined in Chapter 4 and written the corresponding computer programs in FORTRAN. A literature review indicated that there are limited exact or numerical solutions available for crack problems in three-dimensional layered elastic solids. To verify the proposed numerical method, it is necessary to calculate some crack problems for which exact or numerical solutions are available.

Example 1: A penny-shaped crack in a sandwich solid

As is shown in Fig. 5.6, an elastic layer is bonded to two half-spaces and contains a penny-shaped crack parallel to the interfaces between the interlayer and the half-spaces. It is assumed that the two elastic half-spaces have the same elastic properties. The penny-shaped crack is located in the mid-plane of the layer. Using Hankel transforms, Arin and Erdogan (1971) examined this problem where there was a constant pressure on the crack surfaces.

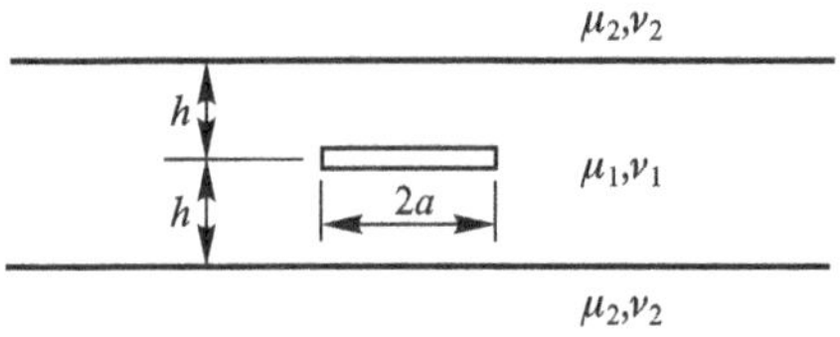

Fig. 5.6 A penny-shaped crack in a sandwich layered system

For convenience, we have chosen the surface of a cylinder to be the boundary surface for the discretization. Consider that the crack is embedded in a finite length cylinder with a circular cross-section that is subjected to a uniform tension loading σ at the far field. Because of symmetry, we use the upper half of the cylinder in the BEM mesh formulation. Figure 5.7 shows the element mesh with 168 boundary elements and 229 nodes. The mesh of the surface containing the crack consists of eight element rings, the crack surface is modeled by four element rings and the traction-singular elements are employed along the crack front.

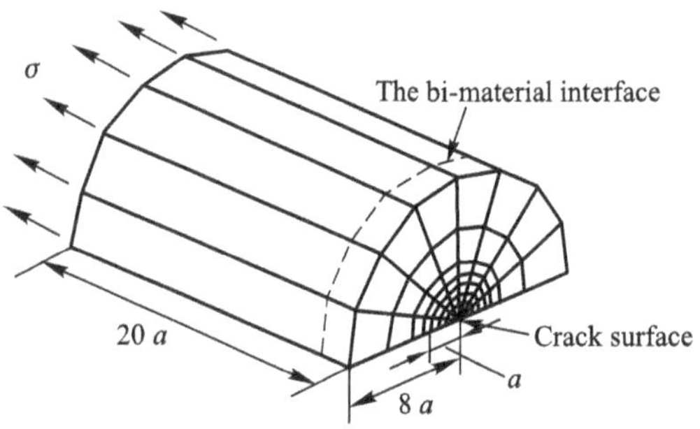

Fig. 5.7 Boundary element mesh for a sandwich solid with a penny-shaped crack

In order to verify the solution, we chose the same elastic parameters for the materials as those used by Arin and Erdogan (1971). Table 5.1 presents the results obtained from the two different methods. The first set of the elastic parameters are given as follows

$$E_1 = 3.15\ \text{GPa}, \quad \nu_1 = 0.35 \ \text{ and } \ E_2 = 70\ \text{GPa}, \quad \nu_2 = 0.22$$

In this case, the differences observed between the two methods are less than 1.49%. The SIF values increase as the layer thickness h/a increases. This is because the lower elastic modulus of material 1 allows the crack to open more easily.

The second set of elastic parameters are given as follows

$$E_1 = 70\ \text{GPa}, \quad \nu_1 = 0.22 \ \text{ and } \ E_2 = 3.15\ \text{GPa}, \quad \nu_2 = 0.35$$

In this case, the difference observed between the results obtained from the two methods is less than 3.75%. The SIF values decrease as the layer thickness h/a increases. This is because the higher elastic modulus of material 1 constrains the crack opening.

Table 5.1 Numerical and exact solutions of stress intensity factors ($K_I/(\sigma\sqrt{\pi a})$) for example 1

	The first set of elastic parameters			The second set of elastic parameters		
h/a	Arin and Erdogan (1971)	This study	Difference (%)	Arin and Erdogan (1971)	This study	Difference (%)
0.2	0.2457	0.2472	–0.25	2.6937	2.7079	–1.42
0.4	0.3205	0.3225	–0.20	1.5757	1.5896	–1.39
0.6	0.3809	0.3790	–0.18	1.1944	1.1569	3.75
0.8	0.4297	0.4247	0.50	0.9884	0.9544	3.40
1.0	0.4713	0.4564	1.49	0.8723	0.8418	3.05
1.2	0.4918	0.4899	1.19	0.8150	0.7789	3.61
1.4	0.5325	0.5209	1.16	0.7214	0.7479	2.35
1.6	0.5531	0.5423	1.08	0.7576	0.7265	3.11
1.8	0.5686	0.5572	1.14	0.7089	0.7155	–0.66
2.0	0.5804	0.5662	1.42	0.6721	0.7054	–1.33

Example 2: A rectangular interface crack in a dissimilar finite cube

It is well known that the stress and/or the displacement fields at the tip of an interface crack exhibit oscillatory behavior. Therefore, the definition of the stress intensity factors for an interface crack is quite different from that for a crack in a homogeneous material. The stress distribution near the crack tip along the interface between two isotropic elastic materials is usually defined as

$$\sigma_y + \mathrm{i}\tau_{xy} = \frac{K_\mathrm{I} + \mathrm{i}K_\mathrm{II}}{\sqrt{2\pi r}}\left(\frac{r}{l}\right)^{\mathrm{i}\varepsilon}, \quad \mathrm{i} = \sqrt{-1} \tag{5.37}$$

where l is the total crack length and is used to normalize the oscillatory term, ε is a bi-material constant and is defined below

$$\varepsilon = \frac{1}{2\pi}\ln\left[\left(\frac{\kappa_1}{\mu_1} + \frac{1}{\mu_2}\right) \Big/ \left(\frac{\kappa_2}{\mu_2} + \frac{1}{\mu_1}\right)\right] \tag{5.38a}$$

$$\kappa_j = \begin{cases} 3 - 4\nu_j, (j = 1,2) & \text{for plane strain} \\ (3 - 4\nu_j)/(1 + \nu_j), (j = 1,2) & \text{for plane stress} \end{cases} \tag{5.38b}$$

The relative displacements between the crack surfaces can be represented as

$$\delta_y + \mathrm{i}\delta_x = \frac{K_\mathrm{I} + \mathrm{i}K_\mathrm{II}}{2(1 + 2\mathrm{i}\varepsilon)\cosh(\varepsilon\pi)}\left[\frac{\kappa_1 + 1}{\mu_1} + \frac{\kappa_2 + 1}{\mu_2}\right]\left(\frac{r}{2\pi}\right)^{1/2}\left(\frac{r}{l}\right)^{\mathrm{i}\varepsilon}, \quad \mathrm{i} = \sqrt{-1} \tag{5.39}$$

where $\delta_j = u_j(r,\pi) - u_j(r,-\pi), j = x, y$.

The SIF values can be evaluated using the extrapolation methods for stresses or displacements. In the present study, the extrapolation method for displacements is applied to evaluate the SIF values. By using Eq. (5.39), the displacement extrapolation method can be expressed as follows

$$\sqrt{K_{\mathrm{I}}^2+K_{\mathrm{II}}^2}=\frac{4\pi\sqrt{1+4\varepsilon^2}\cosh(\pi\varepsilon)}{(\kappa_1+1)/\mu_1+(\kappa_2+1)/\mu_2}\lim_{r\to 0}\sqrt{\frac{\delta_x^2+\delta_y^2}{2\pi r}} \tag{5.40a}$$

$$K_{\mathrm{II}}/K_{\mathrm{I}}=\lim_{r\to 0}\frac{1-(\delta_y/\delta_x)(H_2/H_1)}{\delta_y/\delta_x+H_2/H_1} \tag{5.40b}$$

where

$$H_2/H_1=\frac{\tan(\varepsilon\ln(r/l))-2\varepsilon}{1+2\varepsilon\tan(\varepsilon\ln(r/l))} \tag{5.40c}$$

This method uses the numerical results for the displacements at points that have some distance from the crack tip. Therefore both the numerical errors and the oscillation singularity problem can be avoided.

Figure 5.8 shows a rectangular crack at the interface between two bonded dissimilar cubes. The finite cube is subject to a uniformly distributed load P_0 on the upper and lower plate surfaces. Murakami (1992) and Yuuki and Xu (1994) examined this example using the BEM. For a better comparison, we choose two different dimensions similar to those used in Yuuki and Xu (1994). The dimensions and material properties of the specimen are presented below

$$B=0.5W,\quad H=4W,\quad a/W=0.1$$

and

$$E_1=206\text{ GPa},\quad \nu_1=0.27\text{ and }E_2=304\text{ GPa},\quad \nu_2=0.3$$

The SIF values and the strain energy release rate for the interface crack are defined as follows:

$$\tau_{yz}=\frac{K_{\mathrm{III}}}{\sqrt{2\pi r}} \tag{5.41a}$$

$$F_{\mathrm{I}}=\sqrt{K_{\mathrm{I}}^2+K_{\mathrm{II}}^2} \tag{5.41b}$$

$$\begin{aligned}G_{\mathrm{T}}&=\frac{1}{16\cosh(\varepsilon\pi)}\left[\frac{\kappa_1+1}{\mu_1}+\frac{\kappa_2+1}{\mu_2}\right]F_{\mathrm{I}}^2+\frac{1}{4}\left(\frac{1}{\mu_1}+\frac{1}{\mu_2}\right)K_{\mathrm{III}}^2\\&=G_{\text{in-plane}}+G_{\mathrm{III}}\end{aligned} \tag{5.41c}$$

The displacement extrapolation Eqs. (5.40a~c) can be used to evaluate the SIF values for the interface crack. Furthermore, because the stress τ_{yz} near the crack tip shows a regular singularity of order $r^{-1/2}$, K_{III} can be evaluated by the linear extrapolation expression (5.35c) using the displacements at the midside and corner nodes of the elements on the crack surface.

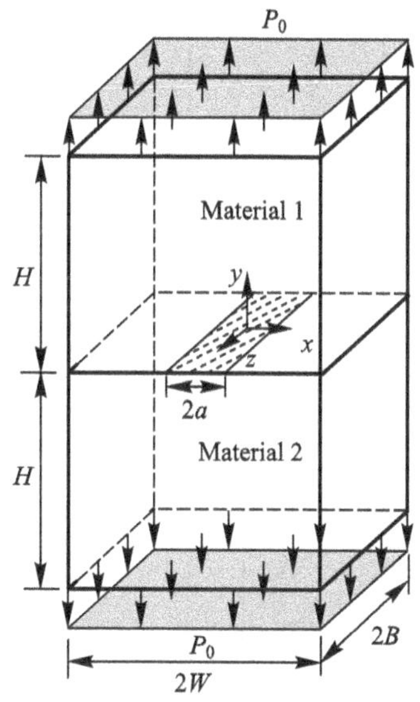

Fig. 5.8 Geometry of a finite cube with an interface crack under a uniform normal stress P_0

As shown in Fig. 5.8, the crack problem is symmetrical about the (x,y) and (y,z) planes. Thus, we can use a quarter of the dissimilar cube in the BEM analysis. To utilize the multi-region method of the BEM, the cracked body is divided into two regions. Region I includes the upper surface of the crack and region II includes the lower surface of the crack. As shown in Figs. 5.9a and b, the surface of region I is discretized into 100 elements and 329 nodes whereas the surface of region II is discretized with 248 elements and 520 nodes. On the common boundary surfaces of the two regions, the nodes and elements of region I corresponds to those of region II one-to-one. We then utilize the fully continuous conditions of tractions and displacements on the common boundary surface to establish the linear equations. The traction-singular elements are used to capture the stress and displacement fields near the crack tip.

Consider first a crack problem in a homogeneous finite cube. Table 5.2 shows the SIF values obtained using two different methods. It can be seen that the absolute differences in the results obtained from the two different methods are 0.74%~1.8%. We now consider the interface crack problem in a finite cube with two dissimilar materials. Table 5.3 gives the SIF values and energy release rates obtained with the two different methods; here there are very small differences between the two different methods. In Table 5.3, the following relations are used

$$\overline{F}_{\mathrm{I}} = F_{\mathrm{I}}/(P_0\sqrt{\pi a}), \quad G_0 = \frac{\kappa_1 + 1}{8\mu_1} P_0^2 \pi a \tag{5.42}$$

From the above analyses, the stress intensity factors obtained with the present formulation are in very good agreement with the existing numerical results.

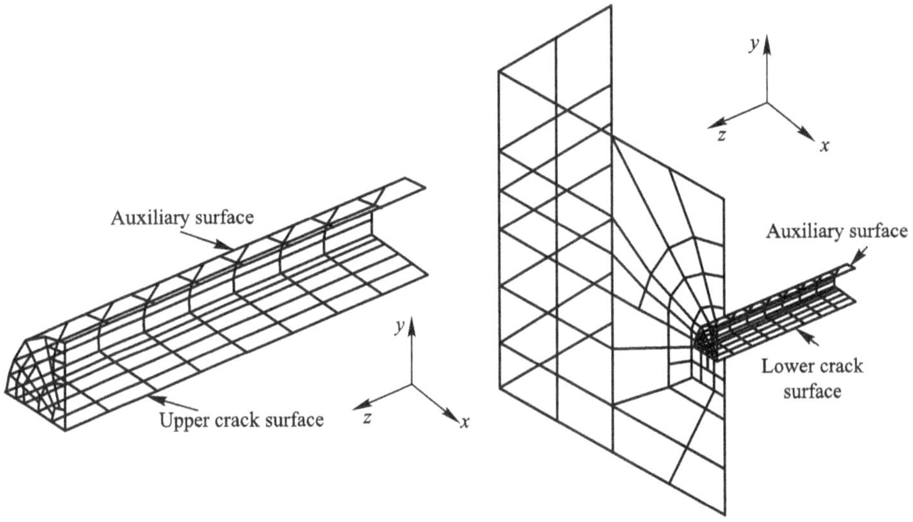

(a) Element mesh with upper surface of a crack for region I

(b) Partial element mesh with the lower surface of a crack for region II

Fig. 5.9 Boundary element mesh for a finite cube with an interface crack

Table 5.2 Numerical and exact solutions of stress intensity factors $(K_{\mathrm{I}}/(\sigma\sqrt{\pi a}))$ for example 2

z/B	Murakami (1992)					This study			
	0.0	0.375	0.75	0.9375	2D	0.0	0.375	0.75	0.9375
$\nu = 0$	1.003	1.003	1.002	1.001		0.9944	0.9915	0.9913	0.9832
$\nu = 0.27$	1.009	1.011	1.025	1.045	1.006	1.0065	1.0062	1.0226	1.0407
$\nu = 0.30$	1.010	1.013	1.031	1.055		1.0076	1.0081	1.0285	1.0514

Table 5.3 Comparison of stress intensity factors and energy release rate for example 2

z/B	Murakami (1992)					This study			
	0.0	0.375	0.75	0.9375	2D	0.0	0.375	0.75	0.9375
$\overline{F}_{\mathrm{I}}/(P_0\sqrt{\pi a})$	1.006	1.009	1.026	1.053	1.008	0.9515	0.9563	0.9856	1.0173
$K_{\mathrm{II}}/K_{\mathrm{I}}$	0.026	0.026	0.023	0.017	0.027	0.0285	0.0291	0.0309	0.0363
$K_{\mathrm{III}}/(P_0\sqrt{\pi a})$	0	−0.017	−0.033	−0.055	0	0	−0.0159	−0.0412	−0.0569
G_{T}/G_0	0.855	0.861	0.891	0.941	–	0.7495	0.7574	0.8062	0.8606
G_{in}/G_0	0.855	0.860	0.890	0.937	0.859	0.7495	0.7571	0.8042	0.8569
G_{III}/G_0	0	0.0003	0.001	0.004	–	0	0.0003	0.0020	0.0038

5.5 Summary

In the initial part of this chapter, the stress and displacement fields near a crack front are considered and different boundary elements, which can capture the singularity near the crack front, are introduced. The traction-singular element is chosen and the corresponding numerical methods are introduced. This special boundary element is then incorporated into Yue's solution-based BEM. By using crack opening displacements, an extrapolation method of the two-point displacements near the crack front is described for the evaluation of the stress intensity factors. Finally, these ideas are combined together to analyze two examples, i.e., a penny-shaped crack in a sandwich solid and a rectangular interface crack in a dissimilar finite cube. These research results have been published in Yue et al. (2002) and cited in many research articles (e.g., Pan et al., 2004; Rangelov et al., 2008; Favata, 2012). In the following chapters, we will employ the proposed method to analyze crack problems in functionally graded materials.

References

Aliabadi MH. Boundary element formulations in fracture mechanics. ASME *Applied Mechanics Reviews*, 1997, 50: 83-96.

Arin K, Erdogan F. Penny-shaped crack in an elastic layer bonded to dissimilar half spaces. *International Journal of Engineering Science*, 1971, 9: 213-232.

Ariza MP, Dominguez J. Boundary element formulation for 3D transversely isotropic cracked bodies. *International Journal for Numerical Methods in Engineering*, 2004, 60: 719-753.

Bains RS. Boundary Element Analysis of Three-dimensional Crack Problems: Weight Function Techniques. Southampton: Wessex Institute of Technology, 1992.

Cruse TA, Wilson RB. Boundary-integral Equation Method for Elastic Fracture Mechanics Analysis, AFOSR-TR-78-0355. Connecticut: United Technologies Corporation, 1977.

Favata A. On the Kelvin problem. *Journal of Elasticity*, 2012, 109: 189-204.

Gao YL, Tan CL, Selvadurai APS. Stress intensity factors for cracks around or penetrating an elliptic inclusion using the boundary element method. *Engineering Analysis with Boundary Elements*, 1992, 10(1): 59-68.

Irwin GR. Analysis of stresses and strains near the end of a crack traversing a plate. *ASME Journal of Applied Mechanics*, 1957, 24: 361-364.

Jia ZH, Shippy DJ, Rizzo FJ. Three-dimensional crack analysis using singular boundary elements. *International Journal for Numerical Methods in Engineering*, 1989, 24: 2257-2273.

Luchi ML, Rizzuti S. Boundary elements of for three-dimensional elastic crack analysis. *International Journal for Numerical Methods in Engineering*, 1987, 24: 2253-2271.

Murakami Y. Stress Intensity Factors Handbook. Oxford: Pergamon Press, 1992.

Pan E, Han F. Green's functions for transversely isotropic piezoelectric multilayered half-spaces. *Journal of Engineering Mathematics*, 2004, 49: 271-288.

Pan E, Yuan FG. Boundary element analysis of three-dimensional crack in anisotropic solids. *International Journal for Numerical Methods in Engineering*, 2000, 48: 211-237.

Rangelov T, Dineva P, Gross D. Effects of material inhomogeneity on the dynamic behavior of cracked piezoelectric solids: a BIEM approach. *ZAMM Journal of Applied Mathematics and Mechanics*, 2008, 88(2): 86-99.

Williams ML. On the stress distribution at the base of a stationary crack. *ASME Journal of Applied Mechanics*, 1957, 24: 109-114.

Xiao HT, Yue ZQ, Tham GL, Lee CF. Analysis of elliptical cracks perpendicular to the interface of two joined transversely isotropic solids. *International Journal of Fracture*, 2005, 133: 329-354.

Yuuki R, Cho SB. Efficient boundary element analysis of stress intensity factors for interface cracks in dissimilar materials. *Engineering Fracture Mechanics*, 1989, 34: 179-188.

Yuuki R, Xu JQ. Boundary element analysis of dissimilar materials and interface crack. *Computational Mechanics*, 1994, 14: 116-127.

Yue ZQ, Xiao HT. Generalized Kelvin solution based boundary element method for crack problems in multilayered solids. *Engineering Analysis with Boundary Elements*, 2002, 26: 691-705.

Yue ZQ, Xiao HT, Pan E. Stress intensity factors of square crack inclined to interface of transversely isotropic bi-material. *Engineering Analysis with Boundary Elements*, 2007, 31: 50-65.

Yue ZQ, Xiao HT, Tham GL, Lee CF, Pan E. Boundary element analysis of three-dimensional crack problems in two joined transversely isotropic solids. *Computational Mechanics*, 2005, 36: 459-474.

Chapter 6

Analysis of Penny-shaped Cracks in Functionally Graded Materials

6.1 Introduction

Analysis of three-dimensional cracks has not developed as fully as its two-dimensional counterpart because of their greater complexities. Consequently, very few exact analyses have been reported, e.g., the ones by Sneddon (1946), Green and Sneddon (1950). Some specialized mathematical tools can be used to obtain these exact solutions. For analyzing the axisymmetric problem related to the indentation of the plane surface of a penny-shaped crack, Selvadurai (2000a) presented a Hankel integral transform development of the governing mixed boundary value problem and its reduction to a single Fredholm integral equation of the second kind and an appropriate consistency condition. The simplest crack problem is that a state of uniform tension in an isotropic homogeneous solid of infinite extent is perturbed by a penny-shaped crack or an elliptical crack. To date exact solutions have not been obtained for the great majority of three-dimensional crack problems.

In most of the existing solutions to crack problems relating to nonhomogeneous solids, it is assumed that the elastic parameters of the composite medium are approximated by simple functions, e.g., an exponential or power form of a space variable. Delale and Erdogan (1983) considered the plane elasticity problem for a nonhomogeneous medium containing a crack and concluded that the stresses around the crack tips have conventional square-root singularities and the effect of the Poisson's ratio is rather negligible. Jin and Noda (1994) drew the conclusion that the singularity and angular distribution of the crack-tip stress field for a nonhomogeneous material are identical to those in a homogeneous material as long as the material properties are continuous. In recent years, investigations into the mechanics of functionally graded materials (FGMs) have grown rapidly. In particular, a special issue of *Engineering Fracture Mechanics, An International Journal,* was devoted to the fracture mechanics of FGMs, see Paulino (2002). Results of such investigations can be used in plane crack problems to predict the initiation and growth of cracks in new materials.

A detailed literature review indicates that there are a limited number of exact or numerical solutions available to solve crack problems in three-dimensional FGMs. Ozturk and Erdogan (1995, 1996) analyzed the problem of a penny-shaped crack which is located in homogeneous dissimilar materials bonded through an interfacial region with graded mechanical properties and are subjected to torsion, compressive and shear tractions. They investigated how the thickness and material non-homogeneity parameter influences the strain energy release rate. Li et al. (1999) investigated the mode I stress intensity factors (SIFs) for functionally graded solid cylinders with an embedded penny-shaped crack or an external circumferential crack using the multiple isoparametric finite element method. Selvadurai (2000b) examined the axisymmetric problem of a penny-shaped crack located at a bonded plane with localized elastic non-homogeneity and generated the crack opening mode SIFs at the crack tip.

In this chapter, we employ the numerical method developed in the previous chapters to analyze the SIFs of penny-shaped cracks and discuss crack growth. In these analyses, the FGM system consists of two homogeneous materials bonded through a non-homogeneous interfacial region of given thickness; the penny-shaped crack surfaces are parallel to or perpendicular to the interfacial layer. The crack surfaces are subjected to uniform compressive or shear loadings. We first provide the SIFs for each case and then discuss the directions and critical loadings of the crack growth by using superposition principles and propagation criteria for fracture mechanics.

6.2 Analysis methods for crack problems in a FGM system of infinite extent

6.2.1 The crack problem in a FGM

In Fig. 6.1, the interfacial layer of the graded material is fully bonded with two homogeneous media of semi-infinite extent and a penny-shaped crack (radius a) is located at the interface of the homogeneous material and the FGMs. The crack surfaces are subjected to a uniform compressive stress p_0. Ozturk and Erdogan (1996) analyzed this crack problem by reducing the related mixed boundary value problem to a system of singular integral equations. In this study, we re-examine this problem to demonstrate the effectiveness and accuracy of the proposed numerical method in dealing with crack problems in FGMs.

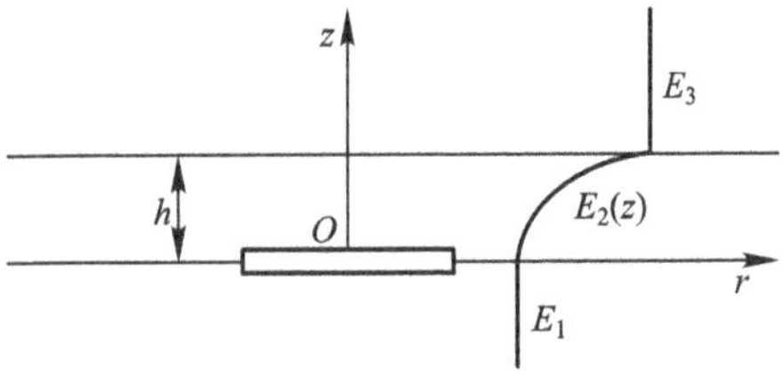

Fig. 6.1 A penny-shaped crack at the interface between a FGM interlayer and a homogeneous medium of semi-infinite extent

It is assumed that the Poisson's ratio of the FGM system is constant, i.e., $\nu_1 = \nu_2 = \nu_3 = \nu$ and the elastic modulus of the FGM is approximated by

$$E_2(z) = E_1 e^{\alpha z} \tag{6.1a}$$

$$\alpha = \frac{1}{h} \ln(E_3/E_1) \tag{6.1b}$$

where h is the interlayer thickness, the constant α can be positive or negative, and E_1, E_2 and E_3 are the elastic moduli of three layers.

6.2.2 The multi-region method for crack problems of infinite extent

It is well known that a straight application of traditional BEM to crack problems leads to a mathematical degeneration if the two crack surfaces are considered co-planar. One method for overcoming this difficulty is to introduce the multi-region method in the BEM.

We consider a FGM system of infinite extent containing a penny-shaped crack as shown in Fig. 6.1. The crack has two open surfaces and two regions for the BEM analysis need to be formed. We can form the first closed curved surface by adding an open imaginary surface in the solid to one of the two crack surfaces. We can form the second closed curved surface by adding the open imaginary surface in the solid to the other crack surface. The two closed curved surfaces divide the entire solid into two regions. We denote the region within the first closed curved surface as region I. Its complement, the region outside the second closed curved surface, is then denoted as region II. Each of the two regions I and II has a crack surface on its boundary. The open imaginary surface serves as a common boundary for the two regions. The two crack surfaces are the actual boundaries. It should also be noted that the two crack surfaces occupy the same area and are separated. There is no open space between the two crack surfaces before the application of the load.

Since the two regions share the same boundary geometry, the boundary integral equations for both the regions are the same except that some signs change and there are slight differences in the $c_{ij}(P)$ terms. Therefore, the coefficient matrix for either region I or II is sufficient to construct the matrix for the entire problem.

By using the discretized boundary integral equations (4.35), the matrix equation for region I can be written as

$$\boldsymbol{A}\boldsymbol{u}_{\mathrm{I}} = \boldsymbol{B}\boldsymbol{t}_{\mathrm{I}} \tag{6.2}$$

where the matrix $\boldsymbol{A}$ is obtained from the integrals containing t_{ij}^{Y}, the matrix $\boldsymbol{B}$ is obtained from the integrals containing u_{ij}^{Y}, and $\boldsymbol{u}_{\mathrm{I}}$ and $\boldsymbol{t}_{\mathrm{I}}$ are the boundary displacements and tractions for region I respectively.

The matrix equation for region II is constructed by taking advantage of Eq. (6.2). It can be expressed as follows

$$-\boldsymbol{A}^{*}\boldsymbol{u}_{\mathrm{II}} = \boldsymbol{B}\boldsymbol{t}_{\mathrm{II}} \tag{6.3}$$

where $\boldsymbol{A}^{*}$ is identical to $\boldsymbol{A}$ except for the diagonal terms, in which the contribution of the boundary at infinity has to be taken into consideration, and $\boldsymbol{B}$ is the same for both regions. We can use the continuity conditions for displacements and tractions over the common boundary

$$u_{\mathrm{I}} = u_{\mathrm{II}}, \quad t_{\mathrm{I}} = -t_{\mathrm{II}} \tag{6.4}$$

By using Eq. (6.4), Eqs. (6.2) and (6.3) can be coupled together. The equations for the entire problem domain can be written as

$$\boldsymbol{G}\boldsymbol{X} = \boldsymbol{R} \tag{6.5}$$

where $\boldsymbol{G}$ is the overall matrix of coefficients; its row and column numbers are equal to two times the row and column numbers of a single domain matrix, $\boldsymbol{X}$ is the vector of the unknowns and consists of the displacements and tractions over the entire boundary except for the applied crack-surface tractions, and $\boldsymbol{R}$ is a known vector containing the effect of the known crack-surface tractions.

For the crack problem in Fig. 6.1, we can choose the surface of a hemisphere in a homogeneous material (material 1) to form the two regions for the BEM analysis. Due to the symmetry of the problem, it is necessary to analyze only a quarter of the hemisphere; only this quarter is discretized, with no elements on the plane of symmetry. Figure 6.2 shows the surface discretization with 88 boundary elements and 289 nodes, in which there are 16 traction-singular elements employed along the crack front.

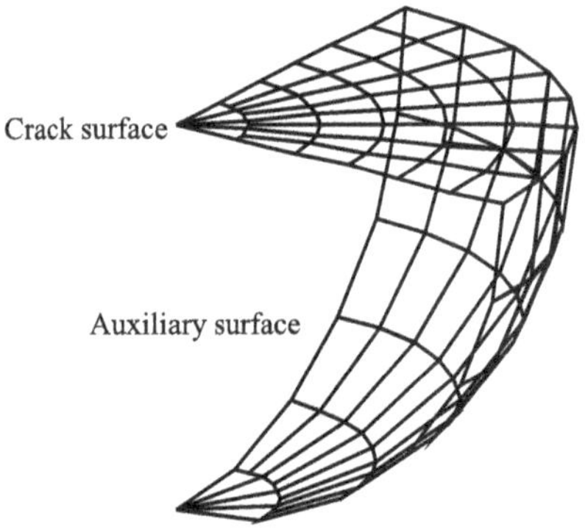

Fig. 6.2 Boundary element mesh for a penny-shaped crack

6.2.3 The layered discretization technique for FGMs

Since the proposed BEM is based on Yue's solution for layered materials, the layered discretization technique can be used to deal with the variation in the elastic parameters of FGMs. For the FGM described in Eq. (6.1), the FGM is closely approximated by n bonded layers of elastic homogeneous media. Each layer has a thickness of h/n and an elastic modulus of $E_2(z)$ at the top of the layer, i.e., for the i-th layer, $z = H_i$, where $H_i =$

ih/n, $(i = 1,2,\cdots,n)$. Two homogeneous materials bonded in an FGM are considered as semi-infinite domains for the layers H_0 and H_{n+1} respectively. For all the layers, the Poisson's ratios are the same and equal to 0.3. Figure 6.3 illustrates an approximation of the continuous depth variation of the elastic modulus when there are a large number of piecewise homogeneous layers, where $n = 20$, $h = 0.5$ and $\alpha a = 3$. It can be observed from Fig. 6.3 that a close approximation of the variation in the elastic modulus can be obtained using a large number of n.

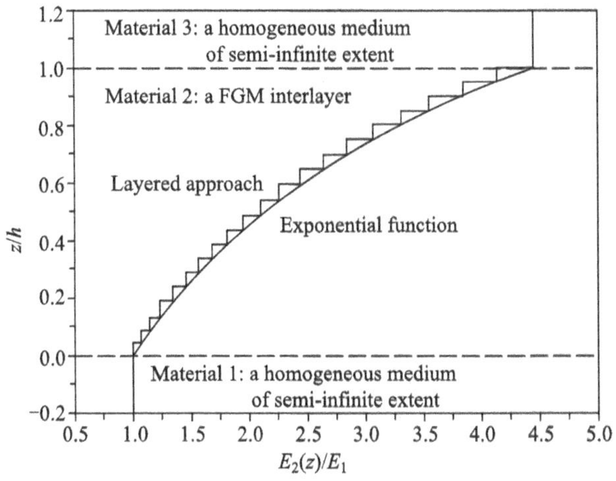

Fig. 6.3 Approximation of the continuous depth variation of the elastic modulus by a layered system of 20 piecewise homogeneous layers for $h = 0.5a$ and $\alpha a = 3$

6.2.4 Numerical verifications

In the following, we will compare the results obtained using the proposed BEM with those given by Ozturk and Erdogan (1996). Four cases, where $\alpha a = 0.4, 0.6, 2$ and 3, are analyzed using the layered approximations and a comparison of the results is shown in Table 6.1. Looking at Table 6.1, it can be observed that a layered approximation ($n = 20$) produces excellent results for the SIF values. For $\alpha a = 0.4$, the absolute errors between the results of the two methods are 1.289% and 0.053% for the normalized SIF K_{I} and K_{II} values, respectively. When $\alpha a = 0.6$, the two absolute errors are 1.3% and 0.12%. For $\alpha a = 2$, the two absolute errors are 0.85% and 0.59% and for $\alpha a = 3$, the two absolute errors are equal to 0.03% and 0.21%.

As the layer discretization number n increases, the SIF values (K_{I} and K_{II}) derived using the proposed BEM approach the exact solutions. The results presented in Table 6.1 demonstrate that accurate SIF values for crack problems in FGMs can be obtained by using the proposed BEM together with a layered discretization technique. In the ensuing, we will present the results obtained by applying the proposed BEM to two particular crack problems in a FGM system. An examination of the relevant literature reveals that currently there is no published research available on the SIF values of these two cracks problems to be investigated.

Table 6.1 Results of the SIF values obtained using the present study and the results of Ozturk and Erdogan (1996) ($h/a = 0.5$ and $\nu = 0.3$)

No. of a layered FGM	$\alpha a = 0.4$		$\alpha a = 0.6$		$\alpha a = 2$		$\alpha a = 3$	
	$\frac{K_I}{p_0\sqrt{\pi a}}$	$\frac{K_{II}}{p_0\sqrt{\pi a}}$	$\frac{K_I}{p_0\sqrt{\pi a}}$	$\frac{K_{II}}{p_0\sqrt{\pi a}}$	$\frac{K_I}{p_0\sqrt{\pi a}}$	$\frac{K_{II}}{p_0\sqrt{\pi a}}$	$\frac{K_I}{p_0\sqrt{\pi a}}$	$\frac{K_{II}}{p_0\sqrt{\pi a}}$
5	0.62253	0.01362	0.61030	0.02185	0.53497	0.06321	0.49312	0.07892
6	0.62340	0.01308	0.61153	0.02102	0.53918	0.06161	0.49810	0.07695
7	0.62395	0.01287	0.61197	0.02056	0.54126	0.06040	0.50265	0.07494
8	0.62462	0.01224	0.61343	0.01985	0.54524	0.05870	0.50740	0.07343
9	0.62487	0.01226	0.61378	0.01988	0.54593	0.05856	0.50799	0.07328
10	0.62547	0.01200	0.61389	0.01919	0.54745	0.05692	0.51013	0.07177
11	0.62549	0.01216	0.61469	0.01975	0.54735	0.05719	0.51021	0.07155
12	0.62584	0.01157	0.61529	0.01891	0.55085	0.05636	0.51509	0.07064
13	0.62536	0.01199	0.61455	0.01949	0.54940	0.05747	0.51317	0.07008
14	0.62617	0.01132	0.61570	0.01852	0.55184	0.05532	0.51592	0.06959
15	0.62651	0.01151	0.61621	0.01882	0.55156	0.05501	0.51480	0.06885
16	0.62618	0.01180	0.61567	0.01926	0.54991	0.05615	0.51425	0.07039
17	0.62668	0.01148	0.61645	0.01877	0.55318	0.05649	0.51760	0.07109
18	0.62671	0.01114	0.61654	0.01827	0.55471	0.05491	0.51739	0.06895
19	0.62601	0.01156	0.61554	0.01887	0.55292	0.05600	0.51836	0.07037
20	0.62619	0.01157	0.61577	0.01889	0.55210	0.05652	0.51747	0.07047
Ozturk and Erdogan (1996)	0.6133	0.0121	0.6026	0.0177	0.5436	0.0506	0.5135	0.0684

6.3 The SIFs for a crack parallel to the FGM interlayer

6.3.1 General

The surfaces of a penny-shaped crack (radius a) shown in Fig. 6.4 are subjected to the following loadings

$$\sigma^{+}_{z'x'} = \sigma^{-}_{z'x'} = q, \quad (0 \leqslant r < a) \tag{6.6a}$$

$$\sigma^{+}_{z'y'} = \sigma^{-}_{z'y'} = 0, \quad (0 \leqslant r < a) \tag{6.6b}$$

$$\sigma^{+}_{z'z'} = \sigma^{-}_{z'z'} = p, \quad (0 \leqslant r < a) \tag{6.6c}$$

where the superscripts + and – correspond to the crack surfaces, which are located at $z' = 0^{+}$ and $z' = 0^{-}$, respectively.

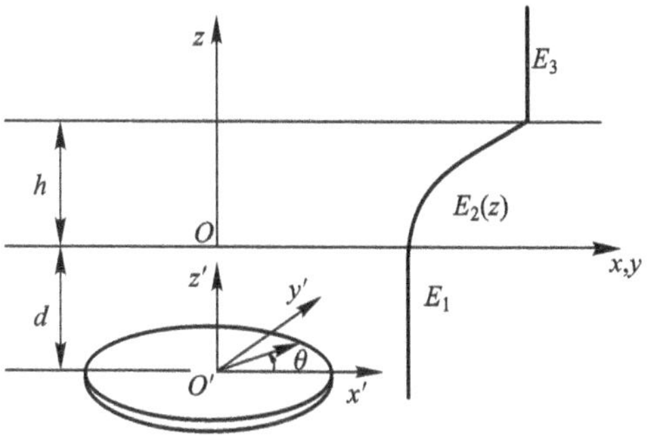

Fig. 6.4 A penny-shaped crack in material 1 and parallel to the FGM interlayer

For convenience, we have chosen a hemisphere (radius a) in a homogeneous material (material 1) that forms the two regions for the BEM analysis. Because of symmetry, it is only necessary to analyze half the hemisphere. We can discretize this half hemisphere and do not need to discretize the plane of the symmetry into the BEM meshes. Figure 6.5 shows the discretized left symmetrical surface containing 176 elements and 553 nodes. We use 32 traction-singular elements along the crack front.

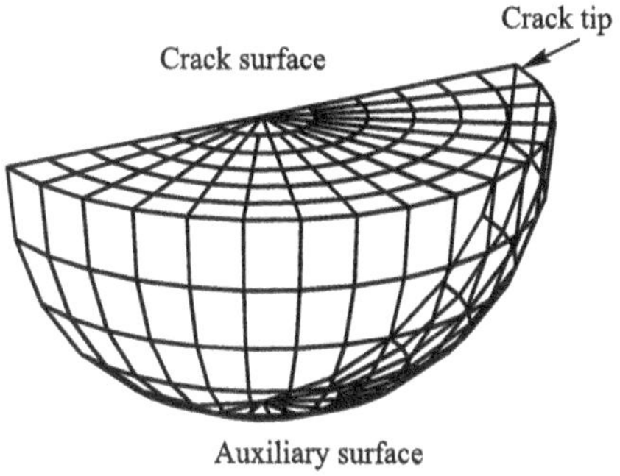

Fig. 6.5 Boundary element mesh of the crack and auxiliary surfaces

6.3.2 A crack subjected to uniform compressive stresses

Let us consider the uniform compressive stresses acting on the crack surfaces, i.e., $q = 0$ and $p = -p_0$ $(p_0 > 0)$ in Eq. (6.6). Figures 6.6 and 6.7 show the mode I and II SIF

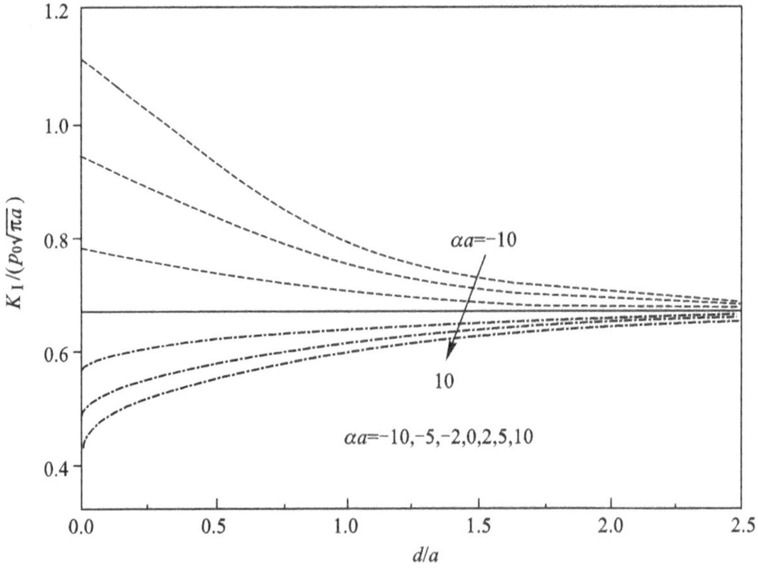

Fig. 6.6 Variations of $K_{\mathrm{I}}/(p_0\sqrt{\pi a})$ with d for different α values ($h = 0.5a$)

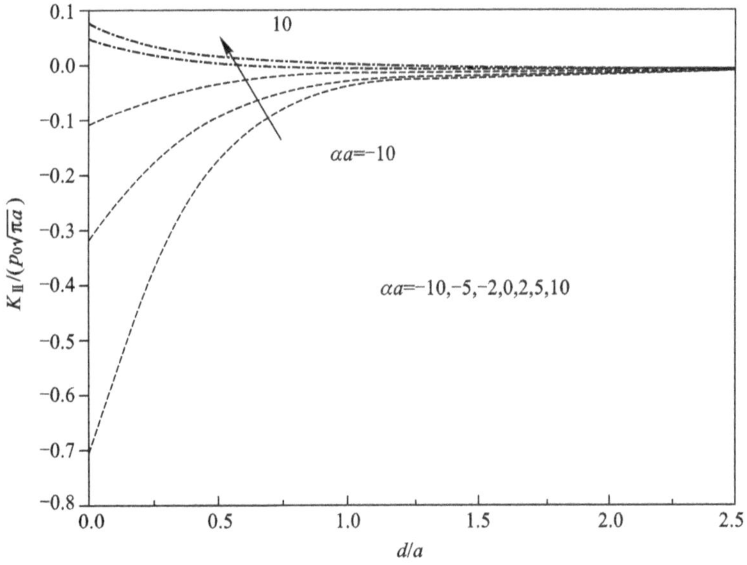

Fig. 6.7 Variations of $K_{\mathrm{II}}/(p_0\sqrt{\pi a})$ with d for different α values ($h = 0.5a$)

values along the crack front for a crack located at a distance d from the interlayer with a non-homogeneity parameter α. If $\alpha \neq 0$, the mode I and II deformations of the crack are coupled together. In this figure, the SIF values vary with different values of α, i.e., $\alpha a =$

$-10, -5, -2, 2, 5, 10$. Notice that the case $\alpha = 0$ corresponds to a homogeneous medium of infinite extent. In this case, the exact solution is $K_I/(p_0\sqrt{\pi a}) = 2/\pi \approx 0.6366$, the numerical solution is 0.6479 and the absolute error between the two solutions is 1.13%.

For a crack in a more compliant body, i.e., where $\alpha > 0$, the crack opening is constrained by the stiffer upper interlayer and the homogeneous half-space, and the constraint becomes more pronounced if the crack is closer to the FGM interlayer. As a result, the mode I SIF values of $\alpha > 0$ are less than those for $\alpha = 0$, and the mode II SIF appears with positive values. On the other hand, for $\alpha < 0$, the crack is located in a stiffer solid, and the compliant upper body tends to increase the crack opening when the crack is closer to the FGM interlayer. As a result, the mode I SIF values of $\alpha < 0$ are more than the ones when $\alpha = 0$, and the mode II SIF appears and has negative values. As the distance d increases, the effect of the FGM interlayer on the crack becomes weak and the SIF values tend to those of a crack problem in a homogeneous medium of infinite extent.

Figure 6.8 shows the variations of the SIF with the thickness h of the FGM interlayer, where $d = 0.2a$ and $\alpha a = -5, 5$. For $\alpha a = -5$, the values of the mode I SIF increase and those for the mode II SIF decrease as the interlayer thickness h increases. For $\alpha a = 5$, the values of the mode I SIF decrease and the ones of the mode II SIF increase as the thickness h increases. In these two cases, variations in the SIF values are not obvious for $h/a > 1.5$.

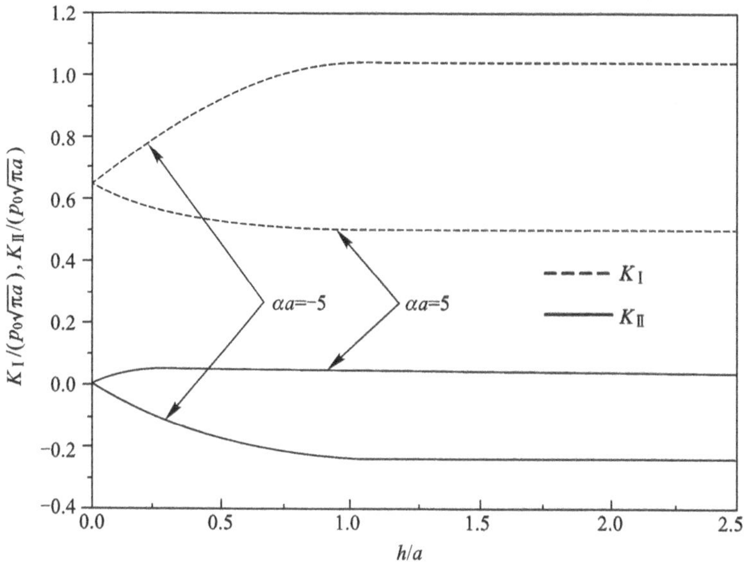

Fig. 6.8 Variations of $K_I/(p_0\sqrt{\pi a})$ and $K_{II}/(p_0\sqrt{\pi a})$ with h for different α values ($d = 0.2a$)

Yue et al. (2003) also analyzed the crack problem for $p = -p_0(r/a)$ and $-p_0(r/a)^2$, and discussed how the SIF varies with the non-homogeneous parameter α, the thickness of the FGM interlayer h and the crack distance d to the interlayer.

6.3.3 A crack subjected to uniform shear stresses

In Eq. (6.6), it is assumed that $q = -q_0 (q_0 > 0)$ and $p = 0$, i.e., the crack surfaces are subjected to uniform shear stresses. Under the action of shear stresses q_0, the mode II and III deformations are coupled together. For this case, the SIFs of the penny-shaped crack in a homogeneous medium of infinite extent are (Tada et al., 2000)

$$K_{\mathrm{I}} = 0 \tag{6.7a}$$

$$K_{\mathrm{II}} = \frac{4}{\pi(2-\nu)} \sqrt{\pi a} q_0 \cos\theta \tag{6.7b}$$

$$K_{\mathrm{III}} = \frac{4(1-\nu)}{\pi(2-\nu)} \sqrt{\pi a} q_0 \sin\theta \tag{6.7c}$$

where ν is Poisson's ratio. It is obvious that the maximum values of K_{II} and K_{III} appear at the crack fronts $\theta = 0^\circ, 90^\circ$, respectively. The exact solution of $(K_{\mathrm{II}})_{\max}/(q_0\sqrt{\pi a})$ is 0.7490, the numerical solution is 0.7544, and the corresponding absolute error is 0.54%. The exact solution of $(K_{\mathrm{III}})_{\max}/(q_0\sqrt{\pi a})$ is 0.5247, the numerical solution is 0.5243, and the corresponding absolute error is 0.04%.

Figures 6.9 and 6.10 show the variations in the mode II and III SIFs with the non-

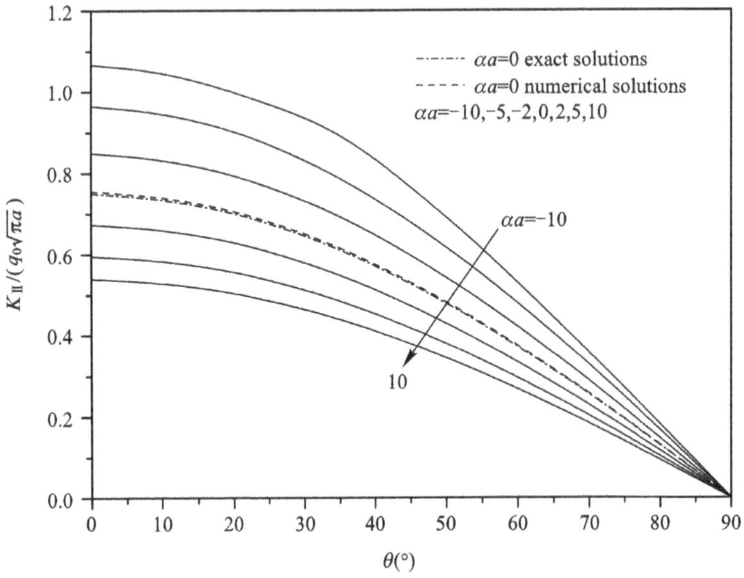

Fig. 6.9 Variations of $K_{\mathrm{II}}/(q_0\sqrt{\pi a})$ with θ for different α values ($d = 0$, $h = 0.5a$)

homogeneous parameter α for a crack located at the interface, i.e., $d = 0$. In comparison with the case where $\alpha = 0$, it is seen that the SIF values decrease for $\alpha > 0$ and increase for $\alpha < 0$. These variations are induced by the upper FGM half-space which constrains or increases the relative sliding of the upper and lower surfaces of the penny-shaped crack.

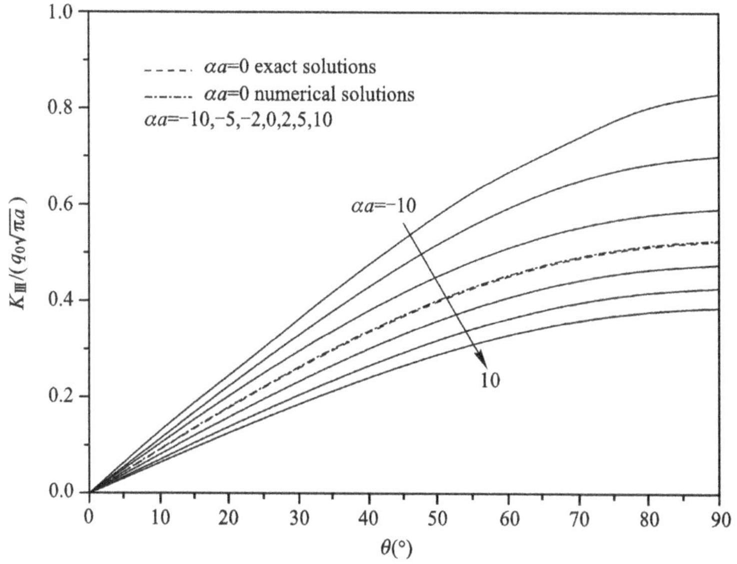

Fig. 6.10 Variations of $K_{III}/(q_0\sqrt{\pi a})$ with θ for different α values ($d = 0$, $h = 0.5a$)

Figure 6.11 shows the variations of the SIF values when the crack is located at a distance d from the interlayer, where $\alpha a = -5,5$ and $h/a = 0.5$. In this figure, K_{II} at the crack front $\theta = 0^\circ$ and K_{III} at $\theta = 90^\circ$ are presented. For $\alpha a = -5$, K_{II} and K_{III} decrease as the distance d increases whereas for $\alpha a = 5$, K_{II} and K_{III} increase as the distance d increases. In these two cases, the values of K_{II} and K_{III} tend towards those for a crack in a homogeneous medium of infinite extent as the distance d approaches infinity.

Figure 6.12 show the variations in the SIF values with an interlayer thickness h where $\alpha a = -5,5$ and $d/a = 0.2$. In this figure, K_{II} at the crack front $\theta = 0^\circ$ and K_{III} at $\theta = 90^\circ$ are presented. For $\alpha a = -5$, K_{II} and K_{III} increase as the thickness h increases, whereas when $\alpha a = 5$, K_{II} and K_{III} decrease with an increase in the thickness h. For $h/a > 1$, the SIF values show no obvious changes when the interlayer thickness increases.

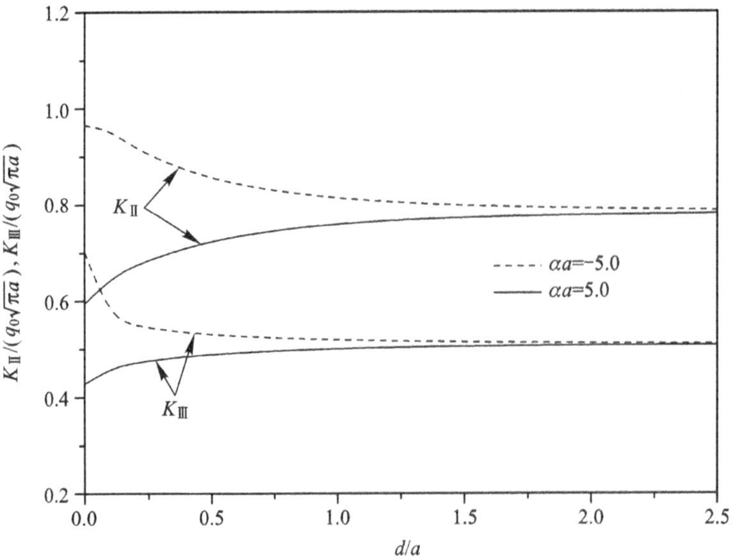

Fig. 6.11 Variations of $K_{\mathrm{II}}/(q_0\sqrt{\pi a})$ and $K_{\mathrm{III}}/(q_0\sqrt{\pi a})$ with d for different α values ($h = 0.5a$)

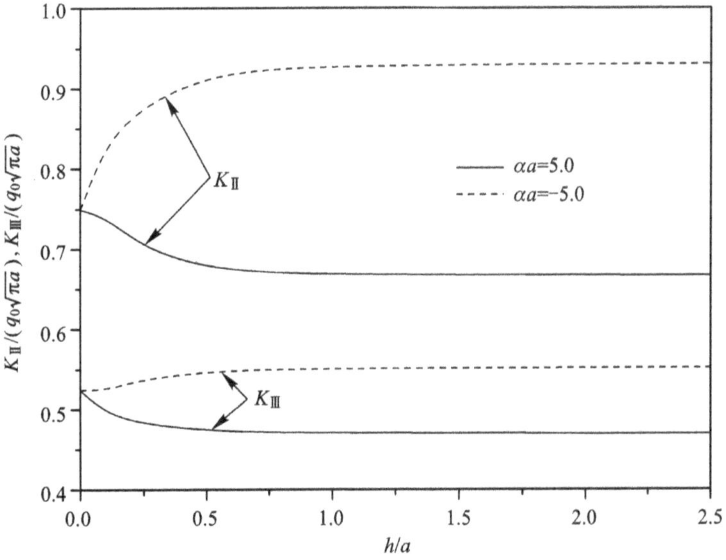

Fig. 6.12 Variations of $K_{\mathrm{II}}/(q_0\sqrt{\pi a})$ and $K_{\mathrm{III}}/(q_0\sqrt{\pi a})$ with h for different α values ($d = 0.2a$)

homogeneous parameter α for a crack located at the interface, i.e., $d = 0$. In comparison with the case where $\alpha = 0$, it is seen that the SIF values decrease for $\alpha > 0$ and increase for $\alpha < 0$. These variations are induced by the upper FGM half-space which constrains or increases the relative sliding of the upper and lower surfaces of the penny-shaped crack.

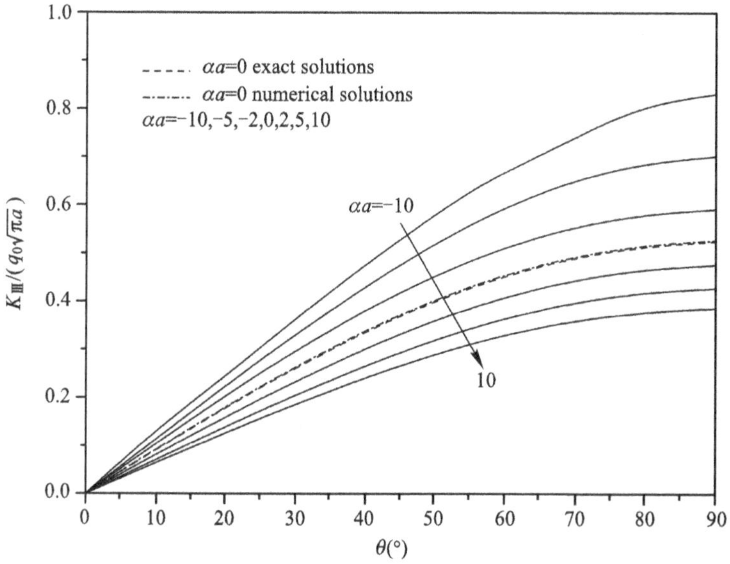

Fig. 6.10 Variations of $K_{\text{III}}/(q_0\sqrt{\pi a})$ with θ for different α values ($d = 0$, $h = 0.5a$)

Figure 6.11 shows the variations of the SIF values when the crack is located at a distance d from the interlayer, where $\alpha a = -5,5$ and $h/a = 0.5$. In this figure, K_{II} at the crack front $\theta = 0°$ and K_{III} at $\theta = 90°$ are presented. For $\alpha a = -5$, K_{II} and K_{III} decrease as the distance d increases whereas for $\alpha a = 5$, K_{II} and K_{III} increase as the distance d increases. In these two cases, the values of K_{II} and K_{III} tend towards those for a crack in a homogeneous medium of infinite extent as the distance d approaches infinity.

Figure 6.12 show the variations in the SIF values with an interlayer thickness h where $\alpha a = -5,5$ and $d/a = 0.2$. In this figure, K_{II} at the crack front $\theta = 0°$ and K_{III} at $\theta = 90°$ are presented. For $\alpha a = -5$, K_{II} and K_{III} increase as the thickness h increases, whereas when $\alpha a = 5$, K_{II} and K_{III} decrease with an increase in the thickness h. For $h/a > 1$, the SIF values show no obvious changes when the interlayer thickness increases.

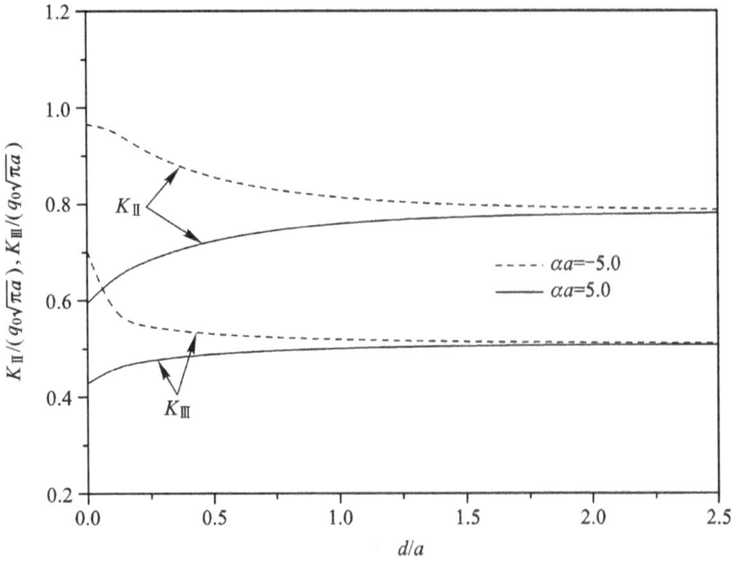

Fig. 6.11 Variations of $K_{II}/(q_0\sqrt{\pi a})$ and $K_{III}/(q_0\sqrt{\pi a})$ with d for different α values ($h = 0.5a$)

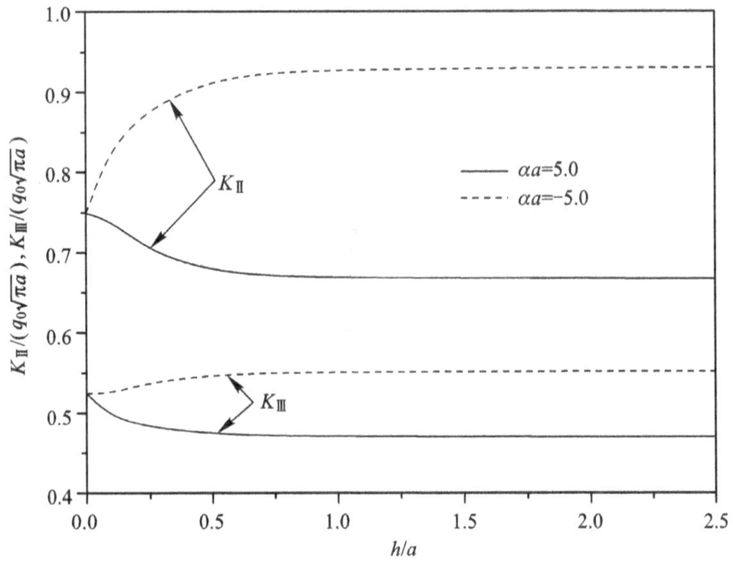

Fig. 6.12 Variations of $K_{II}/(q_0\sqrt{\pi a})$ and $K_{III}/(q_0\sqrt{\pi a})$ with h for different α values ($d = 0.2a$)

6.4 The growth of the crack parallel to the FGM interlayer

6.4.1 The strain energy density factor of an elliptical crack

Using the expressions for the stresses and strains associated with an elliptical crack (Sih and Cha, 1974), Eq. (2.32) can be expressed as

$$\frac{dW}{dV} = \frac{S}{r} + \text{non-singular term} \tag{6.8}$$

where the strain energy density factor S is defined by

$$S = a_{11}K_{\mathrm{I}}^2 + 2a_{12}K_{\mathrm{I}}K_{\mathrm{II}} + a_{22}K_{\mathrm{II}}^2 + a_{33}K_{\mathrm{III}}^2 \tag{6.9}$$

The SIFs K_{I},K_{II} and K_{III} vary with the location along the crack border and the four coefficients a_{11}, a_{12}, a_{22} and a_{33} can be expressed using the spherical angles β and φ at the crack front as follows

$$a_{11} = \frac{\kappa+1}{16\mu\kappa^2\cos\beta}\left[2(1-2\nu) + \frac{\kappa-1}{\kappa}\right] \tag{6.10a}$$

$$a_{12} = \frac{\sqrt{\kappa^2-1}}{8\mu\lambda\kappa^2\cos\beta}\left[\frac{1}{\kappa} - (1-2\nu)\right] \tag{6.10b}$$

$$a_{22} = \frac{1}{16\mu\lambda\kappa^2\cos\beta}\left[4(1-\nu)(\kappa-1) + \frac{1}{\kappa}(\kappa+1)(3-\kappa)\right] \tag{6.10c}$$

$$a_{33} = \frac{1}{4\mu\lambda\kappa\cos\beta} \tag{6.10d}$$

where

$$\kappa = \sqrt{1 + (\tan\beta/\lambda)^2}, \quad \lambda = \cos\varphi \tag{6.10e}$$

The strain energy density factor S is a function of the spherical angles (β, φ). In analyzing the crack growth, the minimum value $S_{\min}$ and the growth direction around the crack front are first determined and then the fracture criterion $S_{\min} = S_{\mathrm{cr}}$ is used to obtain the critical loading σ_{cr} for crack growth.

6.4.2 Crack growth under a remotely inclined tensile loading

In the ensuing, the growth potential of a penny-shaped crack in the two bonded solids with a FGM interlayer under a remotely tensile loading is examined. Using the superposition principle of fracture mechanics, the traction-free crack problem, shown in Fig. 6.13, is examined as a crack loaded by statically self-equilibrium tractions σ. The tractions are needed so that the local crack-tip behavior can be investigated in terms of the SIFs. As a result, the equivalent normal traction p_0 and shear traction q_0 on the crack surface have to be obtained for the remotely inclined tensile loading σ with the loading angle γ as follows

$$p_0 = \sigma \sin^2 \gamma \tag{6.11a}$$

$$q_0 = \sigma \sin \gamma \cos \gamma \tag{6.11b}$$

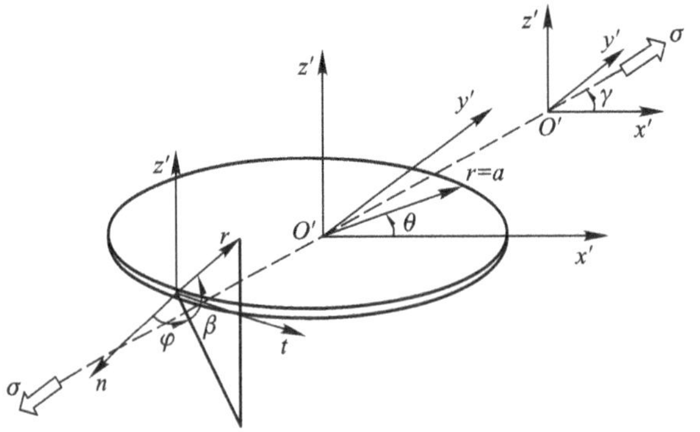

Fig. 6.13 A penny-shaped crack under a remotely inclined tensile loading

Thus, the above SIF values associated with the loadings p_0 and q_0 are used to analyze the crack growth under a remotely inclined loading σ. The strain energy density factor S can be calculated by using the SIF values for p_0 and q_0 and Eq. (6.9). The applied tension σ is always assumed to be in the coordinate plane $O'x'z'$.

(1) The minimum strain energy density factor $S_{\min}$ at the crack front

Let $S_{\min} = S(\beta_0, \varphi_0)$, that is, $S_{\min}$ appears at the position (β_0, φ_0) on the crack front. The $S_{\min}$ of the penny-shaped crack appears in the direction perpendicular to the tangent of the crack front, i.e., $\varphi_0 \equiv 0°$. Figure 6.14 shows the variations of β_0 with α and θ. It can be found that β_0 increases as α increases; however, these changes are different depending upon whether $0° \leqslant \theta \leqslant 90°$ or $90° \leqslant \theta \leqslant 180°$. For $0° \leqslant \theta \leqslant 90°$ and $\alpha < 0$, α has a larger influence on β_0. This phenomena are caused by the different distributions of the mode II SIF values between $0° \leqslant \theta \leqslant 90°$ and $90° \leqslant \theta \leqslant 180°$.

Figure 6.15 shows the $S_{\min}$ variations with α and θ. It can be seen that for the larger angle θ and $\alpha < 0$, the $S_{\min}$ changes are obvious. This is because the K_{II} value induced by p_0 and q_0 is negative for $\alpha < 0$ while the absolute values of K_{II} induced by p_0 are much larger than those induced by q_0. In Fig. 6.15, the maximum $S_{\min}$ values appear at $\theta = 0°$ for $\alpha > 0$ and at $\theta = 180°$ for $\alpha < 0$. In the ensuing, based on the distribution characteristics of $S_{\min}$, the critical loading σ_{cr} to produce crack growth is discussed.

(2) The critical loadings σ_{cr} for crack growth

Figure 6.16 shows the variations of σ_{cr} with α and γ where $d = 0$ and $h = 0.5a$. It can be shown that the σ_{cr} curves for $\alpha > 0$ and $\alpha < 0$ appear on both sides of the curve for $\alpha = 0$ and σ_{cr} increases as α increases. For a given α value, σ_{cr} decreases as γ increases.

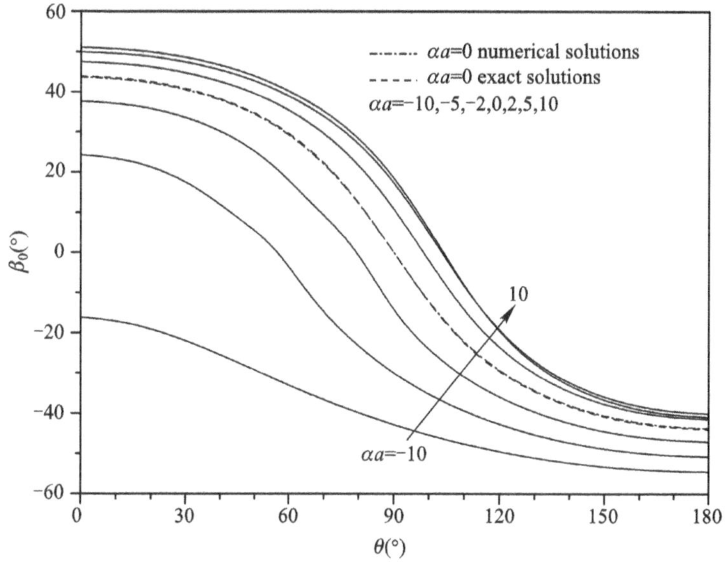

Fig. 6.14 Variations of β_0 with θ for different α values ($d = 0$, $h = 0.5a$ and $\gamma = 60^\circ$)

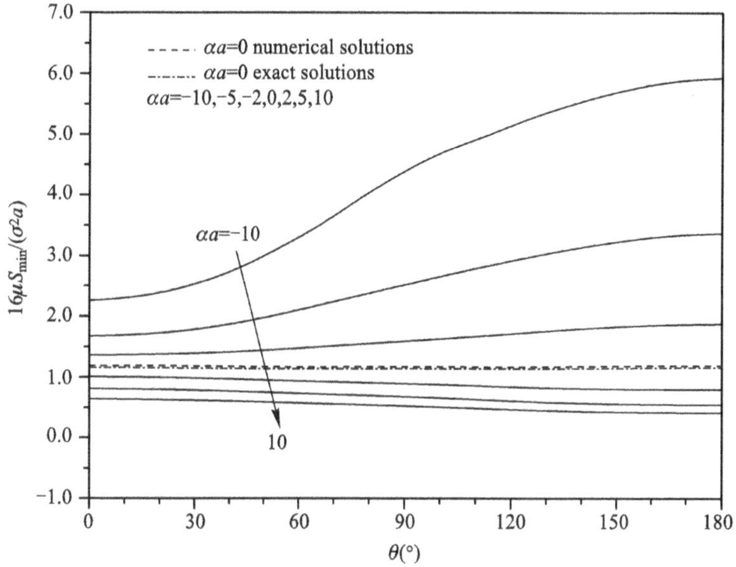

Fig. 6.15 Variations of $S_{\min}$ with θ for different α values ($d = 0$, $h = 0.5a$ and $\gamma = 60^\circ$)

Figure 6.17 shows the variations of σ_{cr} with the distance d where $h = 0.5a$. It can be seen that when the crack is further from the FGM interlayer, σ_{cr} decreases for $\alpha a = 5$ and σ_{cr} increases for $\alpha a = -5$; σ_{cr} tends to behave in the same way as in a homogeneous

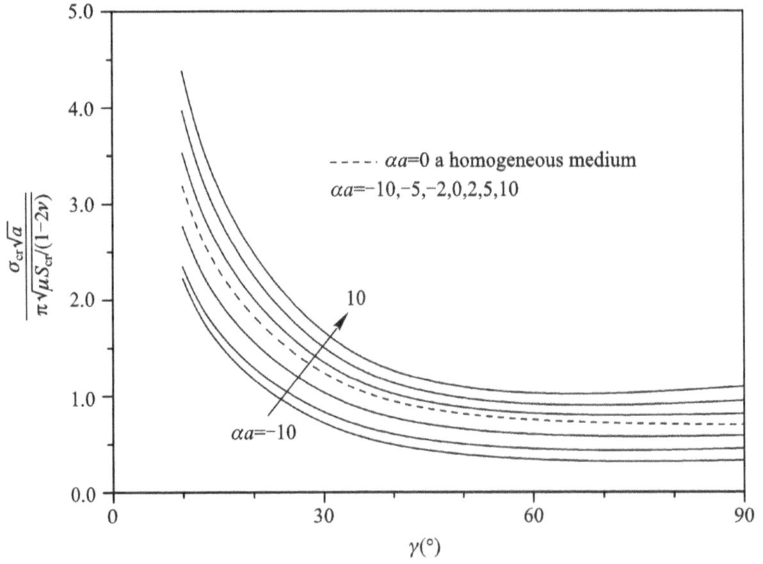

Fig. 6.16 Variations of σ_{cr} with γ for different α values ($d = 0$ and $h = 0.5a$)

medium of infinite extent for the two cases as the distance d approaches infinity. In addition, for a smaller loading angle, the crack distance d to the interlayer has a greater influence on σ_{cr}. For other cases of $\alpha > 0$ and $\alpha < 0$, σ_{cr} has similar variation tendencies.

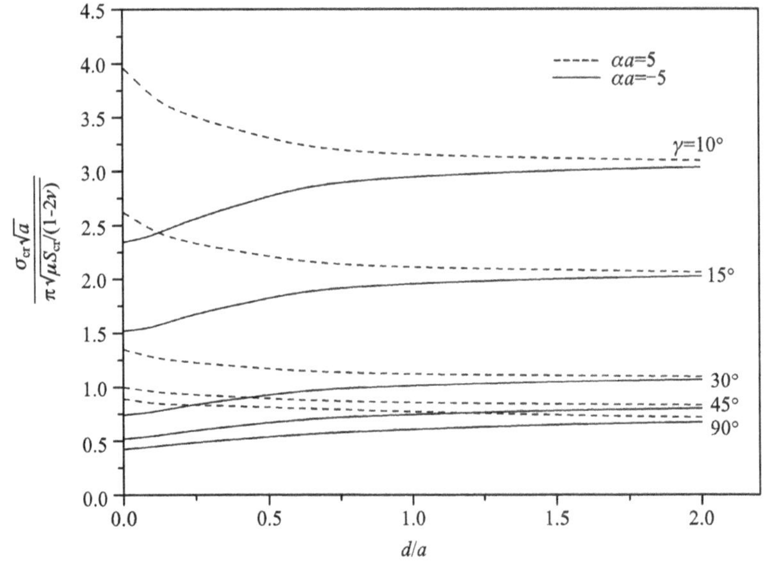

Fig. 6.17 Variations of σ_{cr} with d for different α and γ values ($h = 0.5a$)

Figure 6.18 shows the variations of σ_{cr} with the thickness h of the FGM interlayer where $d = 0.2a$. As the thickness h increases, σ_{cr} increases for $\alpha a = 5$ and σ_{cr} decreases for $\alpha a = -5$. In addition, for a smaller loading angle, the interlayer thickness h has a larger influence on σ_{cr}. For other cases of $\alpha > 0$ and $\alpha < 0$, σ_{cr} has a similar variation tendency.

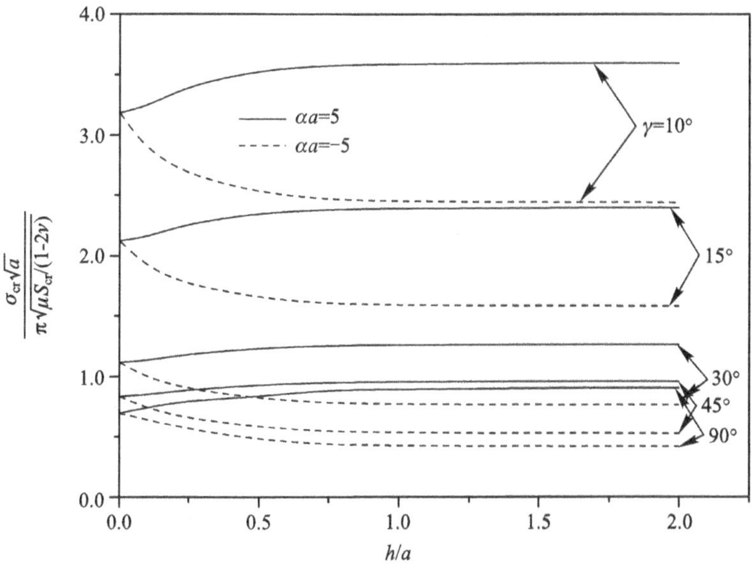

Fig. 6.18 Variations of σ_{cr} with h for different α and γ values ($d = 0.2a$)

6.5 The SIFs for a crack perpendicular to the FGM interlayer

6.5.1 General

In Fig. 6.19a, a FGM interlayer is fully bonded with two homogeneous media of semi-infinite extent to form a three-dimensional FGM system. A penny-shaped crack is located within the homogeneous material 1 and is perpendicular to the graded interfacial zone. The graded interfacial material zone has an elastic modulus according to Eq. (6.1) and the penny-shaped crack surfaces are subject to both normal and shear stresses as follows

$$\sigma^{+}_{x'x'} = \sigma^{-}_{x'x'} = p, \quad (0 \leqslant r < a) \tag{6.12a}$$

$$\sigma^{+}_{x'y'} = \sigma^{-}_{x'y'} = q, \quad (0 \leqslant r < a) \tag{6.12b}$$

$$\sigma^{+}_{x'z'} = \sigma^{-}_{x'z'} = 0, \quad (0 \leqslant r < a) \tag{6.12c}$$

where the superscripts + and – correspond to the crack surfaces, which are located at $x' = 0^+$ and $x' = 0^-$, respectively, and a is the radius of the penny-shaped crack.

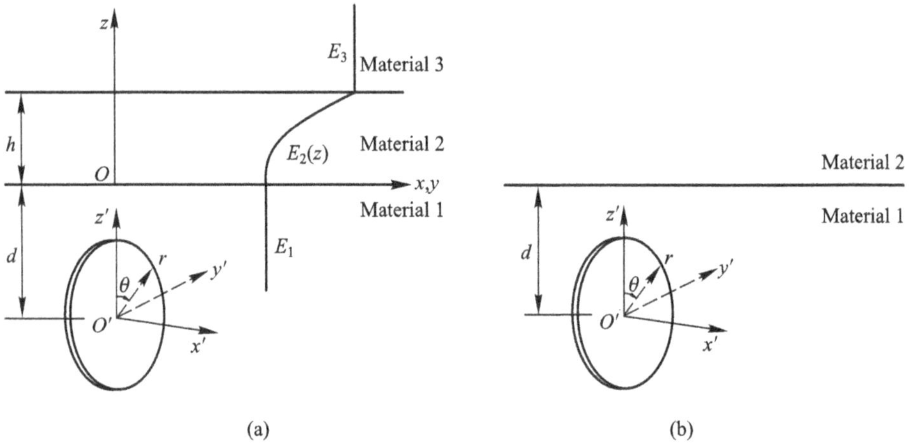

Fig. 6.19 (a) Penny-shaped crack perpendicular to the interface of a bonded bi-material system with a graded interfacial zone, or (b) without a graded interfacial zone

6.5.2 Numerical verifications

Kou and Keer (1995) examined the crack problems shown in Fig. 6.19b using an equivalent body force method. We re-examine this crack problem by using the proposed BEM method. As shown in Fig. 6.19b, the penny-shaped crack surfaces with radius a are perpendicular to the interface of the bi-material system. The bi-material consists of two elastic solids of half-space extent and the two solids are fully bonded together. In Eq. (6.12), let $p = -p_0$ ($p_0 > 0$) and $q = 0$, that is, the crack surfaces are subject to uniform compressive stresses. The elastic modulus and Poisson's ratio are, respectively, denoted by E_1 and ν_1 for material 1 and E_2 and ν_2 for material 2. The ratio of the elastic moduli is $\chi(= E_2/E_1)$ and the Poisson's ratio is chosen to be $\nu_1 = \nu_2 = 0.3$ for both materials. The crack center is located in the lower half-space at a distance d from the bi-material interface.

The cracked body shown in Fig. 6.19b is symmetrical with respect to the $O'x'z'$-plane; therefore, only the left part of the entire crack geometry is examined. For convenience, a hemisphere with radius a in the lower homogeneous material (material 1) is chosen to form the two regions for the BEM analysis. Because of symmetry, only half of the hemisphere needs to be analyzed and this is the section that is discretized into the BEM meshes. Figure 6.20 shows the discretized left symmetrical surface with 176 elements and 553 nodes in which there are 32 traction-singular elements along the crack front. The boundary mesh shown in Fig. 6.20 divides the cracked body into two regions.

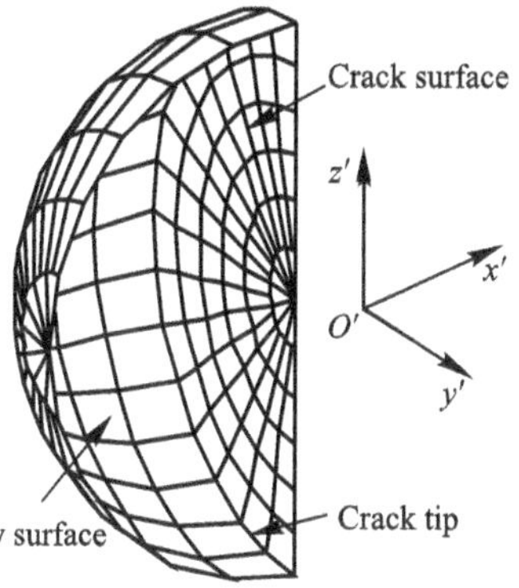

Fig. 6.20 Boundary element meshes for the cracked body

Figure 6.21 shows the mode I SIFs along the crack front for $d = 1.2a$ with different values of χ. The mode I SIF values are normalized with the square root of the crack radius and normal traction on the crack surfaces. It is noted that the curves for $\chi = 1$ and 0 correspond, respectively, to the solutions of a penny-shaped crack in a full-space and a half-space. The case $\chi = \infty$ indicates that the crack is located in the lower half-space bonded to a rigid upper half-space. A comparison of the results from this proposed method with those given by Kou and Keer (1995) is given in Table 6.2.

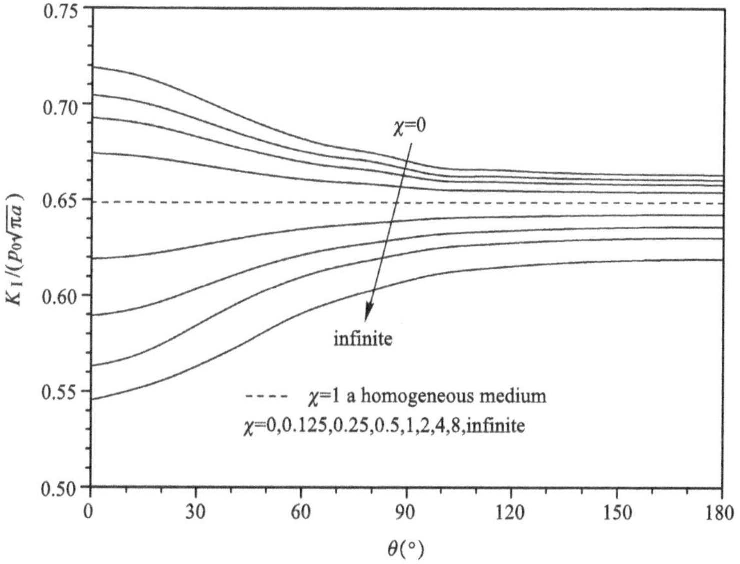

Fig. 6.21 Variations of $K_I/(p_0\sqrt{\pi a})$ with different $\chi(= E_2/E_1)$ values ($d = 1.2a$)

Table 6.2 Comparison of the SIF values ($K_{\mathrm{I}}/(p_0\sqrt{\pi a})$) between the present study and Kou and Keer (1995)

$\chi = E_2/E_1$	θ=11.25°		θ=168.75°	
	Present	Kou and Keer (1995)	Present	Kou and Keer (1995)
∞	0.5504	0.572	0.6191	0.634
8.000	0.5670	0.589	0.6303	0.637
4.000	0.5919	0.602	0.6360	0.639
2.000	0.6203	0.621	0.6424	0.642
1.000	0.6487	0.645	0.6487	0.646
0.500	0.6731	0.672	0.6529	0.650
0.250	0.6910	0.694	0.6564	0.654
0.125	0.7023	0.712	0.6586	0.657
0.000	0.7162	0.735	0.6610	0.661

It should be noted that the results by Kou and Keer are numerical solutions, not analytical ones. The exact solution of the normalized SIF value ($K_{\mathrm{I}}/(p_0\sqrt{\pi a})$) associated with the penny-shaped crack in a homogeneous solid of infinity extent is $2/\pi \approx 0.6366$. For such a crack, the normalized SIF values from the current study and Kou and Keer (1995) are 0.6487 and 0.647, respectively. The numerical results are less than 2% greater than the exact solution $2/\pi$. Furthermore, in Table 6.2, it can be seen that the maximum difference between the present analysis and Kou and Keer's results is 2.1%; this occurs when $\theta = 11.25°$ for $\chi = \infty$. Therefore, it can be concluded that the present method and the element mesh can be further employed for accurate and efficient analyses of penny-shaped crack problems in bi-material systems with a FGM interlayer.

Similar conclusions can be drawn as those given in Kou and Keer (1995): As shown in Fig. 6.21, the SIFs decrease when $\chi < 1$ and increase for $\chi > 1$ as the tip point along the crack front moves away from the interface (i.e., θ increases). For a crack in a more compliant body, i.e., $\chi > 1$, a stiffer upper solid (material 2) constrains the crack opening, whilst for the crack in a less compliant solid, i.e., $\chi < 1$, the softer upper solid tends to increase the crack opening. As the crack front moves away from the interface, the influence of the upper solid becomes weaker and weaker.

6.5.3 The SIFs for a crack subjected to uniform compressive stresses

In the ensuing, we investigate the penny-shaped crack problem as shown in Fig. 6.19a. Consider uniform compressive stresses acting on the crack surfaces, i.e., $p = -p_0$ ($p_0 > 0$) and $q = 0$ in Eqs. (6.12). In this case only the mode I SIF values are not equal to zero and the mode II and III SIF values are always zero.

(1) The variations of the SIF with a non-homogeneity parameter α and the angle θ

Figure 6.22 shows the variations in the SIF values with α and θ where $h = 0.5a$ and $d = 1.2a$, and Table 6.3 presents the corresponding SIF data. In Figure 6.22, the SIF values for positive and negative α values are located below and above the SIF value curve for $\alpha = 0$, respectively, which corresponds to the SIF values of the crack in a homogeneous elastic solid. As the absolute α value increases, the SIF values monotonically move away from the SIF curve for $\alpha = 0$. The SIF values vary with the location of the crack front, which is represented by the angle θ. The location where the crack front is closest to the FGM interfacial layer is at $\theta = 0°$ and the farthest location is at $\theta = 180°$. As the crack front moves from the closest location $\theta = 0°$ to the farthest location $\theta = 180°$, the SIF values monotonically increase and decrease below and above the $\alpha = 0$ curve for positive and negative α values, respectively. The above results indicate the following physical phenomena: For a crack in a compliant body (i.e., $\alpha > 0$ or $E_1 < E_2 < E_3$), the crack opening is constrained by the upper stiffer FGM. Such constraint becomes more pronounced at the crack tip location closest to the interlayer. On the other hand, for a crack in a stiffer solid (i.e., $\alpha < 0$ or $E_1 > E_2 > E_3$), the SIF values become larger and larger as the crack front becomes closer and closer to the interlayer.

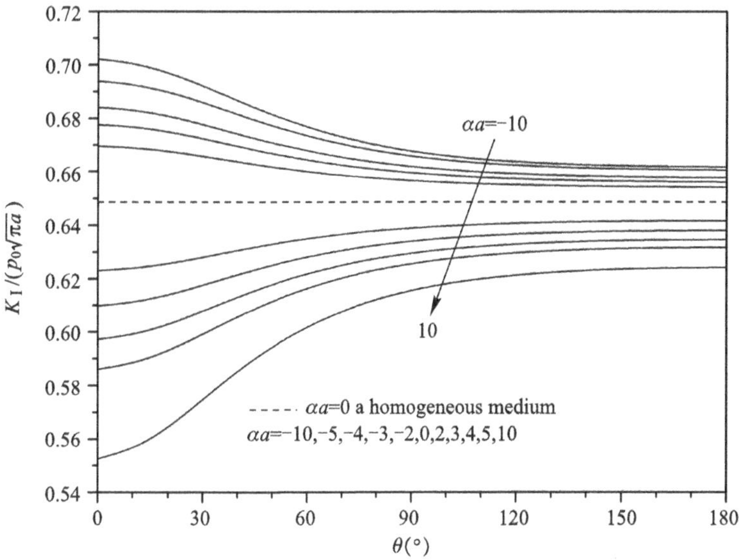

Fig. 6.22 Variations of $K_I/(p_0\sqrt{\pi a})$ with θ for different α values ($d = 1.2a$, $h = 0.5a$)

(2) The variations in the SIF with the distance d of the crack from the FGM interlayer

Figure 6.23 shows how the crack distance d to the FGM interlayer causes variations in the SIF, where $h = 0.5a$ and $\alpha a = -5$ or 5, and Table 6.4 presents the corresponding SIF data. As in the previous observations, we observe that the SIF curves for the four distances considered are located below and above the $\alpha a = 0$ curve for $\alpha a = 5$ and

–5, respectively. For $\alpha a = -5$, the SIF values are highest for $\theta = 0°$ and monotonically decrease to a minimum at $\theta = 180°$. For $\alpha a = 5$, the SIF values are lowest for $\theta = 0°$ and monotonically increase to a maximum at $\theta = 180°$. As d decreases, the SIF values decrease for $\alpha a = 5$ and increase as d increases for $\alpha a = -5$.

Table 6.3 Variations of the SIF values $(K_I/(p_0\sqrt{\pi a}))$ with the parameter α at $\theta = 0°$ and 180° $(d = 1.2a$ and $h = 0.5a)$

αa	$\theta = 0°$	$\theta = 180°$
–10	0.70209	0.66184
–5	0.69385	0.66069
–4	0.68409	0.65788
–3	0.67763	0.65627
–2	0.66958	0.65420
0	0.64862	0.64862
2	0.62301	0.64158
3	0.60982	0.63795
4	0.59728	0.63457
5	0.58601	0.63164
10	0.55242	0.62413

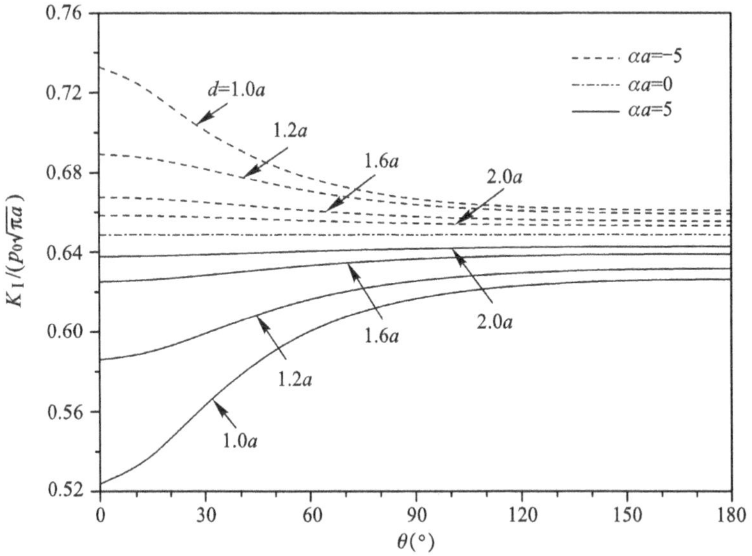

Fig. 6.23 Variations of $K_I/(p_0\sqrt{\pi a})$ with θ for different α and d values $(h = 0.5a)$

Table 6.4 Variations of the SIF values $(K_{\mathrm{I}}/(p_0\sqrt{\pi a}))$ with the parameters α and d ($h = 0.5a$)

d	$\alpha a = 5$		$\alpha a = -5$	
	$\theta = 0°$	$\theta = 180°$	$\theta = 0°$	$\theta = 180°$
1.0a	0.52375	0.62630	0.73285	0.66084
1.2a	0.58601	0.63164	0.68915	0.65911
1.6a	0.62521	0.63898	0.66746	0.65539
2.0a	0.63780	0.64271	0.65840	0.65314

(3) The variations in the SIF with the thickness h of the FGM interlayer

Figure 6.24 shows the variations of the SIF with the thickness h of the FGM interlayer where $\alpha a = -2, 2$ and $d = a$, while Table 6.5 presents the corresponding SIF data. For $d = 1.0a$, the penny-shaped crack is in contact with the FGM interfacial zone at $\theta = 0°$. Here, as h increases, the SIF values decrease and increase for $\alpha a = 2$ and –2, respectively. In particular, it can be seen that the curves indicate that any increase in the thickness after $h > 2.5a$ will not cause visible changes to the SIF values around the crack front $0° \leqslant \theta \leqslant 180°$. This finding is consistent with that given by Choi (2001) for a crack in a FGM bi-material system under plane strain loading.

Table 6.5 Variations of the SIF values $(K_{\mathrm{I}}/(p_0\sqrt{\pi a}))$ with the thickness $h(d = a)$

h	$\alpha a = -2$		$\alpha a = 2$	
	$\theta = 0°$	$\theta = 180°$	$\theta = 0°$	$\theta = 180°$
0.10a	0.66378	0.65058	0.63278	0.64652
0.25a	0.67630	0.65293	0.61792	0.64364
0.50a	0.69106	0.65555	0.59875	0.63963
0.75a	0.69646	0.65703	0.58871	0.63648
1.00a	0.69848	0.65783	0.58304	0.63415
1.25a	0.70004	0.65826	0.57905	0.63253
1.50a	0.70109	0.65850	0.57646	0.63147
2.00a	0.70200	0.65870	0.57407	0.63042
2.50a	0.70252	0.65877	0.57305	0.63004
3.00a	0.70282	0.65897	0.57275	0.62984

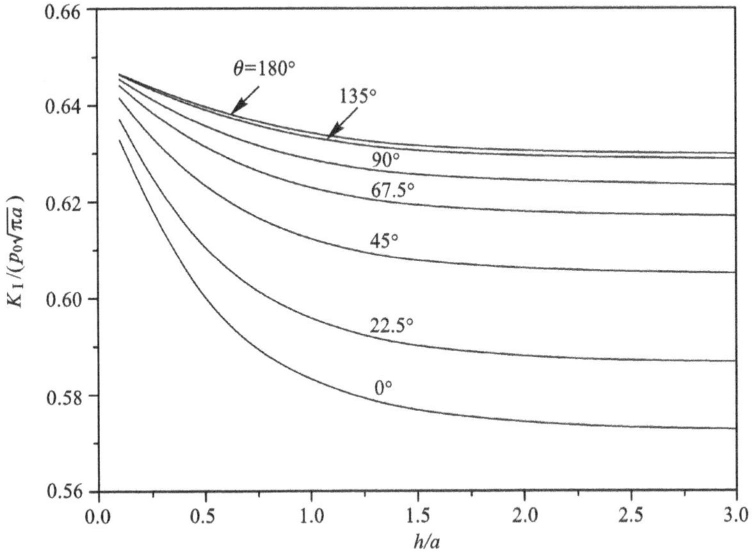

(a) The non-homogeneity parameter αa=2

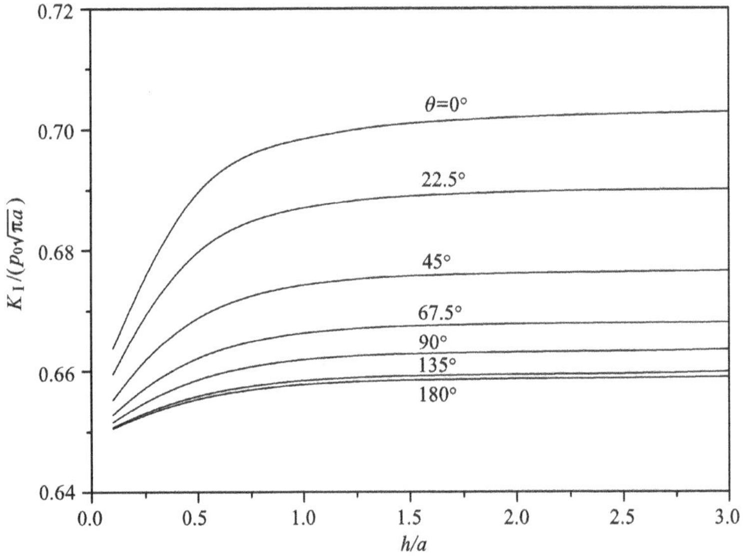

(b) The non-homogeneity parameter αa=−2

Fig. 6.24 Variations of $K_{\mathrm{I}}/(p_0\sqrt{\pi a})$ with h for different α and θ values ($d = a$)

6.5.4 The SIFs for a crack subjected to uniform shear stresses

1. General

In Eq. (6.12), let $p = 0$ and $q = -q_0$ ($q_0 > 0$), that is, the crack surfaces are subjected to uniform shear stresses. In this case the mode I SIF values are zero whereas the mode II and III SIF values are non-zero. The exact solutions for a penny-shaped crack in a homogeneous medium of infinite extent are used to check the numerical results from the present BEM method. For the mode II SIF values $(K_{\mathrm{II}}/(q_0\sqrt{\pi a}))$, the present results give a maximum relative error of 2.1% with respect to the exact solution for different θ angles. This maximum error occurs at the crack front location $\theta = 0°$. For the mode III SIF values $(K_{\mathrm{III}}/(q_0\sqrt{\pi a}))$, the present results give a maximum relative error 2.5% with respect to the exact solution for different θ angles. The crack front location corresponding to this maximum relative error occurs at $\theta = 22.5°$. In the following, we discuss the variations of the mode II and III SIF values with the non-homogeneity parameter α, the crack distance d to the FGM interlayer and the thickness h of the FGM interlayer.

2. The mode II SIF values

(1) The variations of the SIF values with the parameter α

Figure 6.25 shows the variations of SIF values for different α values where $d = 1.2a$ and $h = 0.5a$ and Table 6.6 presents the corresponding SIF data. It can be found that if $\alpha = 0$ the SIF values monotonically decrease from their highest positive values (around 0.77450) at $\theta = 0°$ to their lowest negative values (around –0.77450) at $\theta = 180°$. When $\theta = 90°$, the SIF values are almost zero. Different α values have limited effects on the SIF values and cause the curves move away from the special curve corresponding to $\alpha = 0$. Negative α values slightly increase the SIF values while positive α values cause slight decreases in the SIF values. The larger the absolute α value, the larger the changes in the SIF values. Such effects reach the first maximum at $\theta = 0°$, rapidly reduce to zero at $\theta = 110°$, and then generally increase to the second maximum at $\theta = 180°$.

(2) The variations of the SIF values with the distance d

Figure 6.26 shows the variations of the SIF with the distance d and the angle θ where $h = 0.5a$, $\alpha a = -5$ and 5; Table 6.7 presents the corresponding SIF data. It can be seen that the SIF values monotonically decrease from their highest positive values at $\theta = 0°$ to their lowest negative values at $\theta = 180°$. When $\theta = 90°$, the SIF values are almost zero. Different d values have a limited effect on the SIF values. The larger the d value, the less the change in the SIF values. As d increases, the SIF values increase for positive α, decrease for negative α and approach those corresponding to the special case $\alpha = 0$. Such effects reach the first maximum at $\theta = 0°$, rapidly reduce to zero at $\theta = 110°$, and then generally increase to the second maximum at $\theta = 180°$.

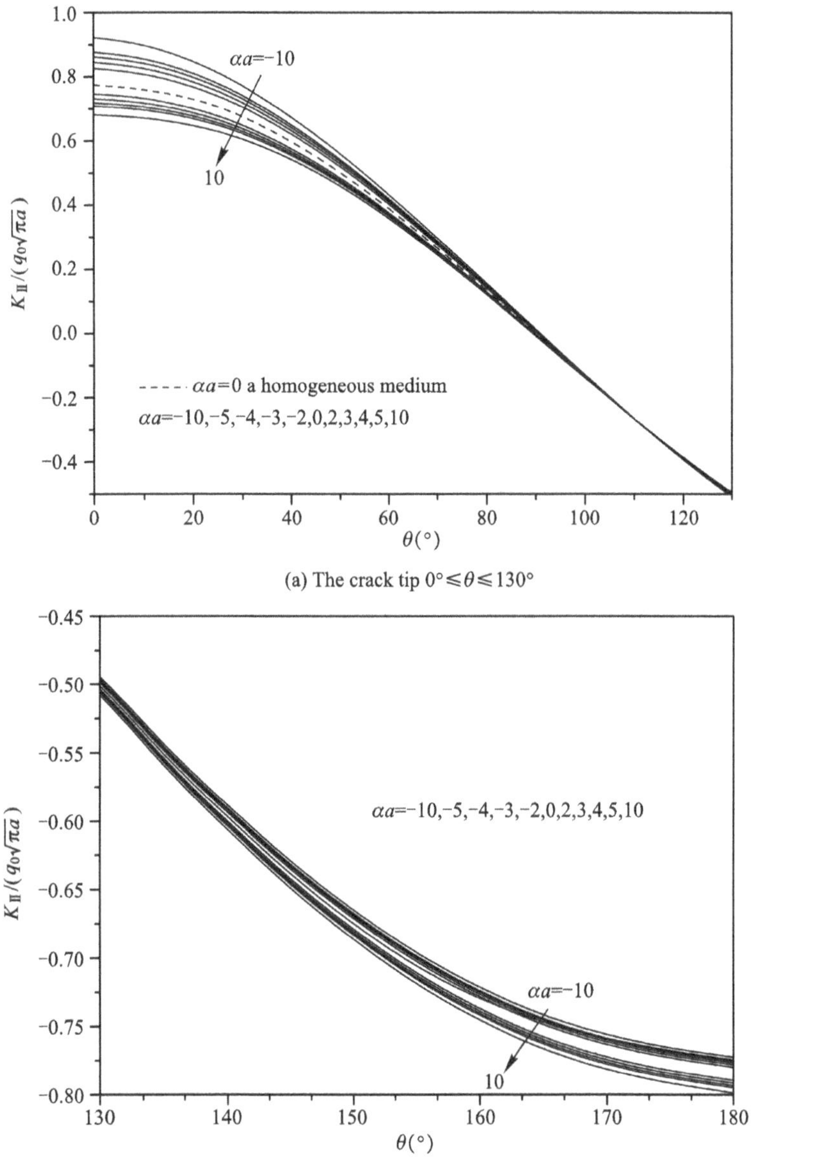

(a) The crack tip 0°≤θ≤130°

(b) The crack tip 130°≤θ≤180°

Fig. 6.25 Variations of $K_{II}/(q_0\sqrt{\pi a})$ with θ for different α values $(d = 1.2a, h = 0.5a)$

Table 6.6 Variations of the SIF values $(K_{\mathrm{II}}/(q_0\sqrt{\pi a}))$ with the parameter α ($d = 1.2a$ and $h = 0.5a$)

$\theta(°)$	$\alpha a = -10$	–5	–4	–3	–2	0	2	3	4	5	10
0.00	0.92177	0.87656	0.86199	0.84507	0.82599	0.77450	0.74646	0.73111	0.71853	0.70847	0.68111
11.25	0.89922	0.85680	0.84302	0.82698	0.80887	0.75947	0.73341	0.71891	0.70707	0.69764	0.67224
22.50	0.83570	0.80015	0.78835	0.77453	0.75888	0.72477	0.69372	0.68134	0.67133	0.66342	0.64272
33.75	0.74045	0.71295	0.70356	0.69247	0.67985	0.65231	0.62742	0.61762	0.60978	0.60367	0.58821
45.00	0.62130	0.60111	0.59400	0.58554	0.57586	0.55472	0.53576	0.52841	0.52259	0.51811	0.50718
56.25	0.48420	0.46992	0.46474	0.45854	0.45142	0.43586	0.42204	0.41675	0.41262	0.40948	0.40207
67.50	0.33383	0.32413	0.32051	0.31616	0.31115	0.30021	0.29062	0.28700	0.28421	0.28213	0.27737
78.75	0.17503	0.16880	0.16643	0.16356	0.16025	0.15306	0.14686	0.14458	0.14285	0.14158	0.13881
90.00	0.01279	0.00922	0.00782	0.00613	0.00418	0.00000	–0.00348	–0.00472	–0.00562	–0.00625	–0.00752

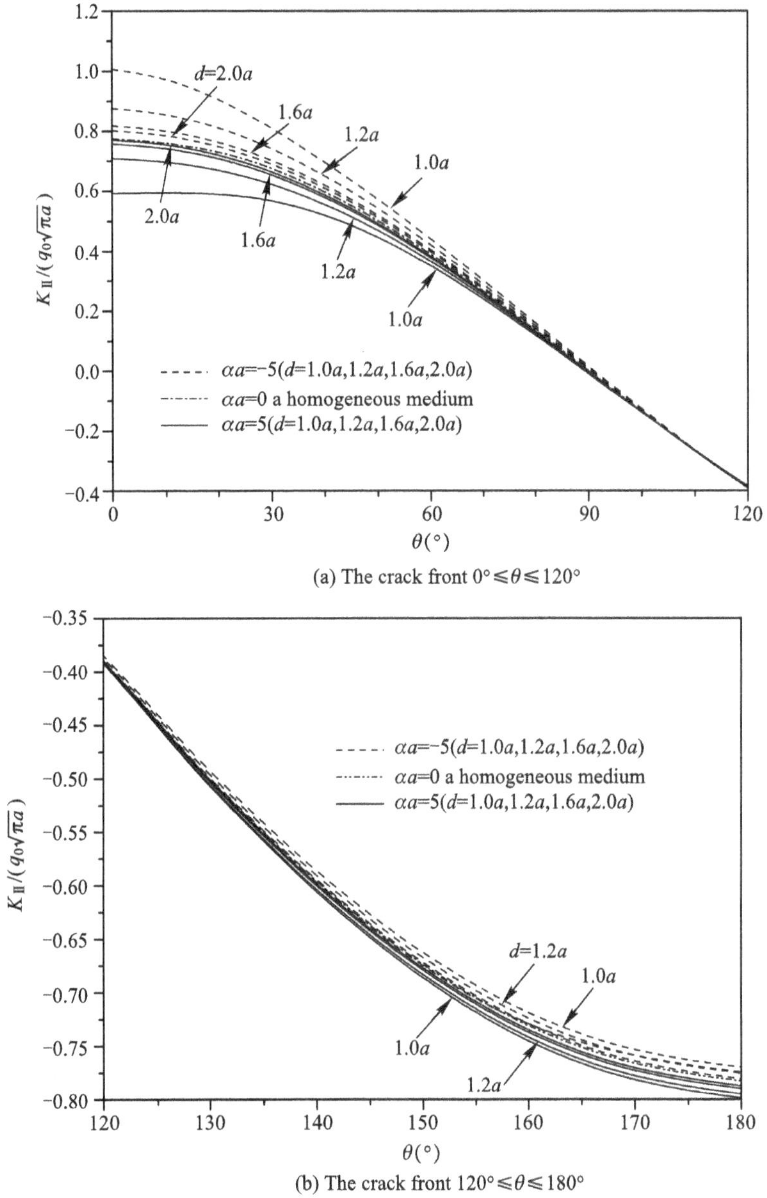

Fig. 6.26 Variations of $K_{\text{II}}/(q_0\sqrt{\pi a})$ with θ for different α and d values ($h = 0.5a$)

Table 6.7 Variations of the SIF values $(K_{\mathrm{II}}/(q_0\sqrt{\pi a}))$ with the parameters α and d $(h = 0.5a)$

$\theta(°)$	$\alpha a = 5$				$\alpha a = -5$			
	d=1.0a	1.2a	1.6a	2.0a	1.0a	1.2a	1.6a	2.0a
0.00	0.59292	0.70847	0.75682	0.77008	1.00636	0.87656	0.81748	0.80131
11.25	0.59545	0.69764	0.74270	0.75541	0.97494	0.85680	0.80142	0.78587
22.50	0.59036	0.66342	0.70050	0.71174	0.88861	0.80015	0.75388	0.74000
33.75	0.55803	0.60367	0.63153	0.64079	0.76998	0.71295	0.67739	0.66580
45.00	0.49049	0.51811	0.53779	0.54497	0.63680	0.60111	0.57525	0.56614
56.25	0.39266	0.40948	0.42271	0.42795	0.49273	0.46992	0.45187	0.44510
67.50	0.27204	0.28213	0.29049	0.29407	0.33889	0.32413	0.31207	0.30733
78.75	0.13593	0.14158	0.14631	0.14852	0.17825	0.16880	0.16125	0.15822
90.00	–0.00880	–0.00625	–0.00423	–0.00312	0.01497	0.00922	0.00506	0.00342

(3) The variations of the SIF values with the thickness h

Figures 6.27 and 6.28 show the effect of the thickness h on the SIF values where $\alpha a = -2$ and 2. In the case $d = a$, the penny-shaped crack is in contact with the FGM interlayer. For $0° \leqslant \theta < 90°$ and as h increases, the SIF values slightly increase for $\alpha a = -2$ and decrease for $\alpha a = 2$. However, for $90° \leqslant \theta < 180°$, the SIF values show almost no change as h increases. In addition, the SIF values hardly change after $h/a > 2.5$.

3. The mode Ⅲ SIF values

(1) The variations of the SIF values with the parameter α

Figure 6.29 shows the mode Ⅲ SIF values with the parameter α and the angle θ where $d = 1.2a$ and $h = 0.5a$ and Table 6.8 presents the corresponding data. In this figure, it can be seen that the SIF values monotonically increase from zero at $\theta = 0°$ to their highest value at $\theta = 90°$ and reduce to zero again at $\theta = 180°$. Different α values have limited effects on the SIF values. In comparison with the corresponding values of the special case $\alpha = 0$, negative α values cause slight increases in the SIF values and positive α values cause slight decreases in the SIF values. The larger the absolute α value, the larger the change in the SIF value. Such effects are mostly evident for $30° \leqslant \theta \leqslant 90°$ and are almost zero for $140° \leqslant \theta \leqslant 180°$.

(2) The variations of the SIF values with the distance d

Figure 6.30 shows the variations of the SIF values with the angle θ and the distance d where $h = 0.5a$ and $\alpha a = -5$ or 5 and Table 6.9 presents the corresponding data. Here the SIF values monotonically increase from zero at $\theta = 0°$ to their highest value at $\theta = 90°$ and reduce to zero again at $\theta = 180°$. Different d values have limited effects on the SIF values. Thus the larger the d value, the less the change in the SIF values. As d increases, the SIF values increase for positive α values and decrease for negative α values, approaching the curve corresponding to the special case $\alpha = 0$. Such effects are mostly evident for $30° \leqslant \theta \leqslant 90°$ and almost zero for $140° \leqslant \theta \leqslant 180°$.

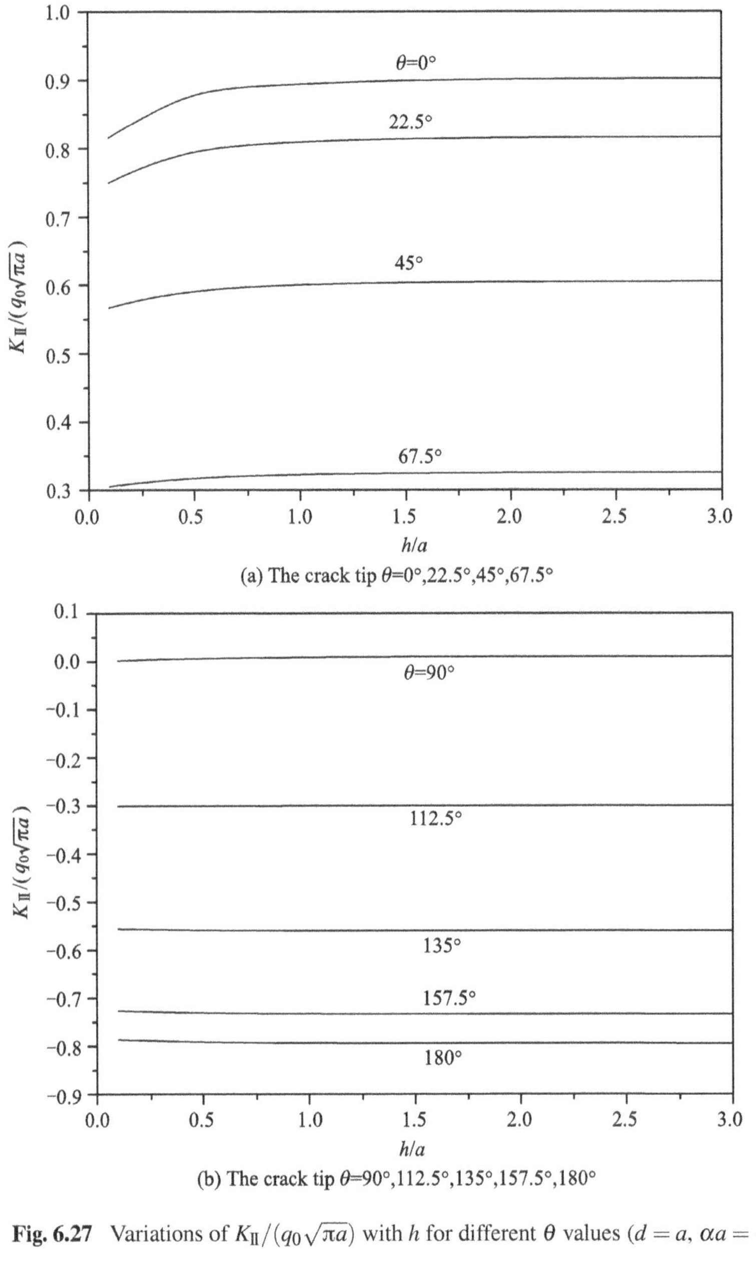

Fig. 6.27 Variations of $K_{II}/(q_0\sqrt{\pi a})$ with h for different θ values ($d = a$, $\alpha a = -2$)

(3) The variations of the SIF values with the thickness h

In Figs. 6.31 and 6.32 we see the variations of the SIF values with the thickness h for $\alpha a = -2$ and 2, respectively, where the distance d is equal to the radius of the penny-shaped crack a. In other words, the penny-shaped crack contacts the FGM interlayer. For

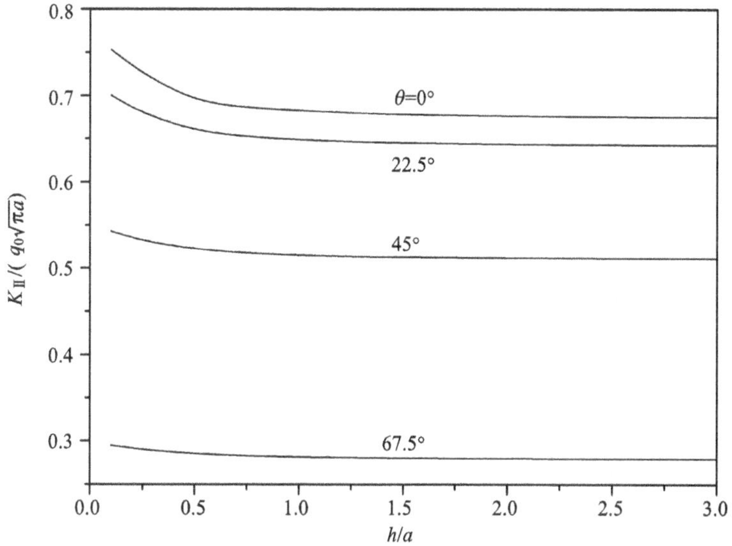

(a) The crack tip θ=0°,22.5°,45°,67.5°

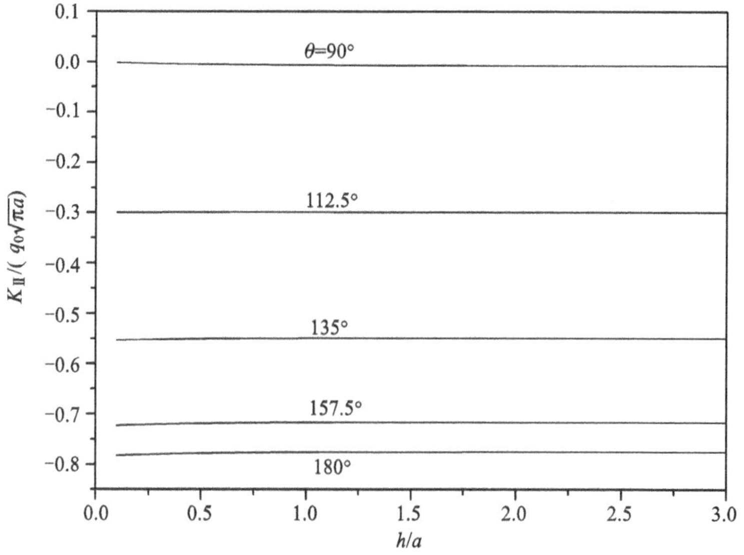

(b) The crack tip θ=90°,112.5°,135°,157.5°,180°

Fig. 6.28 Variations of $K_{\text{II}}/(q_0\sqrt{\pi a})$ with h for different θ values ($d = a$, $\alpha a = 2$)

$0° \leqslant \theta < 90°$, the SIF values increase slightly for $\alpha a = -2$ and decrease for $\alpha a = 2$ as h increases. However, for $90° \leqslant \theta < 180°$, the SIF values show almost no change as h increases. In addition, there is almost no change in the SIF values after $h/a > 2.5$. It

Table 6.8 Variations of the SIF values $(K_{\mathrm{III}}/(q_0\sqrt{\pi a}))$ with the parameter $\alpha(d = 1.2a$ and $h = 0.5a)$

$\theta(°)$	$\alpha a = -10$	–5	–4	–3	–2	0	2	3	4	5	10
0.00	0.00000	0.00000	0.00000	0.00000	0.00000	0.00000	0.00000	0.00000	0.00000	0.00000	0.00000
11.25	0.13916	0.13067	0.12800	0.12493	0.12152	0.11427	0.10790	0.10542	0.10345	0.10190	0.09779
22.50	0.26201	0.24847	0.24405	0.23893	0.23318	0.22085	0.20995	0.20571	0.20233	0.19971	0.19291
33.75	0.36096	0.34600	0.34092	0.33495	0.32818	0.31352	0.30045	0.29537	0.29134	0.28823	0.28040
45.00	0.43491	0.42078	0.41580	0.40988	0.40310	0.38830	0.37504	0.36988	0.36581	0.36268	0.35503
56.25	0.48585	0.47356	0.46910	0.46375	0.45759	0.44402	0.43182	0.42708	0.42334	0.42049	0.41365
67.50	0.51686	0.50671	0.50295	0.49840	0.49314	0.48149	0.47097	0.46688	0.46367	0.46121	0.45542
78.75	0.53140	0.52332	0.52028	0.51659	0.51230	0.50276	0.49410	0.49074	0.48808	0.48606	0.48130
90.00	0.53193	0.52570	0.52333	0.52045	0.51709	0.50957	0.50271	0.50002	0.49790	0.49627	0.49244

should be noted that the crack tips at 67.5° and 112.5°, 45° and 135°, 22.5° and 157.5°, or 11.25° and 168.75° are symmetric pairs with respect to the y'-axis.

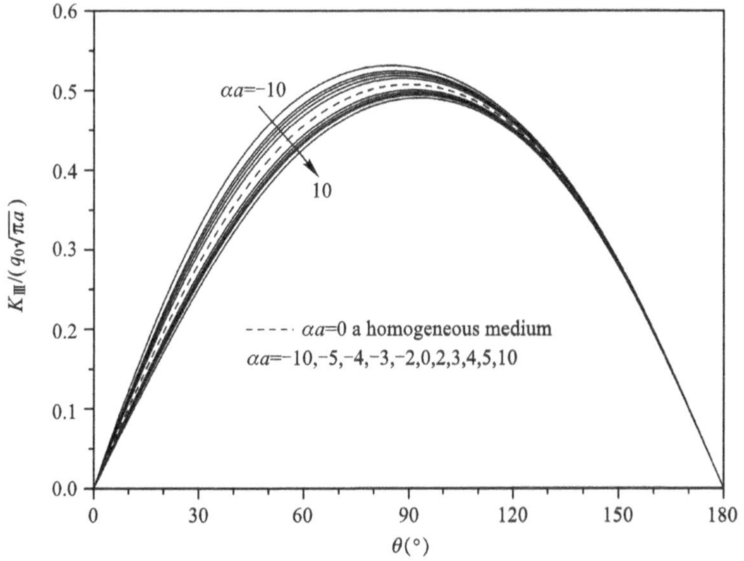

Fig. 6.29 Variations of $K_{III}/(q_0\sqrt{\pi a})$ with θ for different α values ($d = 1.2a$, $h = 0.5a$)

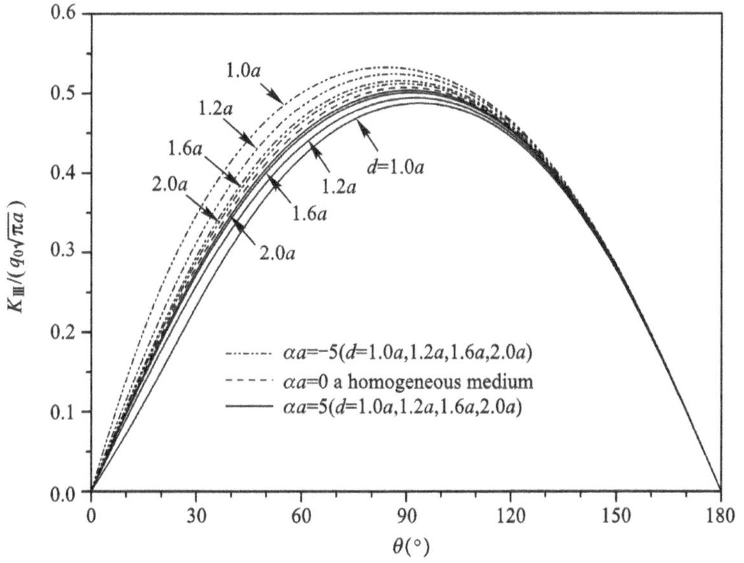

Fig. 6.30 Variations of $K_{III}/(q_0\sqrt{\pi a})$ with θ for different d and α values ($\alpha a = -5, 5$, $h = 0.5a$)

Table 6.9 Variations of the SIF values ($K_{III}/(q_0\sqrt{\pi a})$) with the parameters α and d ($\alpha a = -5, 5$ and $h = 0.5a$)

$\theta(°)$	$\alpha a = 5$				$\alpha a = -5$			
	d=1.0a	1.2a	1.6a	2.0a	d=1.0a	1.2a	1.6a	2.0a
0.00	0.00000	0.00000	0.00000	0.00000	0.00000	0.00000	0.00000	0.00000
11.25	0.07925	0.10190	0.10954	0.11164	0.16197	0.13067	0.12000	0.11731
22.50	0.17122	0.19971	0.21226	0.21603	0.28725	0.24847	0.23125	0.22642
33.75	0.26361	0.28823	0.30251	0.30727	0.37874	0.34600	0.32683	0.32078
45.00	0.34335	0.36268	0.37646	0.38148	0.44576	0.42078	0.40267	0.39629
56.25	0.40560	0.42049	0.43263	0.43739	0.49217	0.47356	0.45792	0.45189
67.50	0.44975	0.46121	0.47141	0.47559	0.52043	0.50671	0.49389	0.48860
78.75	0.47715	0.48606	0.49440	0.49790	0.53329	0.52332	0.51317	0.50874
90.00	0.48926	0.49627	0.50302	0.50585	0.53281	0.52570	0.51785	0.51429

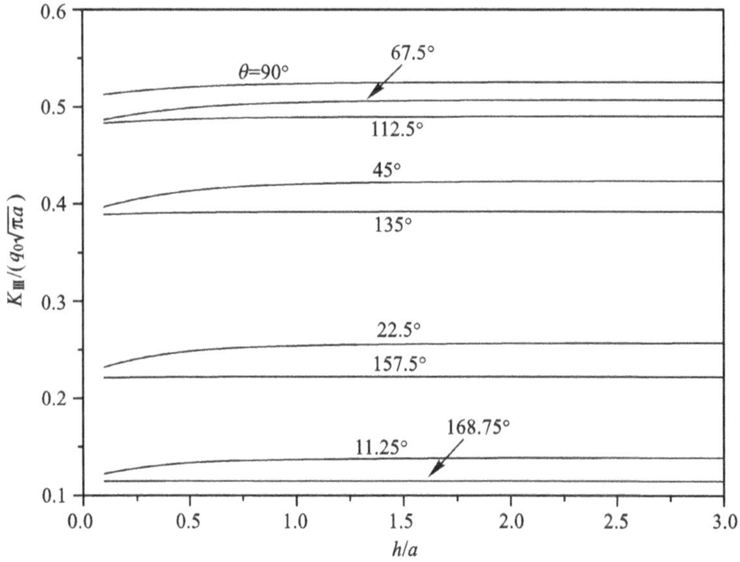

Fig. 6.31 Variations of $K_{III}/(q_0\sqrt{\pi a})$ with h for different θ values ($d = a$, $\alpha a = -2$)

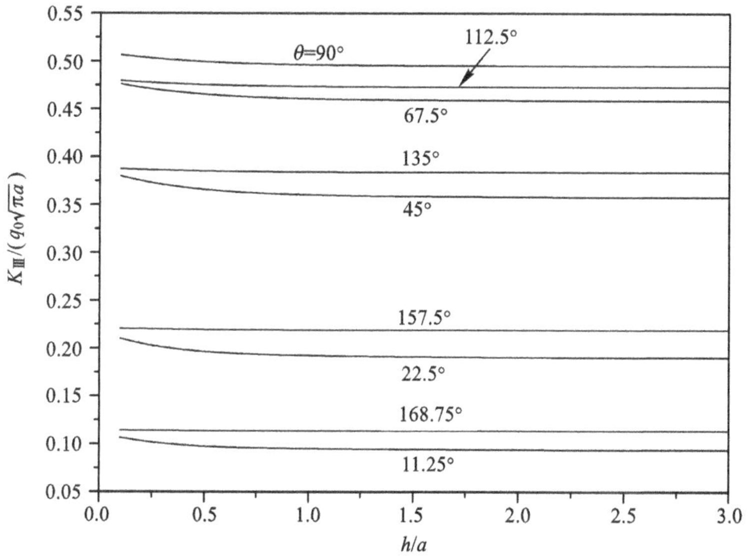

Fig. 6.32 Variations of $K_{\mathrm{III}}/(q_0\sqrt{\pi a})$ with h for different θ values ($d = a$, $\alpha a = 2$)

6.6 The growth of the crack perpendicular to the FGM interlayer

6.6.1 The crack growth under a remotely inclined tensile loading

In the ensuing, we will examine the growth potential of the penny-shaped crack perpendicular to the FGM interlayer. Using the superposition principle, the crack problem under a remotely inclined tensile loading σ, shown in Fig. 6.33, is examined in the same way as that of a crack loaded by statically self-equilibrium tractions σ. Equations (6.11) can be further employed to establish the relationship between the normal traction p_0 and the shear traction q_0 on the crack surfaces and the remotely inclined tensile loading σ. In Section 6.5, we obtained the SIF values of the crack where the surfaces are subject to the loadings p_0 and q_0; these SIF values will be used to calculate the strain energy density factor S and analyze the crack growth. Notice that the tensile loading σ is in the $O'x'z'$-plane, and therefore the crack problem is symmetric.

(1) The minimum strain energy density factor at the crack front

Figure 6.34 shows the influence of the parameter α on $S_{\min}$ when $d = 1.2a$, $h = 0.5a$ and $\gamma = 60°$. The $S_{\min}$ curves of $\alpha > 0$ and $\alpha < 0$ are located on either side of the one for $\alpha = 0$. As the α value increases, the $S_{\min}$ value at a given point decreases. In the vicinity of the crack front $\theta = 0°$, the α values exert a more obvious influence on $S_{\min}$.

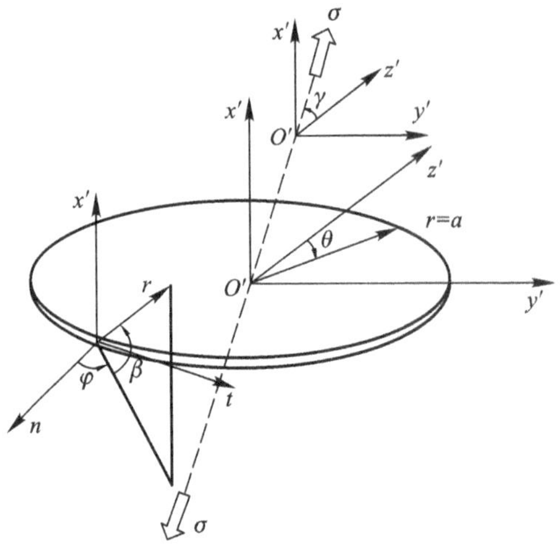

Fig. 6.33 A penny-shaped crack under a remotely inclined tensile loading

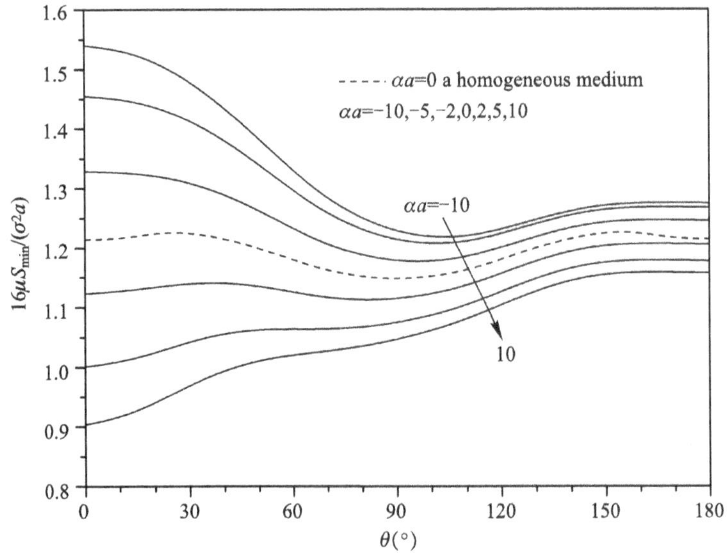

Fig. 6.34 Variations of $S_{\min}$ with θ for different α values ($d = 1.2a$, $h = 0.5a$, $\gamma = 60°$)

(2) The critical loadings of crack growth

In Fig. 6.35 the variations of the critical loading σ_{cr} with the loading angle γ and the parameter α are shown, where $d = 1.2a$ and $h = 0.5a$. For a given α value, σ_{cr} decreases as γ increases. The σ_{cr} curves for $\alpha > 0$ and $\alpha < 0$ are located on either side of the one for $\alpha = 0$. As α increases, σ_{cr} increases. For $\alpha < 0$ the FGM interlayer has an obvious influence on σ_{cr} whereas for $\alpha > 0$ the influence of the FGM interlayer is weak on σ_{cr}. Figure

6.34 can further explain the above phenomenon: It can be seen that when $\alpha > 0$, the S_{min} values of the crack front close to the FGM interlayer decrease; the S_{min} value of the crack front far away from the FGM interlayer can be employed to determine σ_{cr} and, at this location, the α values have a weak influence on S_{min}. If $\alpha > 0$ the results are opposite to those for $\alpha < 0$.

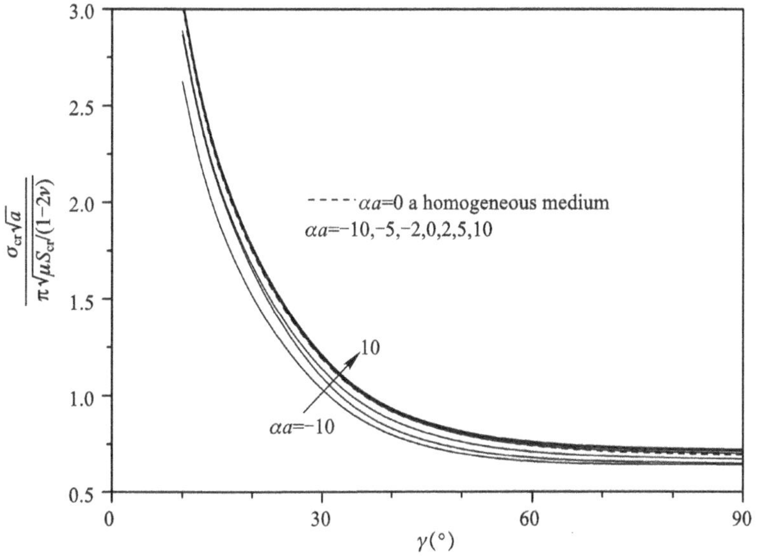

Fig. 6.35 Variations of σ_{cr} with γ for different α values ($d = 1.2a$, $h = 0.5a$)

Under a remotely inclined tensile loading, the crack growth tends towards the direction perpendicular to the external loading. Figure 6.36 shows the variation of σ_{cr} with the loading angle γ where $\alpha a = -5,5$ and $d = a$. If $d = a$, the crack front $\theta = 0°$ is in full contact with the FGM interlayer. For $\alpha a = -5$, σ_{cr} shows obvious variations whereas when $\alpha a = 5$ σ_{cr} varies weakly. For $d > a$, the influence of a given α value on σ_{cr} is not larger than that for $d = a$. That is, Fig. 6.36 shows that the largest influence of the FGM interlayer on σ_{cr} occurs when $\alpha a = -5$ and 5.

Figure 6.37 shows the variations of σ_{cr} with the thickness h of the FGM interlayer where $\alpha a = -2,2$ and $d = 1.2a$. For a smaller loading angle γ, the FGM interlayer has an obvious influence on σ_{cr}. When $\alpha a = -2$, as the thickness h of the FGM interlayer increases, σ_{cr} decreases and varies weakly at $h = a$. If $\alpha a = 2$, as the thickness h of the FGM interlayer increases,σ_{cr} increases weakly. The above variations of σ_{cr} for $\alpha a = -2,2$ are suitable for $\alpha > 0$ and $\alpha < 0$.

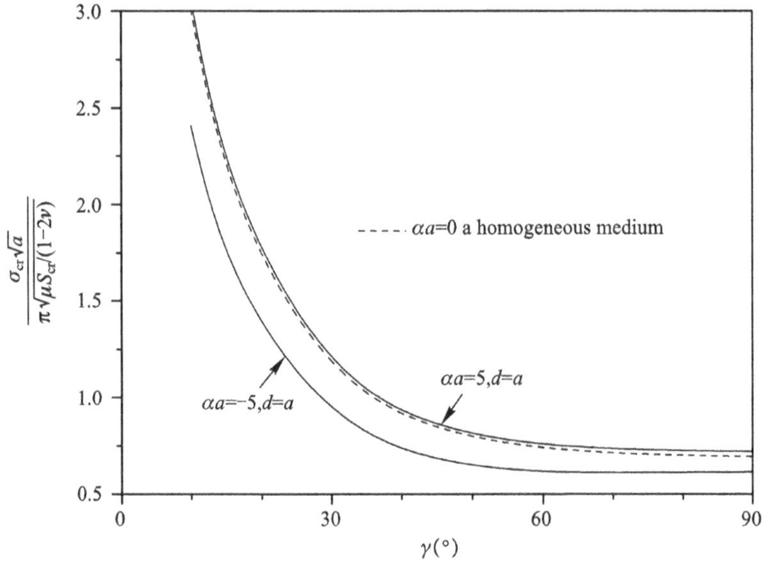

Fig. 6.36 Variations of σ_{cr} with γ for different α values ($d = a$, $h = 0.5a$)

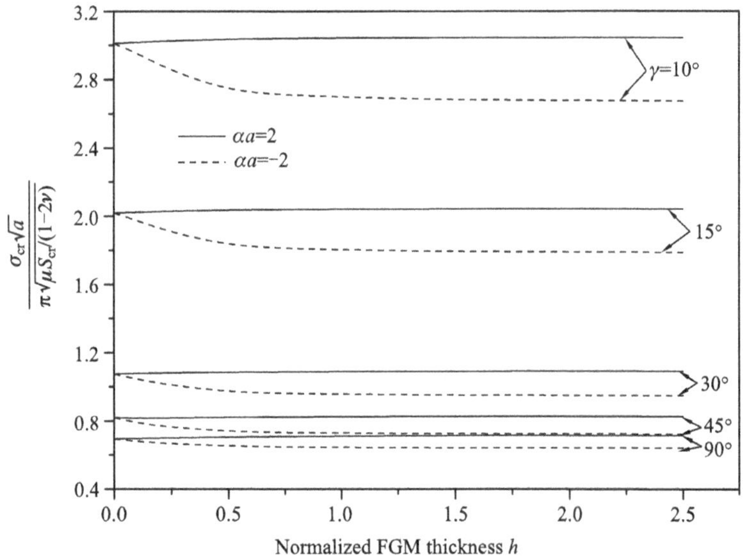

Fig. 6.37 Variations of σ_{cr} with h for different α and γ values ($d = 1.2a$)

6.6.2 The crack growth under a remotely inclined compressive loading

When the direction of the applied loading σ in Fig. 6.33 is reversed, the cracked solid with a FGM interlayer is in a state of compression. The surface of the initiating crack will be different from that associated with the above tensile loading. For such a compressive loading, we need to replace σ in Eqs. (6.11) with $-\sigma$.

(1) The minimum strain energy density factor at the crack front

Figure 6.38 shows the variations of the $S_{\min}$ value at the crack front with the parameter α and the angle θ where $\gamma = 60°$, $d = 1.2a$ and $h = 0.5a$. Looking at this figure we see that the $S_{\min}$ curves of $\alpha > 0$ and $\alpha < 0$ are located on either side of the one for $\alpha = 0$. As α increases, the $S_{\min}$ values at a given point of the crack front decrease. At the crack front $0° \leqslant \theta \leqslant 90°$, the α values have an obvious influence on $S_{\min}$. The crack growth under a remotely inclined compressive loading tends towards the direction of external loading. This is different from the crack growth under a remotely inclined tensile loading.

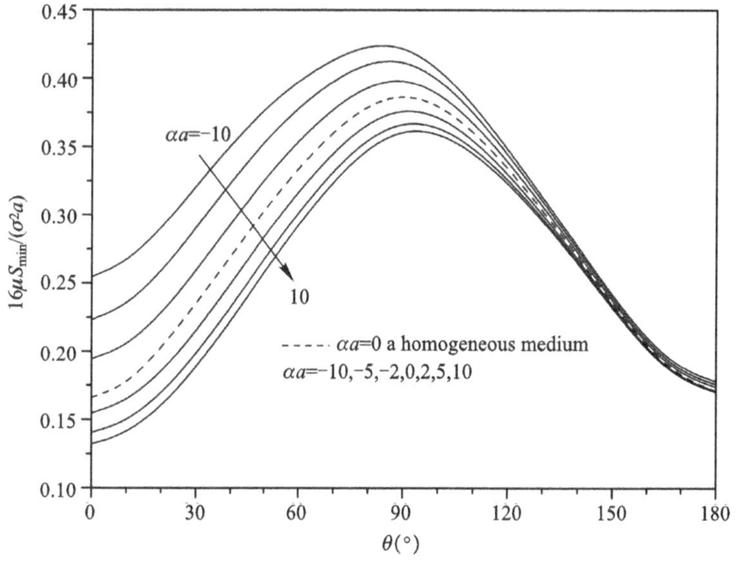

Fig. 6.38 Variations of $S_{\min}$ with θ for different α values ($d = 1.2a$, $h = 0.5a$, $\gamma = 60°$)

(2) The critical loading of crack growth

Figure 6.39 shows the variations of σ_{cr} with α and γ where $\alpha a = -10, 10$, $d = 1.2a$ and $h = 0.5a$. When $\alpha = 0$ and the loading angle γ is close to either $0°$ or $90°$, the σ_{cr} values become larger, and when γ approaches $45°$, σ_{cr} tends towards the minimum values. The σ_{cr} curves for $\alpha > 0$ and $\alpha < 0$ are located on either side of the one for $\alpha = 0$. σ_{cr} varies weakly for $\alpha a = 10$ and has an obvious variation when $\alpha a = -10$ and $\gamma < 45°$. The above variations of σ_{cr} for $\alpha a = -10, 10$ are suitable for $\alpha > 0$ and $\alpha < 0$.

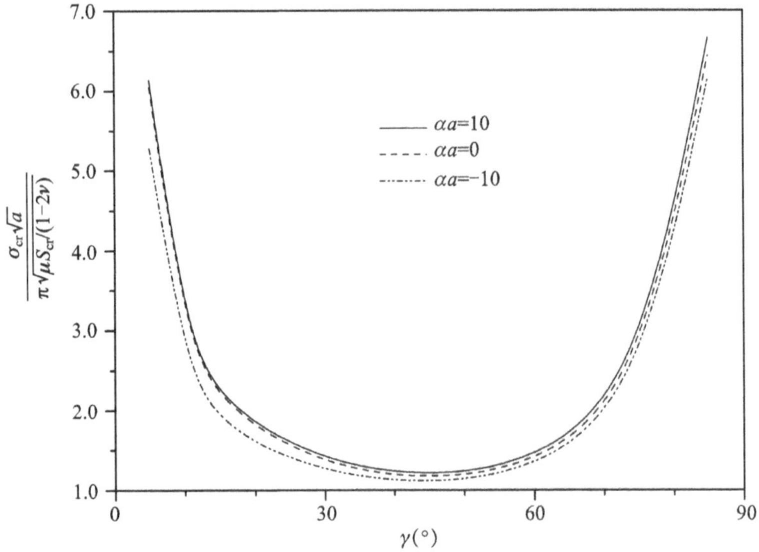

Fig. 6.39 Variations of σ_{cr} with γ for different α values ($d = 1.2a$, $h = 0.5a$)

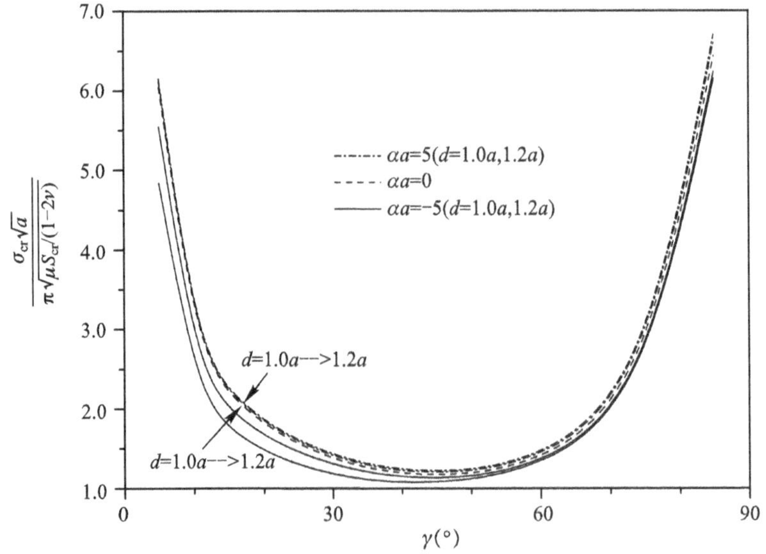

Fig. 6.40 Variations of σ_{cr} with γ for different α and d values ($h = 0.5a$)

In Fig. 6.40 the variations of σ_{cr} with d and γ are shown, where $h = 0.5a$ and $\alpha a = -5, 5$. σ_{cr} varies obviously for $\alpha a = -5$ whereas σ_{cr} varies weakly for $\alpha a = 5$. For $d > a$, the influence of a given α value on σ_{cr} is no larger than the one for $d = a$. That is, in Fig. 6.40 we see that the FGM interlayer has the largest influence on σ_{cr} when $\alpha a = -5$ and 5.

Figure 6.41 shows the variations of σ_{cr} with the thickness h of the FGM interlayer where $d = 1.2a$ and $\alpha a = -2, 2$. Referring to the above variations of σ_{cr} with γ, we have taken $\gamma = 10°, 15°, 30°, 45°$. For $\alpha a = 2$, σ_{cr} increases as h increases but this change is not obvious. For $\alpha a = -2$, σ_{cr} decreases as h increases especially for a smaller loading angle. However, for $h > a$, σ_{cr} varies weakly as h increases.

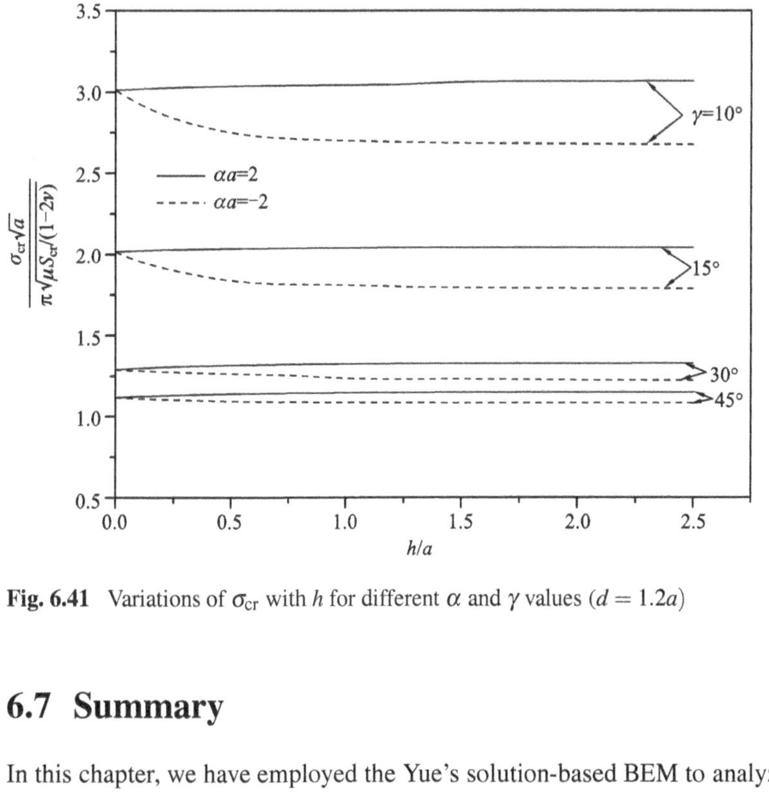

Fig. 6.41 Variations of σ_{cr} with h for different α and γ values ($d = 1.2a$)

6.7 Summary

In this chapter, we have employed the Yue's solution-based BEM to analyze the fracture mechanics of a penny-shaped crack in a FGM system. The crack surfaces are parallel or perpendicular to a graded interlayer of bonded bi-materials. The proposed BEM is characterized by discretizing the FGM interlayer into many layers with each layer having the same elastic properties. It is obvious that the FGM of any distribution of material properties with depth can be analyzed by using the proposed method, which has a high accuracy and efficiency. The proposed numerical method is used to obtain the SIFs of the crack when subjected to uniform compressive and shear tractions. The growth of the penny-shaped crack under remote loadings has been analyzed by using the SIFs and the principles of fracture mechanics.

In a FGM system, the penny-shaped crack subjected to uniform compressive stresses are under mixed mode I and II deformations, while those subjected to uniform shear stresses are under mixed mode II and III deformations. The SIF values are highly affected by the thickness of the FGM interlayer, the non-homogeneity parameter and the crack distance to the FGM interlayer.

In comparison with the crack problem in a homogeneous medium, the critical loadings for crack growth in a FGM system of infinite extent vary considerably. Similar to the SIF value of the crack in a FGM system, the critical loadings for crack growth are highly affected by the thickness of the FGM interlayer, the non-homogeneity parameter and the crack distance to the FGM interlayer. The critical loadings to initiate crack growth increase when the crack is located in a more compliant medium and decrease when the crack is located in a stiffer medium.

These research results have been published in Yue et al. (2003) and Xiao et al. (2005), etc. These results have been cited in many research articles (e.g., Manolis et al., 2004; Kandula et al., 2005; Birman and Byrd, 2007; Dineva et al., 2007; Huang and Chang, 2007; Panzeca et al., 2007) .

References

Birman V, Byrd LW. Modeling and analysis of functionally graded materials and structures. ASME *Applied Mechanics Reviews*, 2007, 60(1-6): 195-216.

Choi HJ. The problem for bonded half-planes containing a crack at an arbitrary angle to the graded interfacial zone. *International Journal of Solids and Structures*, 2001, 38: 6559-6588.

Delale F, Erdogan F. The crack problem for a nonhomogeneous plane. *ASME Journal of Applied Mechanics*, 1983, 50: 609-614.

Dineva PS, Rangelov TV, Manolis GD. Elastic wave propagation in a class of cracked, functionally graded materials by BIEM. *Computational Mechanics*, 2007, 39(3): 293-308.

Green AE, Sneddon IN. The distribution of stress in the neighbourhood of a flat elliptical crack in an elastic solid. *Mathematical Proceedings of the Cambridge Philosophical Society*, 1950, 46: 159-163.

Huang CS, Chang MJ. Corner stress singularities in an FGM thin plate. *International Journal of Solids and Structures*, 2007, 44: 2802-2819.

Kandula SSV, Abanto-Bueno J, Geubelle PH, Lambros J. Cohesive modeling of dynamic fracture in functionally graded materials. *International Journal of Fracture*, 2005, 132: 275-296.

Kou CK, Keer LM. Three-dimensional Analysis of Cracking in a multilayered composite. *ASME Journal of Applied Mechanics*. 1995, 62: 273-281.

Jin Z, Noda N. Crack-tip singular fields in nonhomogeneous materials. *ASME Journal of Applied Mechanics*, 1994, 61: 738-740.

Li C, Zuo Z, Duan ZP. Stress intensity factors for functionally graded solid cylinders. *Engineering Fracture Mechanics*, 1999, 63: 735-749.

Manolis GD, Dineva PS, Rangelov TV. Wave scattering by cracks in inhomogeneous continua. *International Journal of Solids and Structures*, 2004, 41: 3905-3927.

Ozturk M, Erdogan F. An axisymmetric crack in bonded materials with a nonhomogeneous interfacial zone under torsion. *ASME Journal of Applied Mechanics*, 1995, 62: 116-125.

Ozturk M, Erdogan F. Axisymmetric crack problem in bonded materials with a graded interfacial region. *International Journal of Solids and Structures*, 1996, 33:193-219.

Panzeca T, Cucco F, Terravecchia S. Boundary discretization based on the residual energy using the SGBEM. *International Journal of Solids and Structures*, 2007,44: 7239-7260.

Paulino GH. Fracture of functionally graded materials. *Engineering Fracture Mechanics*, 2002, 69: 1519-1520.

Selvadurai APS. The indentation of a precompressed penny-shaped crack. *International Journal of Engineering Science*, 2000a, 38(18): 2095-2111.

Selvadurai APS. The penny-shaped crack at a bonded plane with localized elastic non-homogeneity. *European Journal of Mechanics* - A/Solids, 2000b, 19(3): 525-534.

Sih GC, Cha CK. A fracture criterion for three-dimensional crack problems. *Engineering Fracture Mechanics*, 1974, 6:699-723.

Sneddon IN. The distribution in the neighborhood of a crack in an elastic solid. *Proceedings of the Royal Society*, 1946, A187:229-260.

Tada H, Paris OC, Irwin GR. The Stress Analysis of Cracks Handbook. 3rd ed. London and Bury St Edmunds: Professional Engineering Publishing Limited, 2000.

Xiao HT, Yue ZQ, Tham GL, Chen YR. Stress intensity factors for penny-shaped cracks perpendicular to graded interfacial zone of bonded bi-materials. *Engineering Fracture Mechanics*, 2005, 72: 121-143.

Yue ZQ, Xiao HT, Tham GL. Boundary element analysis of crack problems in functionally graded materials. *International Journal of Solids and Structures*, 2003, 40: 3273-91.

Chapter 7

Analysis of Elliptical Cracks in Functionally Graded Materials

7.1 Introduction

Lin and Smith (1998) pointed out that various internal surface cracks with initially irregular crack fronts evolve into semi-elliptical cracks at the beginning of crack growth. Such semi-elliptical surface cracks in a plate with finite thickness are significant for engineering endeavours (Standford, 2003). Irwin (1962) proposed that the stress intensity factors (SIFs) of these crack-like flaws can be estimated by using those obtained for an elliptical crack in a homogeneous medium of infinite extent and the empirical formulae; and then presented the formulae for calculating the SIFs of a crack that extended partly through a plate. Similarly, the analysis of an elliptical crack of infinite extent is also significantly important.

Green and Sneddon (1950) analyzed the distribution of stresses near a flat elliptical crack in a body of infinite extent under uniform tension at infinity perpendicular to the plane of the crack. Shaha and Kobayashia (1971) derived the SIFs for an elliptical crack embedded in an elastic solid and subjected to an arbitrary internal pressure. In the work by Kassir and Sih (1975), an elliptical crack perpendicular to the surface of a semi-infinite solid and subject to uniform pressure was analyzed. Zheng et al. (1995) calculated the SIFs of surface semi-elliptical cracks in cylinders with an inner radius to wall thickness ratio. Noda et al. (2003) considered an elliptical crack parallel to a bi-material interface by utilizing the body force method and the Green's functions for perfectly bonded elastic half planes. Hachi et al. (2005) proposed a hybridization technique to analyze elliptical cracks in an infinite body and semi-elliptical cracks in cylinders. Zhang et al. (2011) analyzed a semi-infinite elliptical crack in three-dimensional FGMs by employing a boundary-domain-integral equation formulation. Mechab et al. (2011) calculated the J integral for semi-elliptical surface cracks in pipes under bending using three-dimensional finite element analysis.

In this chapter, we will employ the proposed numerical method to analyze an elliptical crack in three-dimensional FGMs. It is assumed that a FGM interlayer with the same thickness is fully bonded with two homogeneous media of semi-infinite extent and an elliptical crack is located parallel or perpendicular to the interlayer. The crack surfaces are subjected to uniform compressive or shear tractions. In the following, we first calculate the SIFs for each loading and then analyze the growth of an elliptical crack under a remotely inclined tensile or compressive loading.

7.2 The SIFs for an elliptical crack parallel to the FGM interlayer

7.2.1 General

As shown in Fig. 7.1a, an elliptical crack is parallel to the FGM interlayer. The FGM elastic modulus is approximated by Eq. (6.1). It is further assumed that the Poisson's ratios of the FGMs are constant and equal to 0.3, i.e., $\nu_1 = \nu_2 = \nu_3 = 0.3$. The ratio of the major and minor axes of the elliptical crack surface is taken as $a/b = 2$. The crack surfaces are subjected to the following loadings

$$\sigma^+_{z'x'} = \sigma^-_{z'x'} = q \tag{7.1a}$$

$$\sigma^+_{z'y'} = \sigma^-_{z'y'} = 0 \tag{7.1b}$$

$$\sigma^+_{z'z'} = \sigma^-_{z'z'} = p \tag{7.1c}$$

where the superscripts + and – correspond to the crack surfaces, which are located at $z' = 0^+$ and $z' = 0^-$, respectively.

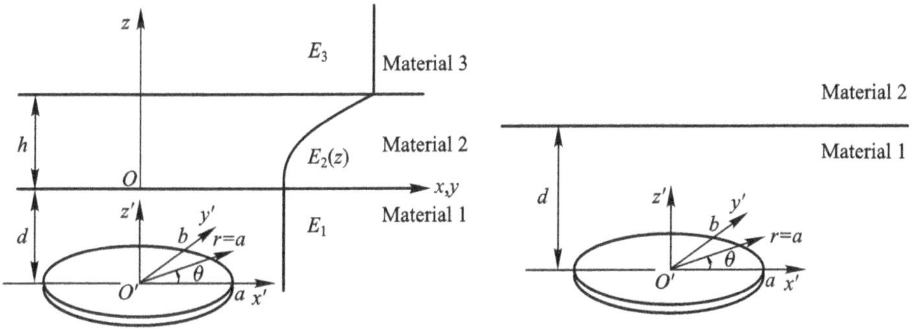

(a) An elliptical crack parallel to the graded interlayer in bonded materials

(b) An elliptical crack parallel to the interface of the bi-material

Fig. 7.1 Elliptical cracks parallel to the graded interlayer or parallel to the interface of the bi-material, $a = 2b$

It is obvious that the cracked body is symmetrical with respect to the $O'x'z'$-plane. Therefore, it is only necessary to consider half of the crack problem. For convenience, half of an ellipsoid in a homogeneous material (material 1) is chosen to form two regions and is discretized for the BEM analysis. Region I is a quarter of the ellipsoid, shown in Fig. 7.2a, and region II is a semi-infinite domain which has the ellipsoid and crack surfaces as an inner boundary. Regions I and II share the ellipsoid surface. We only have to discretize the ellipsoid and crack surfaces and do not need to discretize the plane of symmetry into

BEM meshes. Figure 7.2 shows the discretized surface of region I with 176 elements and 553 nodes. There are 32 traction-singular elements along the crack front. The two sides of the traction-singular elements on the crack surface are perpendicular to the tangent at the crack tip and intersect with the crack front. The inner boundary of region II is discretized into the same mesh as that for region I shown in Fig. 7.2. The auxiliary surfaces of two regions fully coincide and there are one-to-one correspondences among the elements and nodes on the auxiliary surfaces of two regions.

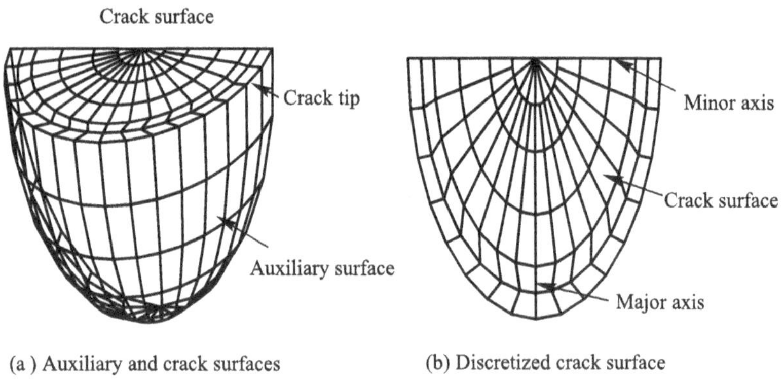

Fig. 7.2 Boundary element meshes of region I for an elliptical crack problem

In Fig. 7.1b, an elliptical crack in a bi-material system is present parallel to the interface between two materials and is subjected to a uniform compressive stress $p = -p_0$ ($p_0 > 0$). Noda et al. (2003) utilized the body force method to obtain the SIFs of the crack problem shown in Fig. 7.1b. In the following, we will employ the proposed numerical method to analyze the crack problem and compare the results with the ones obtained by Noda et al. Table 7.1 presents the SIF values for the crack surfaces located at different distances d from the bi-material interface. It can be seen that the absolute errors of the SIF values ($K_{\mathrm{I}}/(p_0\sqrt{\pi b})$) obtained from the two methods are less than 2.0% and the ones of the SIF values ($K_{\mathrm{II}}/(p_0\sqrt{\pi b})$) are about 0.2%.

Table 7.1 The SIF values for an elliptical crack parallel to the interface of two half-spaces ($E_2/E_1 = 0.5, a = 2b$)

$d/(2b)$	$K_{\mathrm{I}}/(p_0\sqrt{\pi b})$				$K_{\mathrm{II}}/(p_0\sqrt{\pi b})$			
	A	B	$A-B$	$(A-B)/A\%$	A	B	$A-B$	$(A-B)/A\%$
0.1	0.9756	0.95585	0.01975	2.0244	0.0735	0.07223	0.00127	1.7279
0.2	0.9554	0.93578	0.01962	2.0536	0.0577	0.06018	0.00248	–4.2981
0.3	0.9347	0.92124	0.01346	1.4400	0.0442	0.04289	0.00131	2.9638
0.4	0.9143	0.90957	0.00473	0.5173	0.0322	0.02951	0.00269	8.3540
0.5	0.8963	0.89591	0.00390	0.0435	0.0227	0.01801	0.00047	20.6608
1.0	0.84817	0.84974	–0.00157	–0.1851	0.00414	0.00539	–0.00125	–30.1932
2.0	0.82982	0.82878	0.00104	0.1253	0.00038	0.00047	–0.00009	–23.6842

Note: A and B correspond, respectively, to the data given in Noda et al. (2003) and by utilizing the present BEM method.

7.2.2 Elliptical crack under a uniform compressive stress

Consider a uniform compressive stress acting on the crack surfaces, i.e., $p = -p_0$ ($p_0 > 0$) and $q = 0$ in Eq. (7.1). In this case, the exact solutions for the SIFs at the front of the elliptical crack in a homogeneous solid of infinite extent are as follows (Tada et al., 2000)

$$K_{\mathrm{I}} = \frac{p_0\sqrt{\pi b}}{E(k)}\left\{\sin^2\theta + \frac{b^2}{a^2}\cos^2\theta\right\}^{1/4} \tag{7.2a}$$

$$K_{\mathrm{II}} = 0 \tag{7.2b}$$

$$K_{\mathrm{III}} = 0 \tag{7.2c}$$

where

$$E(k) = \int_0^{\pi/2}\sqrt{1-k^2\sin^2\varphi}\,\mathrm{d}\varphi, \quad k^2 = 1 - b^2/a^2$$

in which a and b are the lengths of the major and minor axes of the elliptical crack surface, respectively. For $a = 2b$, the exact solution of the mode I SIF ($K_{\mathrm{I}}/(p_0\sqrt{\pi b})$) at the crack front of the minor axis is $E(k) \approx 0.8242$ and the numerical solution obtained using the proposed method is 0.8257. The difference in the results of the two methods is about 0.15%.

In the following, we will analyze the elliptical crack in a FGM system shown in Fig. 7.1a. Figures 7.3 ~ 7.14 present the variations of the SIFs for the elliptical crack subjected to a uniform compressive stress with different influential factors, including the non-homogeneity parameter α, the thickness h of the FGM interlayer and the crack distance d to the FGM interlayer.

Due to the crack geometry and the influence of the FGM interlayer, the mode I, Ⅱ and Ⅲ deformations of the elliptical crack under the uniform compressive stress are coupled together. By considering the symmetry of the cracked body, it can be found that the three modes of deformations are symmetrical with respect to the $O'x'z'$-plane. For simplification, the SIF values at the crack front $0^\circ \leqslant \theta \leqslant 90^\circ$ will be presented in the figures.

(1) Variations of the SIF values with the non-homogeneity parameter α

Figures 7.3 ~ 7.5 present the variations of the mode I, Ⅱ and Ⅲ SIFs with the parameter α, where $h = 0.5a$ and $d = 0$. Tables 7.2 ~ 7.4 present the corresponding data for the SIFs. Notice that when $d = 0$ the crack is located at the interface between the FGM interlayer and the homogeneous medium of semi-infinite extent and when $\alpha = 0$ the FGM system degenerates into a homogeneous medium.

In these figures, we can see that the SIF values for both positive and negative values of α are located, respectively, at the two sides of the SIF curve for $\alpha = 0$. In Fig. 7.3, the mode I SIF values increase as the values of α decrease. This is because a compliant graded interlayer increases the crack opening. In Figs. 7.4 and 7.5, the mode Ⅱ and Ⅲ SIF values are presented. For $\alpha > 0, K_{\text{II}}$ and K_{III} values are always positive whilst for $\alpha < 0$ K_{II} and K_{III} values are always negative. The variations in the SIF values are more evident for $\alpha < 0$ than for $\alpha > 0$.

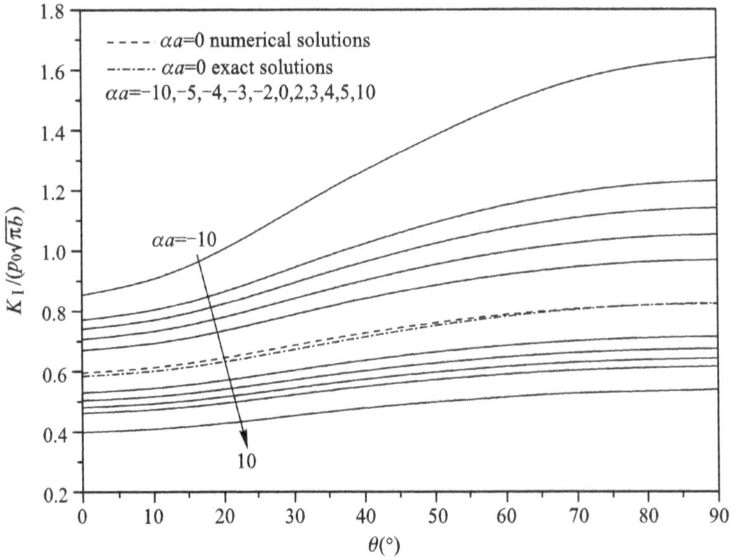

Fig. 7.3 Variations of $K_{\text{I}}/(p_0\sqrt{\pi b})$ with θ for different α values under p_0 ($h = 0.5a, d = 0$)

(2) Variations of the SIF values with the parameters α and d

Figures 7.6 ~ 7.8 illustrate the variations of the mode I, Ⅱ and Ⅲ SIF values as the parameter α changes, where $h = 0.5a$ and $d = 0.2a$. When compared with the above case ($d = 0$), it can be observed that there are weaker variations in the corresponding SIF

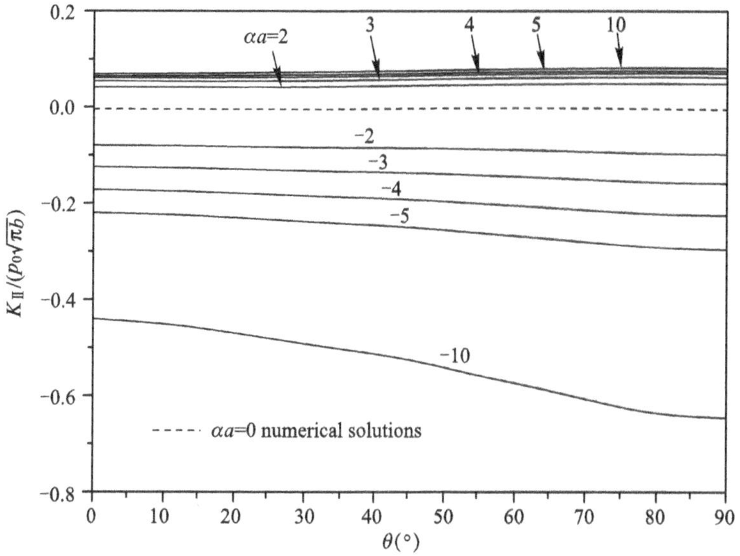

Fig. 7.4 Variations of $K_{II}/(p_0\sqrt{\pi b})$ with θ for different α values under p_0 $(h = 0.5a, d = 0)$

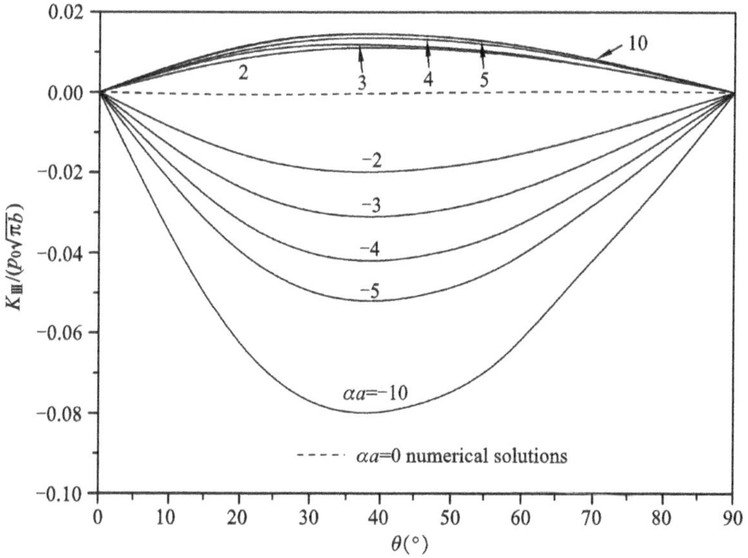

Fig. 7.5 Variations of $K_{III}/(p_0\sqrt{\pi b})$ with θ for different α values under p_0 $(h = 0.5a, d = 0)$

Table 7.2 The SIF values $(K_I/(p_0\sqrt{\pi b}))$ for an elliptical crack under p_0 for different α values $(h = 0.5a, d = 0)$

$\theta(°)$	$\alpha a = 10$	5	4	3	2	0	–2	–3	–4	–5	–10
0.00	0.39925	0.46173	0.48081	0.50332	0.52963	0.59506	0.66998	0.70653	0.74034	0.77035	0.85407
11.25	0.40862	0.47263	0.49234	0.51567	0.54304	0.61165	0.69137	0.73080	0.76786	0.80133	0.90365
22.50	0.43301	0.50036	0.52154	0.54685	0.57682	0.65334	0.74529	0.79215	0.83759	0.88010	1.03064
33.75	0.46388	0.53378	0.55645	0.58383	0.61679	0.70245	0.80940	0.86580	0.92207	0.97633	1.18895
45.00	0.49069	0.56325	0.58726	0.61651	0.65222	0.74576	0.86560	0.93074	0.99661	1.06126	1.32575
56.25	0.50896	0.58495	0.61034	0.64138	0.67948	0.7802	0.91162	0.98494	1.05996	1.13476	1.45407
67.50	0.52768	0.60242	0.62800	0.65955	0.69847	0.80308	0.94255	1.02173	1.10407	1.18745	1.55628
78.75	0.53290	0.61147	0.63813	0.67091	0.71129	0.81966	0.96484	1.04765	1.13430	1.22248	1.61745
90.00	0.53971	0.61617	0.64247	0.67492	0.71508	0.82347	0.96983	1.05383	1.14213	1.23238	1.64125

Table 7.3 The SIF values $(K_{\mathrm{II}}/(p_0\sqrt{\pi b}))$ for an elliptical crack under p_0 for different α values $(h = 0.5a, d = 0)$

$\theta(°)$	$\alpha a = 10$	5	4	3	2	0	–2	–3	–4	–5	–10
0.00	0.06928	0.06535	0.06128	0.05381	0.04141	–0.00440	–0.07937	–0.12428	–0.17165	–0.21953	–0.44052
11.25	0.06992	0.06553	0.06137	0.05381	0.04136	–0.00467	–0.08025	–0.12572	–0.17389	–0.22278	–0.45038
22.50	0.07190	0.06648	0.06210	0.05432	0.04165	–0.00512	–0.08255	–0.12958	–0.17987	–0.23133	–0.47470
33.75	0.07422	0.06818	0.06368	0.05579	0.04295	–0.00459	–0.08431	–0.13334	–0.18619	–0.24068	–0.50031
45.00	0.07576	0.07006	0.06570	0.05792	0.04501	–0.00297	–0.08470	–0.13582	–0.19109	–0.24840	–0.52176
56.25	0.07928	0.07340	0.06902	0.06114	0.04789	–0.00166	–0.08739	–0.14187	–0.20111	–0.26305	–0.56362
67.50	0.08047	0.07464	0.07028	0.06235	0.04894	–0.00203	–0.09128	–0.14832	–0.21076	–0.27633	–0.59670
78.75	0.08252	0.07620	0.07168	0.06348	0.04962	–0.00330	–0.09664	–0.15667	–0.22279	–0.29263	–0.63998
90.00	0.08188	0.07563	0.07114	0.06297	0.04911	–0.00396	–0.09782	–0.15827	–0.22490	–0.29533	–0.64568

Table 7.4 The SIF values $(K_{\mathrm{III}}/(p_0\sqrt{\pi b}))$ for an elliptical crack under p_0 for different α values ($h = 0.5a, d = 0$)

$\theta(^\circ)$	$\alpha a = 10$	5	4	3	2	0	−2	−3	−4	−5	−10
0.00	0.00000	0.00000	0.00000	0.00000	0.00000	0.00000	0.00000	0.00000	0.00000	0.00000	0.00000
11.25	0.00638	0.00745	0.00731	0.00668	0.00533	−0.00049	−0.01012	−0.01553	−0.02088	−0.02581	−0.04108
22.50	0.01073	0.01271	0.01253	0.01152	0.00928	−0.00067	−0.01732	−0.02672	−0.03604	−0.04462	−0.07051
33.75	0.01220	0.01488	0.01480	0.01373	0.01117	−0.00045	−0.02023	−0.03149	−0.04259	−0.05273	−0.08132
45.00	0.01143	0.01441	0.01444	0.01352	0.01105	−0.00019	−0.01960	−0.03083	−0.04176	−0.05170	−0.07895
56.25	0.01032	0.01277	0.01279	0.01199	0.00984	0.00004	−0.01700	−0.02689	−0.03654	−0.04537	−0.07027
67.50	0.00724	0.00914	0.00919	0.00865	0.00718	0.00016	−0.01203	−0.01893	−0.02564	−0.03164	−0.04618
78.75	0.00371	0.00477	0.00483	0.00457	0.00383	0.00020	−0.00616	−0.00978	−0.01333	−0.01654	−0.02538
90.00	0.00000	0.00000	0.00000	0.00000	0.00000	0.00000	0.00000	0.00000	0.00000	0.00000	0.00000

values. In Figs. 7.9 to 7.11, this phenomenon is more; these figures show the variations of the mode I, II and III SIF values with d, showing that the effects of the FGM interlayer rapidly become weaker and weaker as d increases. For $d/a > 2$, this influence may be neglected.

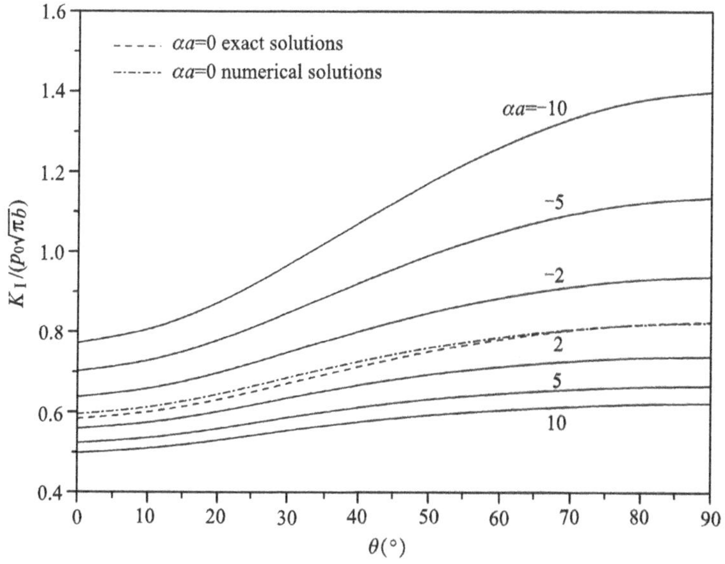

Fig. 7.6 Variations of $K_{\mathrm{I}}/(p_0\sqrt{\pi b})$ with θ for different α values under p_0 $(h = 0.5a, d = 0.2a)$

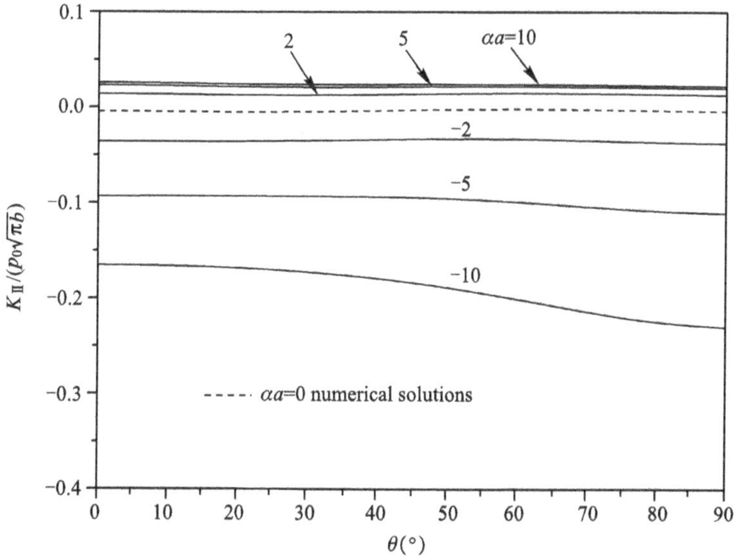

Fig. 7.7 Variations of $K_{\mathrm{II}}/(p_0\sqrt{\pi b})$ with θ for different α values under p_0 $(h = 0.5a,\ d = 0.2a)$

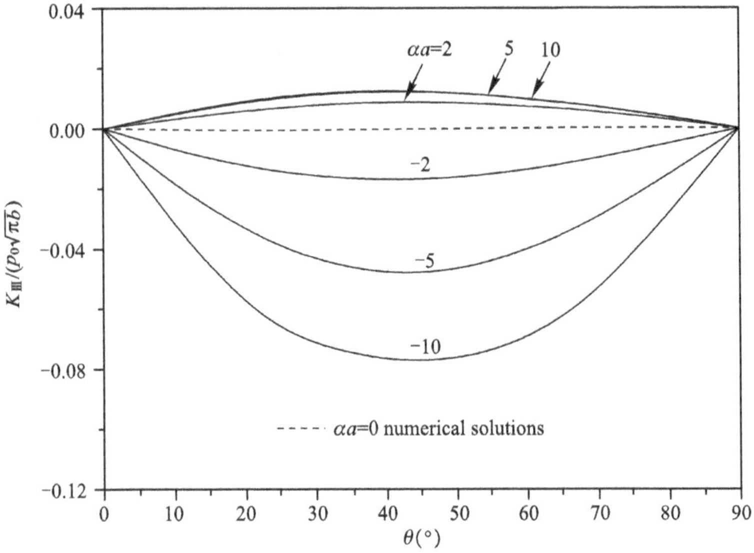

Fig. 7.8 Variations of $K_{\text{III}}/(p_0\sqrt{\pi b})$ with θ for different α values under p_0 ($h = 0.5a$, $d = 0.2a$)

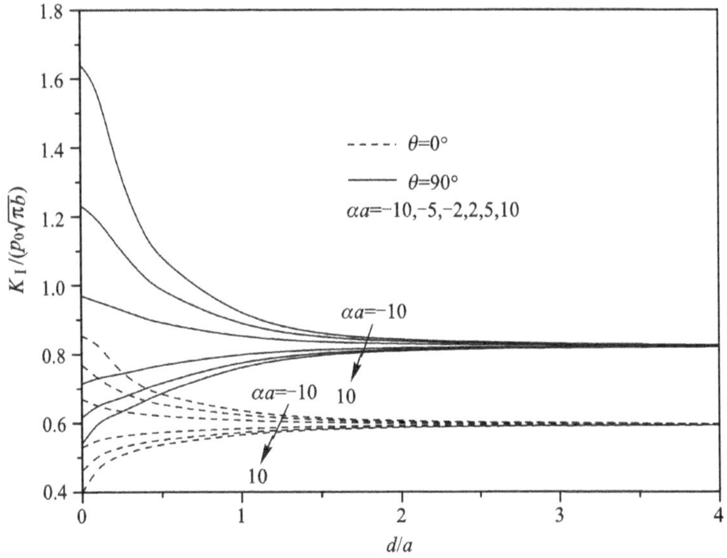

Fig. 7.9 Variations of $K_{\text{I}}/(p_0\sqrt{\pi b})$ with d for different α and θ values under p_0 ($h = 0.5a$)

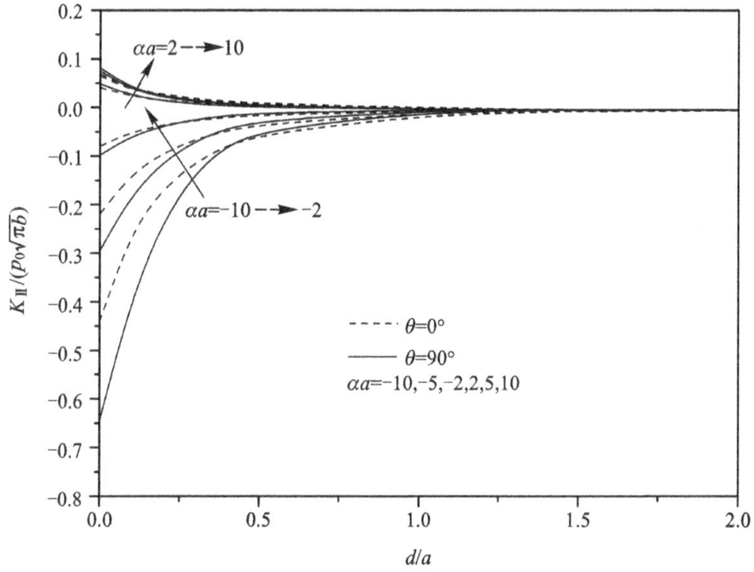

Fig. 7.10 Variations of $K_{\text{II}}/(p_0\sqrt{\pi b})$ with d for different α and θ values under p_0 $(h = 0.5a)$

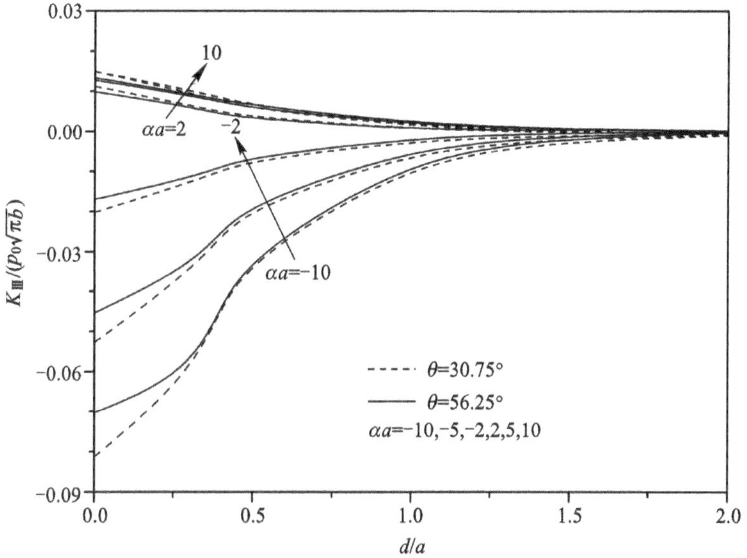

Fig. 7.11 Variations of $K_{\text{III}}/(p_0\sqrt{\pi b})$ with d for different α and θ values under p_0 $(h = 0.5a)$

(3) Variations of the SIF values with the thickness h of the FGM interlayer

In Figs. 7.12 ∼ 7.14, the variations of the SIF values with h for $\alpha a = -5, 5$ and $d = 0.2a$ are shown. As the thickness h increases, the mode I SIF values increase for $\alpha a = -5$ whilst they decrease for $\alpha a = 5$. For $\alpha a = -5$ and 5, the absolute values of the mode

Ⅱ and Ⅲ SIFs increase as the FGM thickness increases. For these two different material parameters ($\alpha a = -5, 5$), the three SIF modes at different locations on the crack front show no obvious changes for $h > a$.

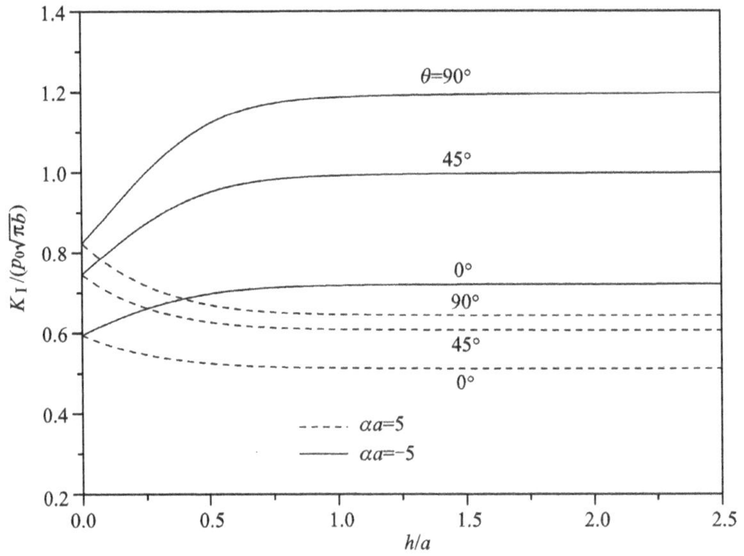

Fig. 7.12 Variations of $K_{\mathrm{I}}/(p_0\sqrt{\pi b})$ with h for different α and θ values under p_0 ($d = 0.2a$)

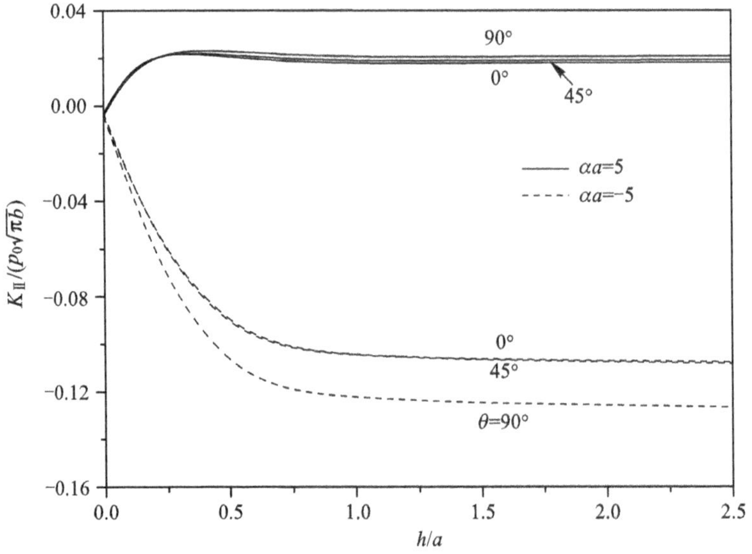

Fig. 7.13 Variations of $K_{\mathrm{II}}/(p_0\sqrt{\pi b})$ with h for different α and θ values under p_0 ($d = 0.2a$)

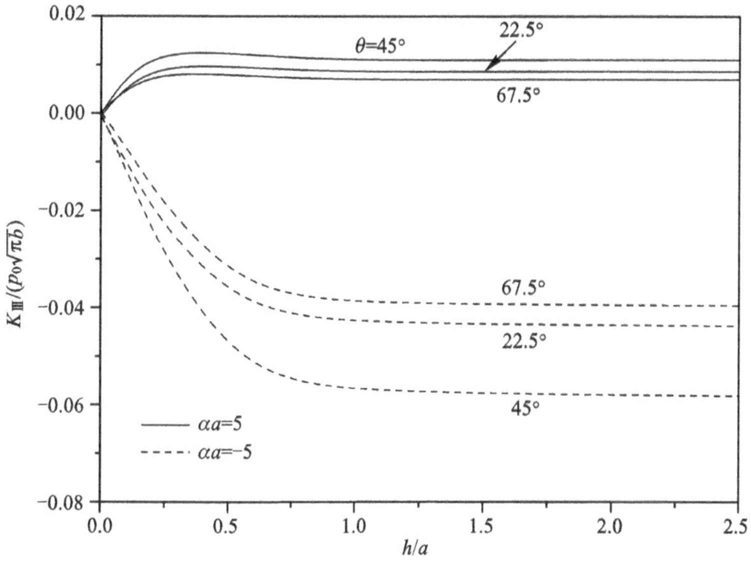

Fig. 7.14 Variations of $K_{\mathrm{III}}/(p_0\sqrt{\pi b})$ with h for different α and θ values under p_0 $(d = 0.2a)$

7.2.3 Elliptical crack under a uniform shear stress

Consider a uniform shear stress acting on the crack surfaces, i.e., $p = 0$ and $q = -q_0$ $(q_0 > 0)$ in Eq. (7.1). For this case, the exact solutions of the SIFs at the front of the elliptical crack in a homogeneous solid of infinite extent are given as follows (Tada et al., 2000)

$$K_{\mathrm{I}} = 0 \tag{7.3a}$$

$$K_{\mathrm{II}} = q_0\sqrt{\pi b}k^2[\sin^2\theta + k'^2\cos^2\theta]^{-1/4}\frac{\sin\theta}{C} \tag{7.3b}$$

$$K_{\mathrm{III}} = q_0\sqrt{\pi b}(1-\nu)k^2k'[\sin^2\theta + k'^2\cos^2\theta]^{-1/4}\frac{\cos\theta}{C} \tag{7.3c}$$

where

$$C = (k^2 + \nu k'^2)E(k) - \nu k'^2 K(k), \quad K(k) = \int_0^{\pi/2}\frac{\mathrm{d}\varphi}{\sqrt{1-k^2\sin^2\varphi}},$$

$$E(k) = \int_0^{\pi/2}\sqrt{1-k^2\sin^2\varphi}\,\mathrm{d}\varphi, \quad k^2 = 1-(b/a)^2, \quad k' = b/a$$

In this example, we take $a/b = 2$, that is, the ratio of the major and minor axes of the elliptical crack surface is equal to 2. Since the elliptical crack is still in material 1 and has the same dimensions as the ones given in Section 7.2.2, the boundary element meshes shown in Fig. 7.2 can be further utilized to analyze the crack problem. From Eqs. (7.3a) ~ (7.3c), we can see that the mode II and III SIFs are always coupled together for the uniform shear traction on the crack surfaces. The exact solution of the mode Ⅱ SIF ($K_{\rm II}/(q_0\sqrt{\pi b})$) at the crack front of the minor axis is 0.89561 and the numerical solution is 0.90687. The largest error in results obtained using the two methods is approximately 1.0% for the mode Ⅱ SIF values. The exact solution of the SIF values ($K_{\rm III}/(q_0\sqrt{\pi b})$) at the crack tip of the major axis is 0.44330 while the numerical solution is 0.43819; thus the largest error of the mode Ⅲ SIF values is 0.5% and occurs at the crack tip of the major axis. In the following, we will discuss the variations of the mode Ⅱ and Ⅲ SIFs with different influential factors.

By considering the symmetry of the cracked body, the mode Ⅱ deformation is antisymmetrical with respect to the $O'x'z'$-plane and the mode Ⅲ deformation is symmetrical with respect to the $O'x'z'$-plane. For simplification, the SIF values at the crack front $0^\circ \leqslant \theta \leqslant 90^\circ$ will be presented in the following figures.

1. The mode Ⅱ stress intensity factors

Figures 7.15 to 7.18 illustrate the variations of the SIF values with the nonhomogeneity parameter α, the crack distance d to the interlayer, and the thickness h of the FGM interlayer. Table 7.5 lists the corresponding SIF values for different α values when $h = 0.5a$ and $d = 0$.

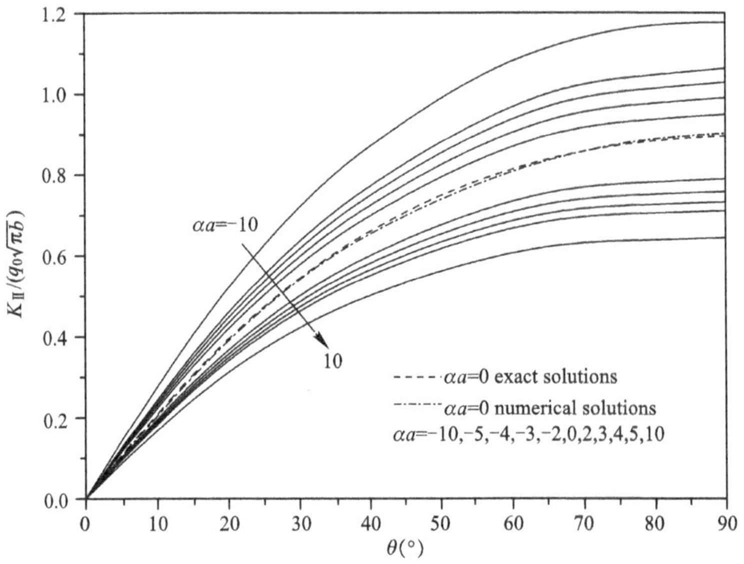

Fig. 7.15 Variations of $K_{\rm II}/(q_0\sqrt{\pi b})$ with α for different θ values under q_0 ($h = 0.5a, d = 0$)

(1) Variations of the SIF values with the parameter α

The variations of the SIF values with the parameter α where $h = 0.5a$ and $d = 0$ are shown in Fig. 7.15, where it can be seen that the SIF values for positive and negative values of α are located, respectively, on either side of the SIF curve for $\alpha = 0$. As the α values decrease, the SIF values increase. This is because a compliant graded interlayer increases the crack sliding. Figure 7.16 shows the variations of the SIF values with α for $h = 0.5a$ and $d = 0.2a$; the effect of different α values becomes weaker as the crack distance to the FGM interlayer increases. This can be more clearly observed in Fig. 7.17.

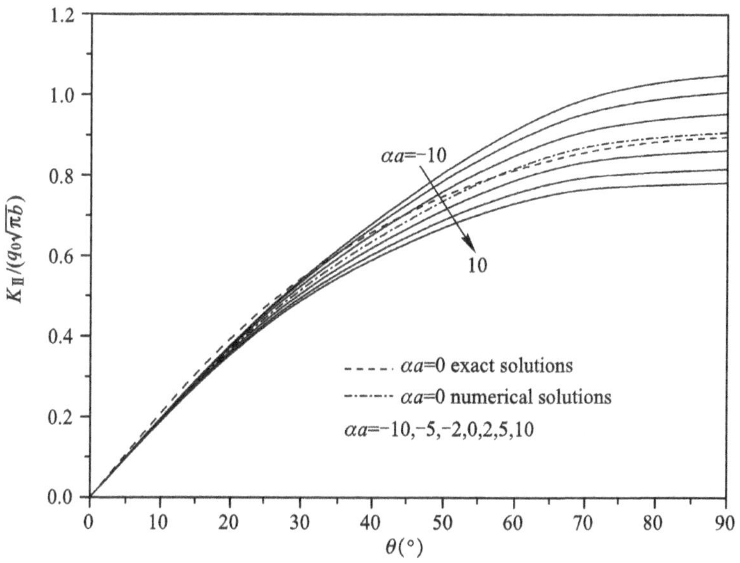

Fig. 7.16 Variations of $K_{\mathrm{II}}/(q_0\sqrt{\pi b})$ with α for different θ values under q_0 $(h = 0.5a, d = 0.2a)$

(2) Variations of the SIF values with the crack distance d to the FGM interlayer

The variations of the SIF values with d to the FGM interlayer for $\alpha a = -10, -5, -2, 2, 5, 10$ and $\theta = 45°, 90°$ are shown in Fig. 7.17. In this figure, the effects of the inhomogeneity parameter α rapidly weaken as the distance of the crack from the FGM interlayer becomes larger. When $d/a > 2$, the effects of the parameter α on the SIF values may be neglected.

(3) Variations of the SIF values with the thickness h of the FGM interlayer

Figure 7.18 illustrates the variations of the SIF values with h for $\alpha a = -5, 5, d = 0.2a$ and $\theta = 11.25°, 22.5°, 45°, 90°$. When $d = 0.2a$ the crack is located away from the interface between the homogeneous medium and the FGM interlayer. The SIF values increase when $\alpha a = -5$ and decrease for $\alpha a = 5$ as the interlayer thickness h increases. For $\alpha a = -5$ and 5, the SIF values at different locations of the crack front show no obvious changes for $h > a$.

Table 7.5 The SIF values ($K_{\text{II}}/(q_0\sqrt{\pi b})$) for an elliptical crack under q_0 for different α values ($h = 0.5a, d = 0$)

$\theta(°)$	$\alpha a = 10$	5	4	3	2	0	−2	−3	−4	−5	−10
0.00	0.00061	0.00059	0.00058	0.00057	0.00055	0.00072	0.00058	0.00063	0.0007	0.00078	0.00113
11.25	0.19448	0.21150	0.21584	0.22087	0.22672	0.23456	0.25703	0.26511	0.2729	0.28038	0.31681
22.50	0.34998	0.38159	0.39006	0.39988	0.41126	0.43877	0.46998	0.48559	0.50065	0.51503	0.58384
33.75	0.46056	0.50537	0.51761	0.53190	0.54855	0.59607	0.63408	0.65663	0.67825	0.69862	0.79158
45.00	0.53642	0.58896	0.60419	0.62217	0.64317	0.70123	0.75087	0.77903	0.80573	0.83061	0.93429
56.25	0.59474	0.65420	0.67208	0.69341	0.71847	0.78680	0.84771	0.88143	0.91327	0.94262	1.05820
67.50	0.63300	0.69699	0.71690	0.74076	0.76903	0.85079	0.91569	0.95402	0.99009	1.02310	1.13294
78.75	0.63849	0.70527	0.72614	0.75141	0.78149	0.88975	0.93631	0.97604	1.01263	1.04528	1.17334
90.00	0.64480	0.71126	0.73267	0.75865	0.78960	0.90142	0.94969	0.99078	1.02876	1.06262	1.17656

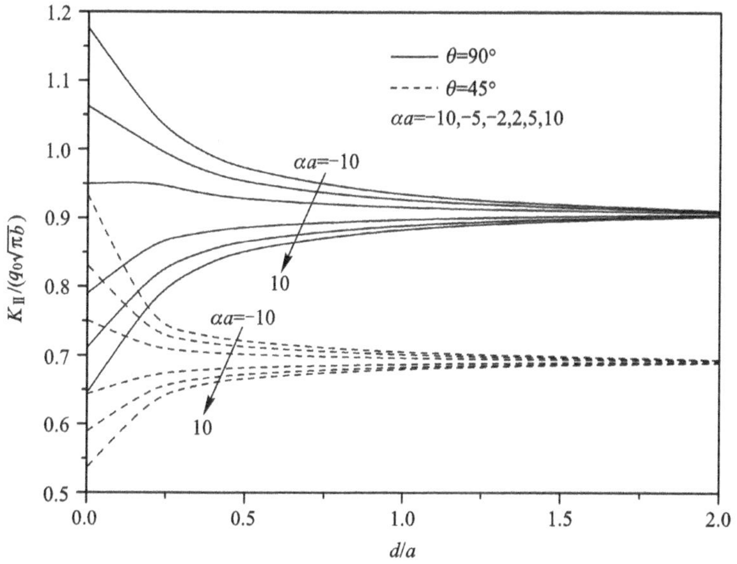

Fig. 7.17 Variations of $K_{II}/(q_0\sqrt{\pi b})$ with d for different α and θ values under q_0 $(h = 0.5a)$

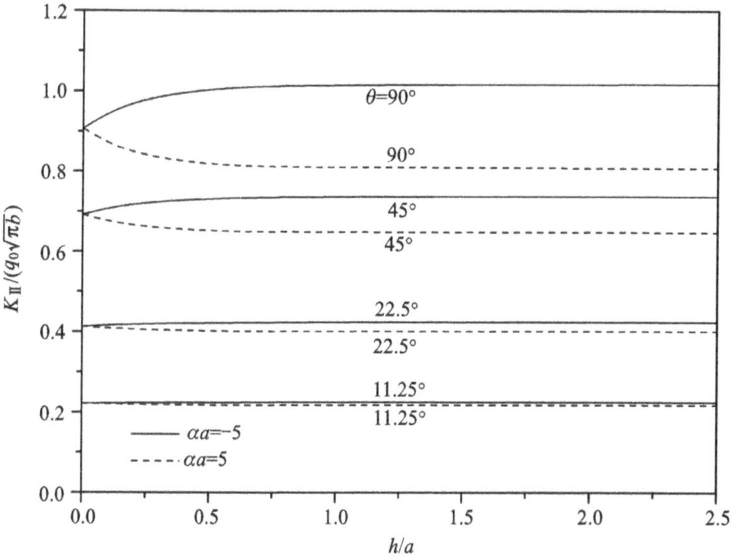

Fig. 7.18 Variations of $K_{II}/(q_0\sqrt{\pi b})$ with h for different α and θ values under q_0 $(d = 0.2a)$

2. The mode III stress intensity factors

Figures 7.19 to 7.22 illustrate the variations in the mode III SIF values as the nonhomogeneity parameter α of the FGM, the crack distance d to the interlayer, and the thickness h

of the FGM interlayer change. Table 7.6 lists the corresponding SIF values with different α values for $h = 0.5a$ and $d = 0$.

(1) Variations of the SIF values with the parameter α for $d = 0$

In Fig. 7.19 the SIF values for different α values with $h = 0.5a$ and $d = 0$ are shown. When $d = 0$, the crack is located at the interface between the homogeneous medium and the FGM interlayer. In this figure, it can be observed that the SIF values for positive and negative values of α are located on either side of the SIF curve for $\alpha = 0$. As the α values decrease, the SIF values increase because a compliant graded interlayer increases the crack sliding.

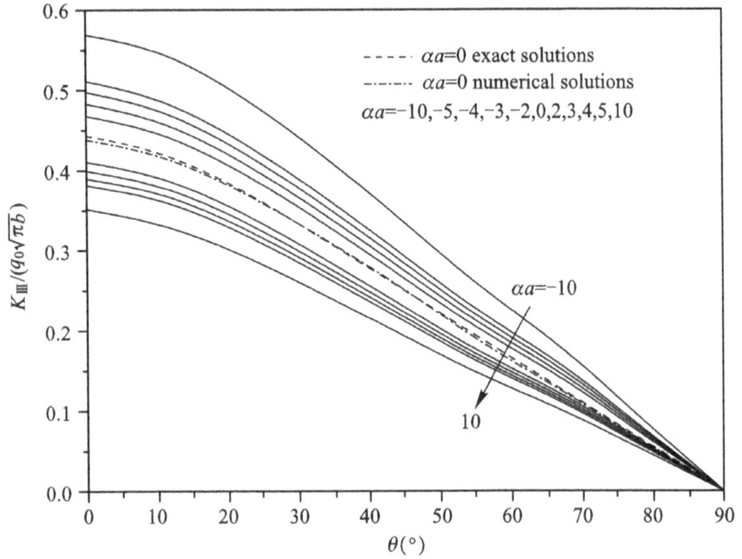

Fig. 7.19 Variations of $K_{\mathrm{III}}/(q\sqrt{\pi b})$ with α for different θ values under q_0 ($h = 0.5a, d = 0$)

(2) Variations of the SIF values with the parameter α for $d = 0.2a$

Figure 7.20 shows the SIF values for $h = 0.5a$ and $d = 0.2a$. If $d = 0.2a$, the crack is located at a distance from the interface between the homogeneous medium and the FGM interlayer. It can be found that the effects of different α values become weaker as the crack distance d increases. This can be more clearly observed in Fig. 7.21, which shows how the effects of the inhomogeneity parameters α rapidly become weak as the crack distance d increases. When $d > 2a$, the effects of the parameter α on the SIF values may be neglected.

(3) Variations of the SIF values with the thickness h of the FGM interlayer

Figure 7.22 illustrates the variations of the SIF values with h for $\alpha a = -5, 5$ and $d = 0.2a$; here the SIF values increase for $\alpha a = -5$ and decrease for $\alpha a = 5$ as the FGM thickness increases. It can be found that for $h > a$ the SIF values at different locations on the crack front display no obvious changes.

Table 7.6 The SIF values ($K_{\mathrm{III}}/(q_0\sqrt{\pi b})$) for an elliptical crack under q_0 for different α values ($h = 0.5a, d = 0$)

$\theta(°)$	$\alpha a = 10$	5	4	3	2	0	−2	−3	−4	−5	−10
0.00	0.35161	0.38138	0.38967	0.39934	0.41055	0.43819	0.46773	0.48278	0.49724	0.51101	0.56922
11.25	0.33286	0.36383	0.37184	0.38116	0.39196	0.41901	0.44714	0.46167	0.47561	0.48887	0.54922
22.50	0.29368	0.32035	0.32762	0.33607	0.34586	0.37174	0.39609	0.40938	0.42218	0.43433	0.49097
33.75	0.24426	0.26772	0.27396	0.28123	0.28968	0.31433	0.33356	0.34528	0.35659	0.36736	0.41804
45.00	0.19242	0.21096	0.21599	0.22183	0.22863	0.25145	0.26415	0.27371	0.28294	0.29176	0.33402
56.25	0.14083	0.15441	0.15819	0.16259	0.16772	0.17988	0.19492	0.20233	0.20953	0.21648	0.24768
67.50	0.09981	0.10996	0.11275	0.11596	0.11972	0.12243	0.13959	0.14526	0.15086	0.15638	0.17724
78.75	0.04942	0.05358	0.05492	0.05649	0.05832	0.05999	0.06799	0.07061	0.07317	0.07560	0.08592
90.00	0.00000	0.00000	0.00000	0.00000	0.00000	0.00000	0.00000	0.00000	0.00000	0.00000	0.00000

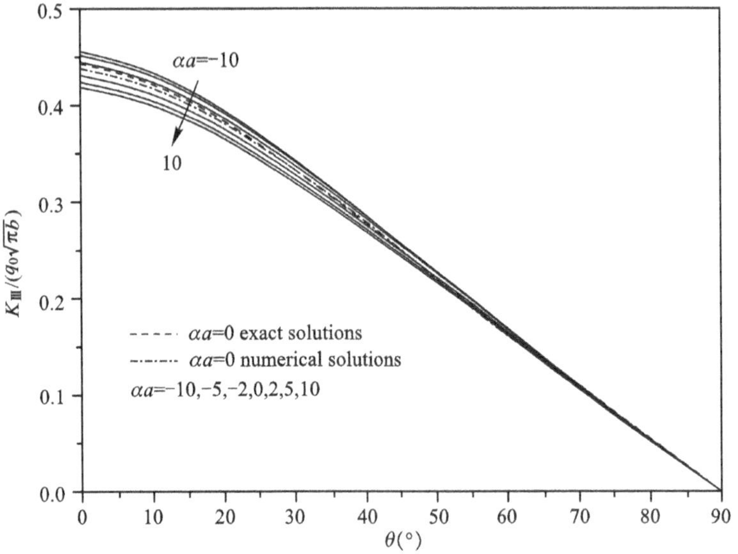

Fig. 7.20 Variations of $K_{\text{III}}/(q\sqrt{\pi b})$ with α for different θ values under q_0 $(h = 0.5a, d = 0.2a)$

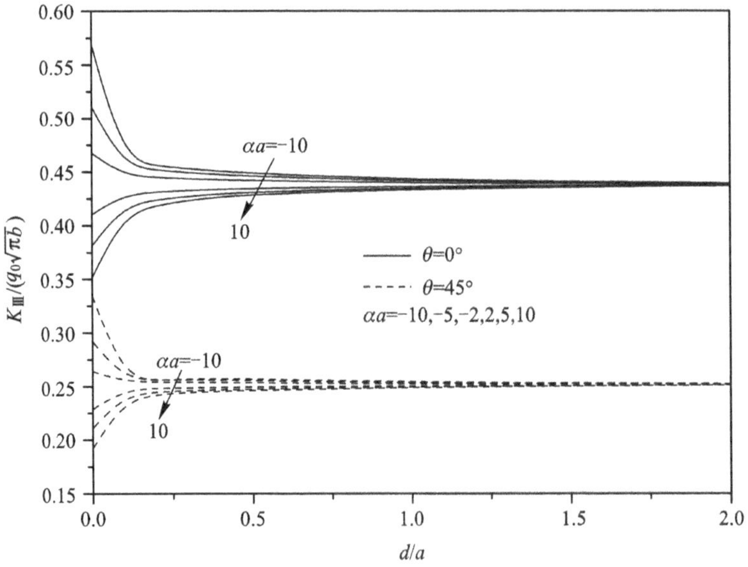

Fig. 7.21 Variations of $K_{\text{III}}/(q_0\sqrt{\pi b})$ with d for different α and θ values under q_0 $(h = 0.5a)$

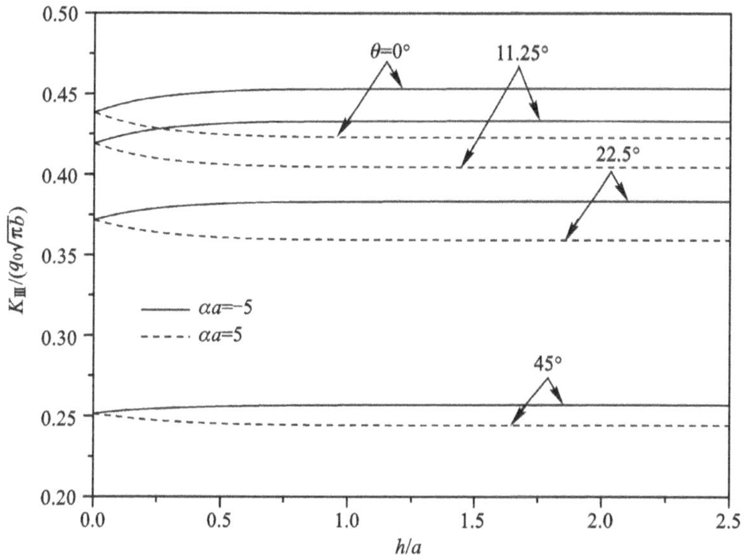

Fig. 7.22 Variations of $K_{\text{III}}/(q_0\sqrt{\pi b})$ with h for different α and θ values under $q_0 (d = 0.2a)$

7.3 The growth of an elliptical crack parallel to the FGM interlayer

In this section, we analyze the growth potential of an elliptical crack in the two bonded solids with a FGM interlayer under a remotely tensile loading as shown in Figs. 7.1a and 7.23. The elliptical crack is in material 1 of a FGM system and is parallel to the FGM

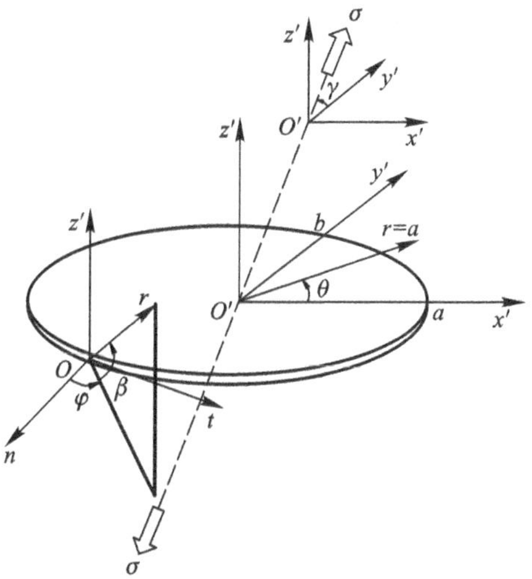

Fig. 7.23 An elliptical crack subject to a remotely inclined loading

interlayer. The SIF values given above and those derived from the similar method outlined in Chapter 6 will be employed. The superposition principle (6.11) of fracture mechanics is first used to calculate the SIF values of an elliptical crack under the remotely tensile loading and subsequently Eq. (6.8) are utilized to calculate the strain energy density factor at the crack front. Finally, the fracture criterion proposed by Sih and Cha (1974) are employed to obtain the directions and critical loadings of crack growth.

1. The minimum strain energy density factor at the crack front

Figure 7.23 illustrates an elliptical crack subjected to a remotely tensile loading σ in the $O'y'z'$- plane; a local spherical coordinate system (r, β, φ) is attached to the crack front. The minimum strain energy density factor is taken as $S_{\min} = S(\beta_0, \varphi_0)$. Table 7.7 presents the variations of the spherical angles (β_0, φ_0) with θ around the crack periphery and the parameter α, where $h = 0.5a$, $d = 0$ and $\gamma = 60°$. As pointed out by Sih and Cha (1974), for an elliptical crack in a homogeneous medium the φ_0 values at any crack-tip point are related to the crack geometry whilst the corresponding β_0 values in the φ_0-plane depend on the crack geometry, the load position and the Poisson's ratio. The results obtained from this calculation also show that φ_0 is not related to the parameter α. Figure 7.24 illustrates the variations of the spherical angles β_0 with the α values; the β_0 values increase as the α values decrease because the SIF values show an increase when α decreases. The mode II and III SIF values obtained from the shear stress q_0 and compressive stress p_0 have different distribution characteristics when $-90° \leqslant \theta \leqslant 0°$ and $0° \leqslant \theta \leqslant 90°$. As a result, the β_0 values are different in the two ranges for any given θ around the crack periphery.

Table 7.7 Values of φ_0 and β_0 for different α and θ values ($d = 0$)

$\theta(°)$	$\varphi_0(°)$	$\beta_0(°)$						
		$\alpha a = 10$	5	2	0	–2	–5	–10
–90.00	0.00000	39.1314	39.3463	40.2774	42.3123	43.5490	45.6570	47.3492
–78.75	16.0141	38.7093	38.8845	39.7840	41.8677	43.1440	45.3776	47.6235
–67.50	27.9384	37.8534	38.0323	38.9321	40.7460	42.4157	44.8211	47.0250
–56.25	34.7185	35.6553	35.7541	36.6844	38.6197	40.5783	43.4736	46.6175
–45.00	36.8699	32.6312	32.8511	33.9568	35.8388	38.4559	41.9364	45.9272
–33.75	34.7185	29.2539	29.7113	31.1579	33.0112	36.6835	40.9725	46.0485
–22.50	27.9384	23.6894	24.3028	26.4392	28.9403	33.9423	39.4967	45.9427
–11.25	16.0141	10.3310	11.8926	15.9956	20.4650	28.1221	35.8955	44.1011
0.00	0.0000	–18.1005	–15.1872	–8.7679	0.0000	12.9456	26.5570	38.2637
11.25	–16.0141	–34.0428	–31.6566	–27.1810	–20.4650	–10.0159	7.8445	26.2554
22.50	–27.9384	–40.2854	–38.0128	–34.1738	–28.9403	–20.9919	–6.7683	11.8825
33.75	–34.7185	–43.2959	–41.2156	–37.6754	–33.0112	–25.9008	–13.4814	2.9747
45.00	–36.8699	–45.2146	–43.2860	–39.9933	–35.8388	–28.9034	–16.9941	0.5860
56.25	–34.7185	–47.1932	–45.3928	–42.3596	–38.6197	–31.8507	–19.8381	–1.8248
67.50	–27.9384	–48.1303	–46.6571	–44.0288	–40.7460	–34.4094	–22.6205	–2.4847
78.75	–16.0141	–48.4398	–46.9868	–44.5147	–41.8677	–35.4397	–23.6459	–2.2712
90.00	0.0000	–48.3858	–47.0614	–44.7582	–42.3123	–36.1265	–24.7437	–1.2857

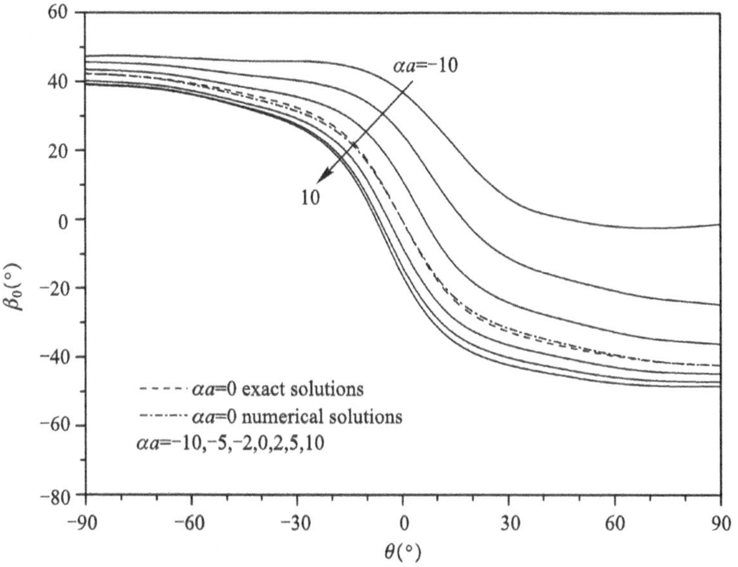

Fig. 7.24 Variations of β_0 with θ for different α values $(d = 0, h = 0.5a, \gamma = 60°)$

The variations of $S_{\min}$ with θ around the crack periphery and the parameter α where $h = 0.5b$, $d = 0$ and $\gamma = 60°$ are illustrated in Fig. 7.25. For an elliptical crack in a homogeneous solid, the maximum $S_{\min}$ value around the whole crack tip appears at the intersection of the minor axis with the crack border. It should be noted that the $S_{\min}$ curves, shown in Fig. 7.25, correspond to those SIF values obtained from the exact and numerical solutions. It is evident that the parameter α influences the $S_{\min}$ value at the crack tip, especially for $\alpha < 0$. For any given θ value, the $S_{\min}$ value increases as the α value decreases. For $\alpha > 0$, the maximum value of $S_{\min}$ appears at the crack front $\theta = 90°$ whereas when $\alpha < 0$ the maximum value of $S_{\min}$ appears at $\theta = -90°$. Thus, the σ_{cr} values of the elliptical crack in the FGM can be calculated by setting $S_{\min} = S_{cr}$ at $\theta = -90°$ for $\alpha < 0$ and at $\theta = 90°$ for $\alpha > 0$, respectively.

Figure 7.26 shows the variations of $S_{\min}$ with the loading angle γ at the crack front $\theta = -90°, 90°$. It can be observed that for all the loading angles γ (except $\gamma = 0°$) the $S_{\min}$ values at $\theta = -90°$ and $90°$ increase as the α value decreases. The $S_{\min}$ values at $\theta = -90°$ show more evident variations than those at $\theta = 90°$. A discussion of this phenomenon will be presented in the following.

Considering the crack geometry, the mode III SIF values at $\theta = -90°$ and $90°$ are zero. For $\alpha > 0$, the mode II SIF values at $\theta = 90°$ by q_0 are positive and increase and those by p_0 are positive and decrease as α increases. Thus α affects the mode II SIF values and accordingly the $S_{\min}$ value become weak under tensile loading. For $\alpha > 0$, the mode II SIF values at $\theta = -90°$ by p_0 are positive and increase and those by q_0 are negative and increase as α increases. Therefore α affects the mode II SIF values so that

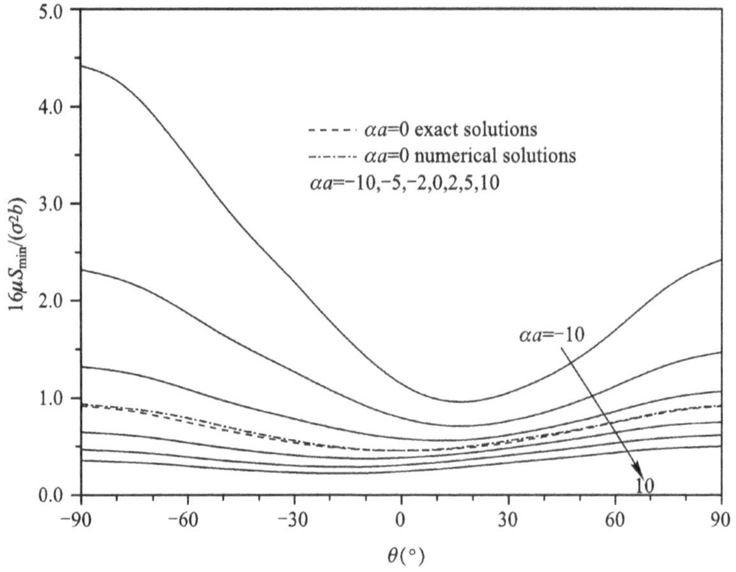

Fig. 7.25 Variations of $S_{\min}$ with θ for different α values $(d = 0, h = 0.5a, \gamma = 60^\circ)$

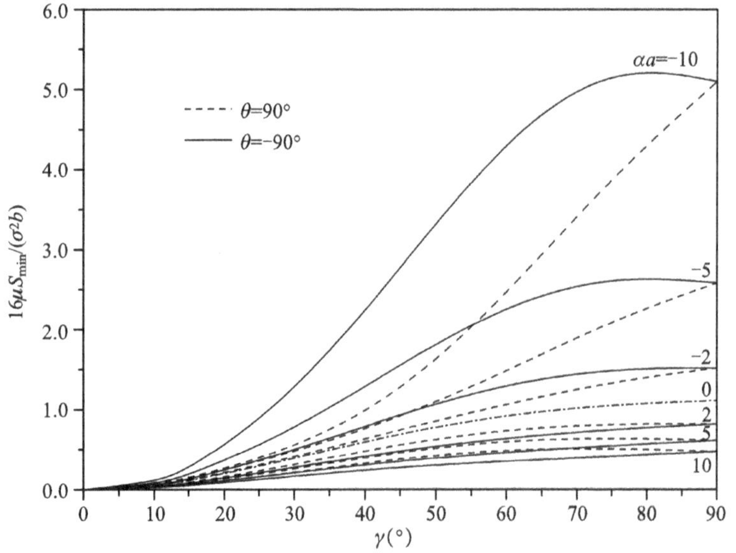

Fig. 7.26 Variations of $S_{\min}$ with γ for different α and θ values $(d = 0, h = 0.5a)$

the $S_{\min}$ value becomes greater under tensile loading. Similar analyses are suitable for $\alpha < 0$; this also results in different variations of the spherical angles β_0 in the two ranges

$-90° \leqslant \theta \leqslant 0°$ and $0° \leqslant \theta \leqslant 90°$. For $\gamma = 90°$, the effects of q_0 on the mode II SIF values disappear; as a result, the S_{min} values at $\theta = -90°$ and $90°$ are equal.

2. The critical loadings for crack growth

Figure 7.27 presents the variations of the critical loading σ_{cr} with the loading angle γ and the parameter α, where $h = 0.5a$ and $d = 0$. Here, the σ_{cr} value decreases monotonically as the loading angle γ increases and decreases as the parameter α decreases under a given γ value.

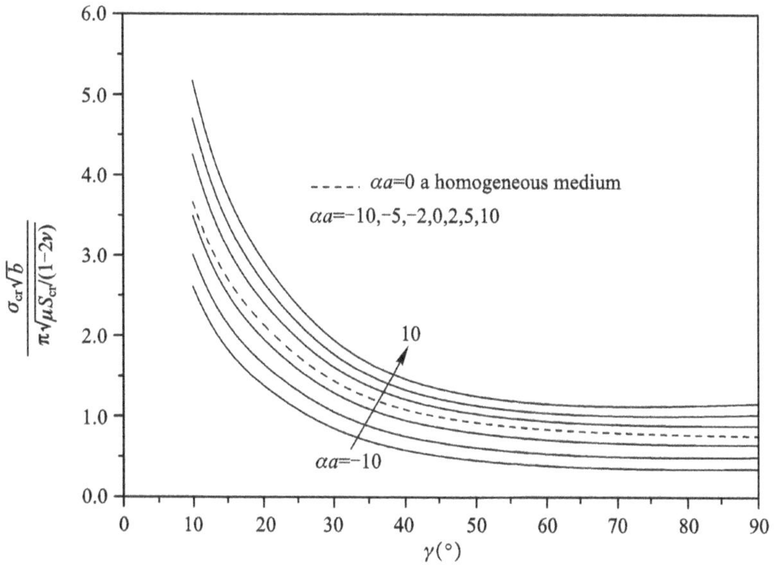

Fig. 7.27 Variations of σ_{cr} with γ for different α values $(d = 0, h = 0.5a)$

In Fig. 7.28 the variations of σ_{cr} with the crack distance d to the FGM interlayer are shown, where $h = 0.5a$ and $\alpha a = -5, 5$. It is evident that, for a given loading angle γ, the σ_{cr} value decreases for $\alpha a = 5$ and increases for $\alpha a = -5$ as the crack distance d from the FGM interlayer increases. In the general case, as the crack distance d increases, the σ_{cr} value decreases for $\alpha > 0$ and increases for $\alpha < 0$.

Figure 7.29 shows the variations in the σ_{cr} values with the thickness h for different loading angles γ (45°, 60° and 90°) and $\alpha a = 5$ or -5. The σ_{cr} values for $h = 0$ represent the results for a homogeneous solid. It can be observed that the σ_{cr} value increases as the thickness h increases for $\alpha a = 5$ whilst the σ_{cr} value decreases for $\alpha a = -5$. Such variations of the σ_{cr} value are evident for $h \leqslant a$ and relatively weak for $h > a$. In the general case, the σ_{cr} value decreases as the thickness h increases for $\alpha < 0$ whilst it increases for $\alpha > 0$.

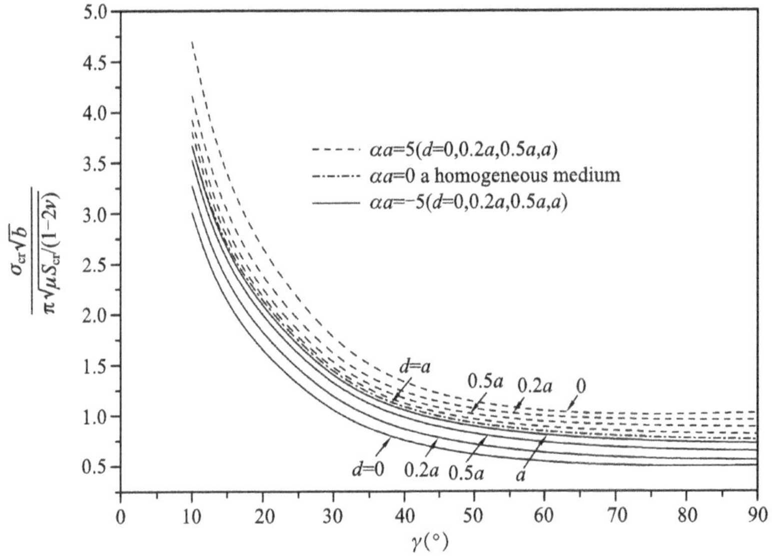

Fig. 7.28 Variations of σ_{cr} with d for different γ and α values $(h = 0.5a)$

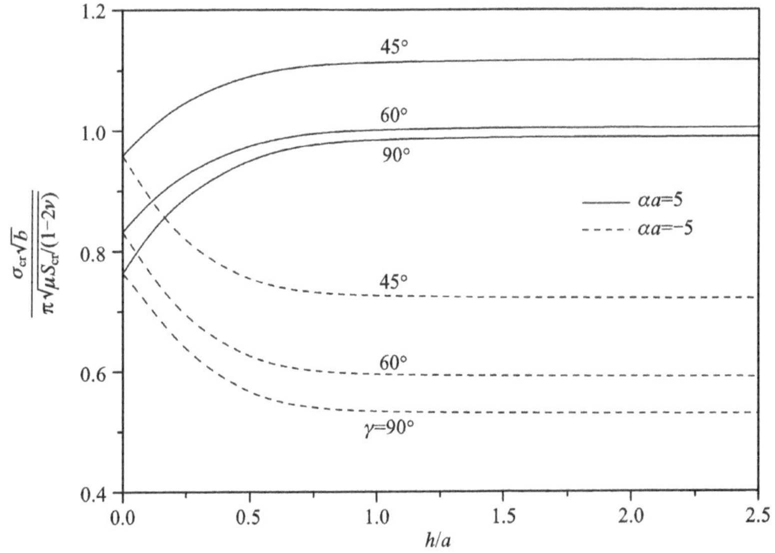

Fig. 7.29 Variations of σ_{cr} with h for different γ and α values $(d = 0.2a)$

7.4 The SIFs for an elliptical crack perpendicular to the FGM interlayer

7.4.1 General

The elastic modulus of a FGM system shown in Fig. 7.30 is approximated by Eq. (6.1). It is further assumed that the Poisson's ratios of the FGMs are constant and equal to 0.3, i.e., $\nu_1 = \nu_2 = \nu_3 = 0.3$. The ratio of the major and minor axes of the elliptical crack surface is taken as $a/b = 2$. The minor axis of the elliptical crack in Fig. 7.30 is perpendicular to the FGM interlayer. The crack surfaces are subjected to the following loadings

$$\sigma^{+}_{x'x'} = \sigma^{-}_{x'x'} = p \tag{7.4a}$$

$$\sigma^{+}_{x'y'} = \sigma^{-}_{x'y'} = 0 \tag{7.4b}$$

$$\sigma^{+}_{x'z'} = \sigma^{-}_{x'z'} = q \tag{7.4c}$$

where the superscripts + and – correspond to the crack surfaces, which are located at $x' = 0^+$ and $x' = 0^-$, respectively.

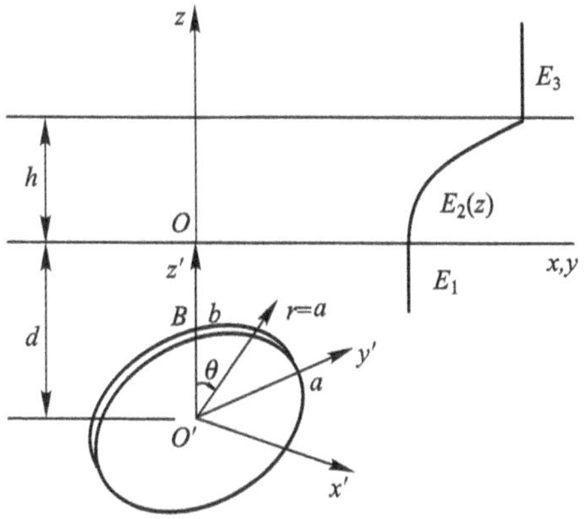

Fig. 7.30 Coordinate systems for an elliptical crack perpendicular to a FGM interfacial zone in two fully bonded solids and subject to tension and shear $(a = 2b)$

In Fig. 7.30, the cracked body is symmetrical with respect to the $O'x'z'$-plane. Therefore, we only need to consider half of the crack problem and employ a similar discretizing method to that outlined in Section 7.2. Figure 7.31 shows the discretized surface of region I with 176 elements and 553 nodes. There are 32 traction-singular elements along the crack front. The two sides of the traction-singular elements on the crack surface are perpendicular to the tangent at the crack front and intersect with the crack front. The boundary of region II is discretized into the same mesh as the one of region I shown in Fig. 7.31.

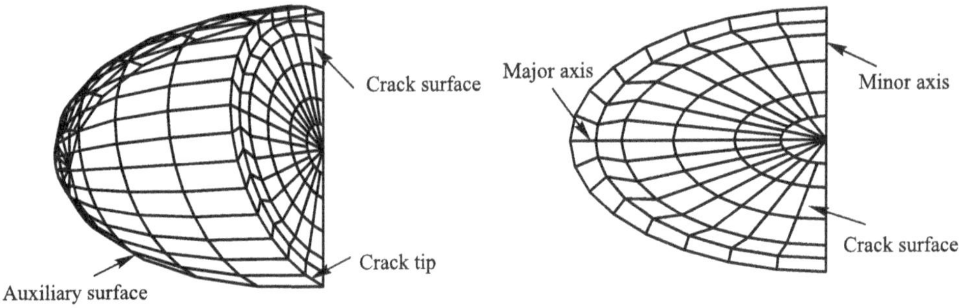

(a) Auxiliary and crack surfaces

(b) Crack surface

Fig. 7.31 Boundary element meshes for an elliptical crack problem

7.4.2 Elliptical crack under a uniform compressive stress

The crack surfaces are subjected to a uniform compressive stress, i.e., $p = -p_0$ ($p_0 > 0$) and $q = 0$ in Eq. (7.4). In order to verify the element mesh for an elliptical crack in a homogeneous medium of infinite extent, we obtain the exact solution of $K_{\mathrm{I}}/(p_0\sqrt{\pi b})$ at the crack front of the minor axis as $1/E(k) = 0.8257$. Our numerical solution from the proposed method is 0.8342; therefore, the absolute errors of results from the two methods are about 1.0%.

In the following, we will examine the SIFs of an elliptical crack in the FGMs. Figures 7.32 ~ 7.34 show the variations of the SIF values with the nonhomogeneity parameter α, the crack distance d to the FGM interlayer and the interlayer thickness h. Notice that the elliptical crack in the FGM system subjected to a uniform compressive stress has mixed deformation modes and the cases $\alpha = 0$ and $h = 0$ correspond to a homogeneous medium of infinite extent.

(1) Variations of the SIF values with the nonhomogeneity parameter α

Figure 7.32 illustrates the SIF values with α and Table 7.8 presents the corresponding data. In this figure, the SIF values for positive and negative values of α are located, respectively, at either side of the curve of the SIF values for $\alpha = 0$. As the values of α decrease, the SIF values increase. This is because a compliant graded interlayer increases the crack opening. The crack tip $\theta = 0°$ is located at the minor axis of the elliptical crack and is closest to the interlayer. Evidently, the maximum influence of the graded interlayer on the SIFs occurs at the crack tip $\theta = 0°$. For $\alpha < 0$, the increments of the SIF values at the elliptical crack tip decrease as θ approaches 90° for the crack tip $0° \leqslant \theta \leqslant 90°$ and increase weakly as θ approaches 180° for the crack tip $90° \leqslant \theta \leqslant 180°$. This phenomenon of weak increases is different from that for a penny-shaped crack outlined in Chapter 6. For $\alpha > 0$, a similar influence can be observed, indicating that the influence of the graded interlayer on the SIF values for the elliptical crack may not only be dependent on the crack-tip distances to the interlayer but also on the elliptical crack shape.

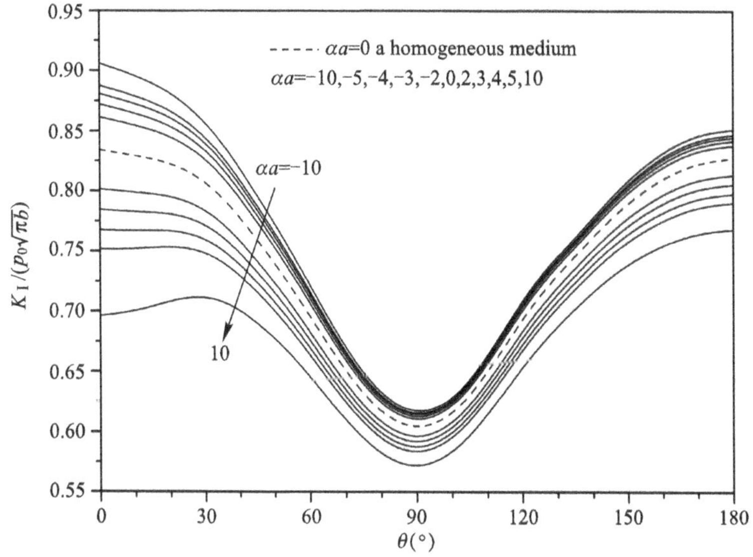

Fig. 7.32 Variations of $K_{\mathrm{I}}/(p_0\sqrt{\pi b})$ with α for different θ values under p_0 $(d = 1.2b, h = 0.5a)$

(2) Variations of the SIF values with the crack distance d to the FGM interlayer

Figure 7.33 shows the SIF values of the elliptical crack tip for different d where $\alpha a = -5$ and 5 and Table 7.9 presents the corresponding data. From this figure, a trend can be observed where the mode I SIF decreases as d decreases for $\alpha a = 5$. This implies that more pronounced constraints are induced by the stiffer solid as the crack is located closer to the FGM interface. The opposite behavior can be observed for $\alpha a = -5$; when the crack is located farther away from the interface, the SIF values become less sensitive to the variations in the material parameters and the locations of the crack front, and the solutions for different α values approach those of an elliptical crack in a homogeneous elastic solid of infinite extent.

(3) Variations of the SIF values with the thickness h of the FGM interlayer

The effect of h on the SIF values, where $\alpha a = -5$ and 5, is illustrated in Fig. 7.34 and the corresponding data are tabulated in Table 7.10. At different locations of the crack front, the SIF values decrease for $\alpha a = 5$ as h increases and the opposite behavior can be observed for $\alpha a = -5$. When h/a approaches 2.5 there are no obvious changes for the SIF values at different locations of the crack front for $\alpha a = -5$ and 5.

Table 7.8 The SIF values ($K_I/(p_0\sqrt{\pi b})$) for an elliptical crack under p_0 with different α values ($h = 0.5a, d = 1.2b$)

$\theta(°)$	$\alpha a = -10$	−5	−4	−3	−2	0	2	3	4	5	10
0.00	0.90610	0.88766	0.88080	0.87215	0.86146	0.83419	0.80151	0.78442	0.76760	0.75167	0.69613
11.25	0.89518	0.87834	0.87202	0.86404	0.85418	0.82890	0.79848	0.78251	0.76676	0.75184	0.69999
22.50	0.87859	0.86446	0.85904	0.85215	0.84360	0.82166	0.79526	0.78144	0.76787	0.75509	0.71155
33.75	0.84709	0.83594	0.83154	0.82590	0.81887	0.80080	0.77910	0.76781	0.75682	0.74656	0.71274
45.00	0.79674	0.78836	0.78491	0.78045	0.77484	0.76032	0.74289	0.73389	0.72522	0.71721	0.69182
56.25	0.74269	0.73653	0.73385	0.73032	0.72584	0.71409	0.69992	0.69265	0.68569	0.67934	0.65993
67.50	0.67918	0.67525	0.67339	0.67090	0.66767	0.65898	0.64817	0.64253	0.63709	0.63212	0.61722
78.75	0.63139	0.62874	0.62736	0.62546	0.62293	0.61593	0.60697	0 .60221	0.59761	0.59340	0.58095
90.00	0.61171	0.60982	0.60871	0.60714	0.60500	0.59890	0.59086	0.58652	0.58228	0.57839	0.56698

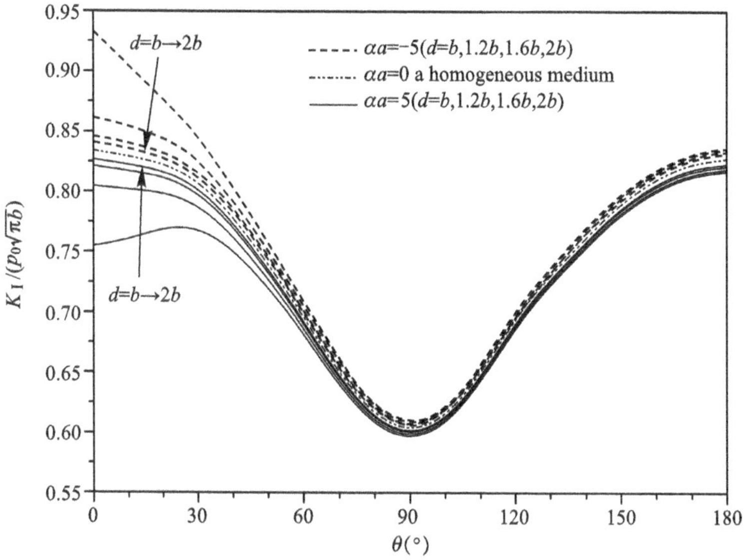

Fig. 7.33 Variations of $K_I/(p_0\sqrt{\pi b})$ with d for different α and θ values under p_0 $(h = 0.5a)$

Table 7.9 The SIF values $(K_I/(p_0\sqrt{\pi b}))$ for an elliptical crack under p_0 with different d values $(h = 0.5a)$

$\theta(°)$	$\alpha a = -5$				$\alpha a = 5$			
	d = 2.0b	1.6b	1.2b	1.0b	d = 1.0b	1.2b	1.6b	2.0b
0.00	0.84101	0.84583	0.86140	0.93305	0.75495	0.80467	0.82143	0.82698
11.25	0.83521	0.83976	0.85398	0.90040	0.76065	0.80155	0.81699	0.82222
22.50	0.82732	0.83133	0.84274	0.87014	0.77247	0.79851	0.81111	0.81567
33.75	0.80575	0.80904	0.81735	0.83255	0.76726	0.78256	0.79185	0.79556
45.00	0.76462	0.76718	0.77288	0.78108	0.73772	0.74644	0.75288	0.75576
56.25	0.71785	0.71982	0.72371	0.72828	0.69826	0.70341	0.70787	0.71006
67.50	0.66192	0.66332	0.66588	0.66852	0.64801	0.65118	0.65418	0.65578
78.75	0.61843	0.61951	0.62133	0.62302	0.60753	0.60972	0.61192	0.61317
90.00	0.60119	0.60208	0.60351	0.60472	0.59182	0.59352	0.59529	0.59635

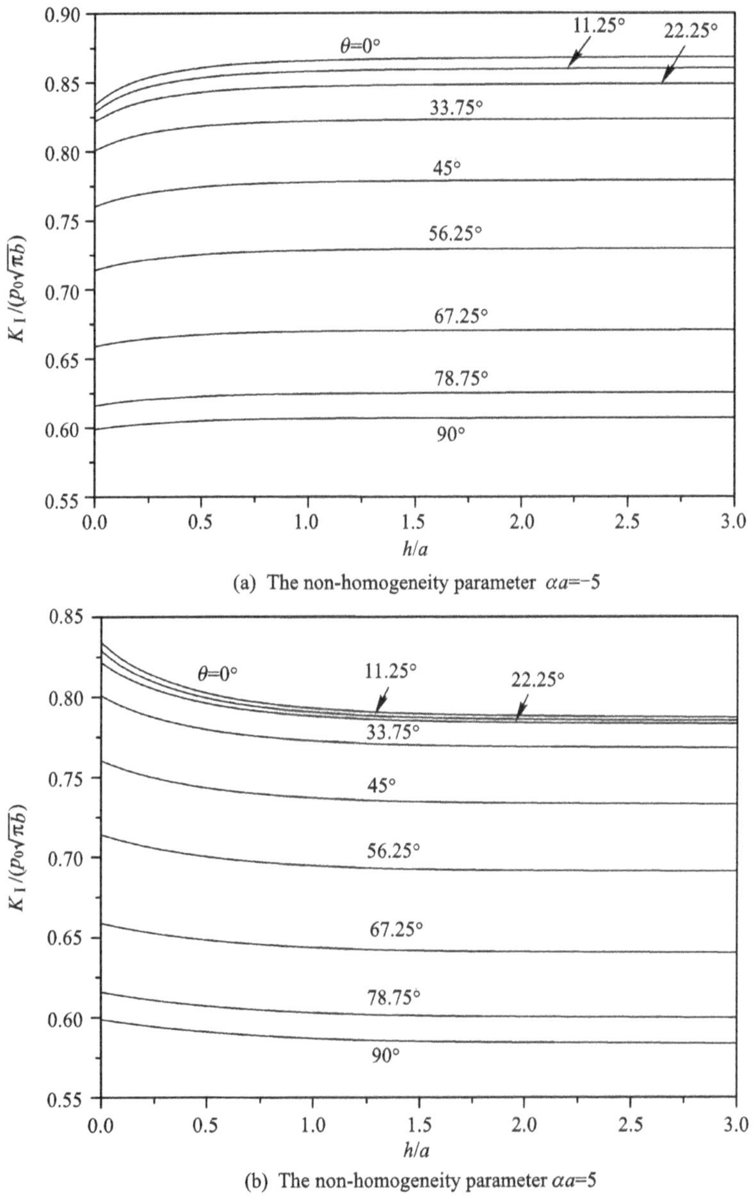

Fig. 7.34 Variations of $K_I/(p_0\sqrt{\pi b})$ with h for different α and θ values $(d = 1.2b)$

Table 7.10 The SIF values ($K_I/(p_0\sqrt{\pi b})$) for an elliptical crack under p_0 with different h values ($d = 1.2b$)

$h/(2b)$	$\alpha a = -5$					$\alpha a = 5$				
	$\theta = 0.0°$	11.25°	22.50°	33.75°	90.00°	$\theta = 0.0°$	11.25°	22.50°	33.75°	90.00°
0.0	0.83419	0.82890	0.82166	0.80080	0.59890	0.83419	0.82890	0.82166	0.80080	0.59890
0.10	0.84559	0.83942	0.83051	0.80774	0.60087	0.82243	0.81802	0.81248	0.79358	0.59681
0.25	0.85466	0.84786	0.83793	0.81393	0.60297	0.81171	0.80801	0.80371	0.78633	0.59419
0.50	0.86146	0.85418	0.84360	0.81887	0.60500	0.80151	0.79848	0.79526	0.77910	0.59086
0.75	0.86434	0.85679	0.84598	0.82101	0.60603	0.79586	0.79319	0.79050	0.77490	0.58845
1.00	0.86564	0.85795	0.84703	0.82199	0.60654	0.79266	0.79019	0.78774	0.77241	0.58677
1.25	0.86648	0.85868	0.84767	0.82256	0.60682	0.79057	0.78823	0.78596	0.77080	0.58564
1.50	0.86701	0.85915	0.84807	0.82289	0.60698	0.78924	0.78699	0.78484	0.76979	0.58491
2.00	0.86751	0.85958	0.84843	0.82320	0.60711	0.78801	0.78584	0.78380	0.76885	0.58421
2.50	0.86778	0.85981	0.84862	0.82335	0.60716	0.78747	0.78535	0.78336	0.76848	0.58396
3.00	0.86783	0.85990	0.84871	0.82344	0.60725	0.78704	0.78504	0.78305	0.76803	0.58352

7.4.3 Elliptical crack under a uniform shear loading

Consider that the crack surfaces are subjected to a uniform shear loading, i.e., $p = 0$ and $q = -q_0$ ($q_0 > 0$) in Eq. (7.4). In this case, the mode II and III deformations are coupled together. The boundary element meshes shown in Fig. 7.31 are further utilized. An elliptical crack in a homogeneous medium of infinite extent is analyzed to verify the element mesh. The exact solution of $K_{II}/(q_0\sqrt{\pi b})$ at the crack front of the minor axis is about 0.89561, the numerical solution is 0.90687 and the absolute error is 1.0%. The exact solution of $K_{III}/(q_0\sqrt{\pi b})$ at the crack front of the major axis is about 0.44330, the numerical solution is 0.41819 and the absolute error is 2.5%. In the following, we will discuss, respectively, the variations of the mode II and III SIF values due to the presence of the FGM interlayer.

1. Results of the mode II SIF values

(1) Variations of the SIF values with the nonhomogeneity parameter α

Figure 7.35 illustrate the variations of the mode II SIF with α and Table 7.11 lists the corresponding SIF values. In Fig. 7.35, it can be seen that the first maximum influence of α on the mode II SIF values occur in the neighborhood of $\theta = 0°$ and the second maximum influence of α on the SIF value occurs in the neighborhood of $\alpha = 180°$. In the neighborhood of $\alpha = 105°$, small variations in the mode II SIF values appear; these observations are closely related to the shear loading conditions. When compared with the results of an elliptical crack in a homogeneous elastic solid of infinite extent , the mode

Ⅱ SIF values increase for $\alpha < 0$ and decrease for $\alpha > 0$. It can also be found that the influence of α on the mode Ⅱ SIF values under q_0 are similar to those under p_0.

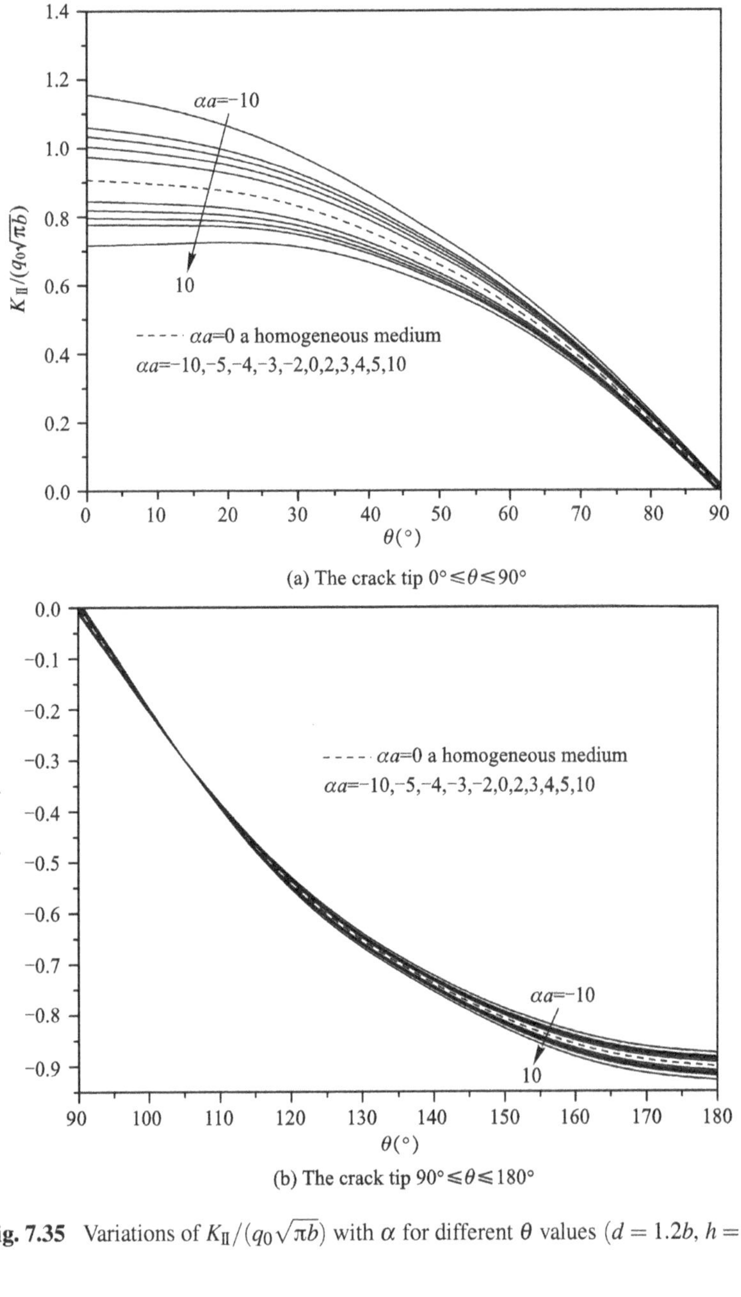

Fig. 7.35 Variations of $K_{\mathrm{II}}/(q_0\sqrt{\pi b})$ with α for different θ values $(d = 1.2b,\ h = 0.5a)$

Table 7.11 The SIF values $(K_{\mathrm{II}}/(q_0\sqrt{\pi b}))$ for an elliptical crack under q_0 with different α values $(h = 0.5a, d = 1.2b)$

$\theta(°)$	$\alpha a = -10$	–5	–4	–3	–2	0	2	3	4	5	10
0.00	1.15575	1.06046	1.03421	1.00515	0.97359	0.90687	0.84472	0.81830	0.79555	0.77628	0.71583
11.25	1.11883	1.03276	1.00899	0.98264	0.95402	0.89339	0.83681	0.81274	0.79201	0.77445	0.71948
22.50	1.05169	0.98462	0.96553	0.94415	0.92070	0.87062	0.82377	0.80395	0.78700	0.77277	0.72936
33.75	0.94615	0.89886	0.88473	0.86867	0.85082	0.81224	0.77611	0.76097	0.74818	0.73759	0.70661
45.00	0.80808	0.77590	0.76575	0.75401	0.74079	0.71188	0.68483	0.67365	0.66433	0.65675	0.63565
56.25	0.66225	0.64026	0.63292	0.62430	0.61447	0.59277	0.57257	0.56435	0.55759	0.55220	0.53792
67.50	0.39782	0.40672	0.41020	0.41461	0.42004	0.43356	0.44815	0.45473	0.46047	0.46530	0.47926
78.75	0.21119	0.21637	0.21847	0.22117	0.22456	0.23316	0.24258	0.24682	0.25050	0.25357	0.26210
90.00	-0.01162	-0.00934	-0.00834	-0.00701	-0.00529	-0.00072	0.00445	0.00680	0.00881	0.01048	0.01485

(2) Variations of the SIF values with the crack distance d to the FGM interlayer
The mode II SIF values with d where $h = 0.5a$ and $\alpha a = -5$ or 5 are shown in Fig. 7.36

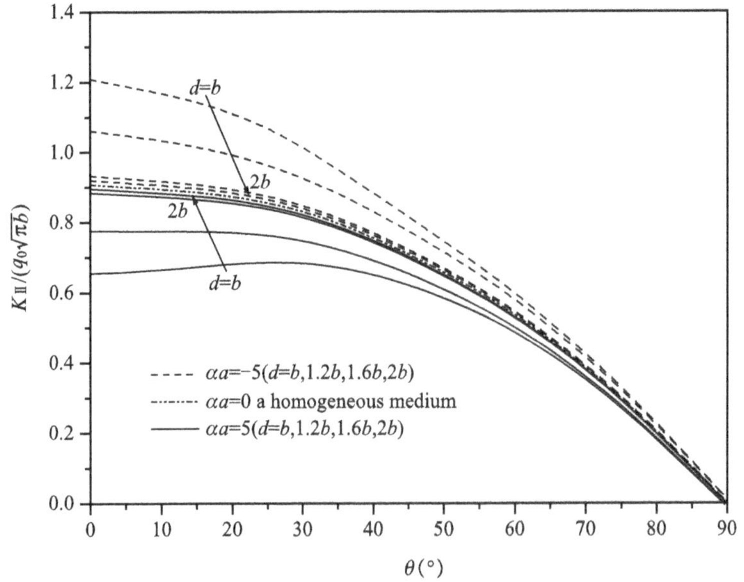

(a) The crack tip $0° \leqslant \theta \leqslant 90°$

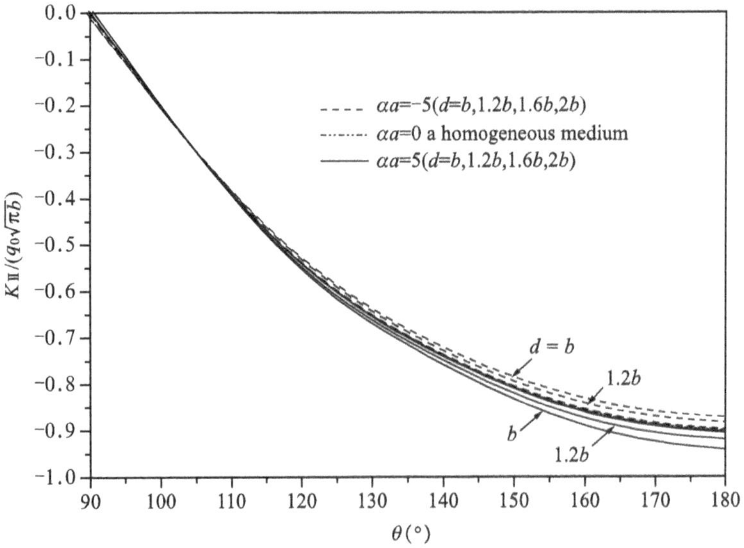

(b) The crack tip $90° \leqslant \theta \leqslant 180°$

Fig. 7.36 Variations of $K_{\text{II}}/(q_0\sqrt{\pi b})$ with d for different α and θ values under q_0 $(h = 0.5a)$

and Table 7.12 lists the corresponding SIF values. As with the uniform compressive loading, the mode Ⅱ SIF values decrease as d decreases for $\alpha a = 5$, which implies a more pronounced constraint by the constituent materials as the crack is located closer to the FGM interface. An opposite behavior is encountered for $\alpha a = -5$. When the crack is located farther away from the FGM interlayer, the FGM exerts a weak influence on the SIF values.

Table 7.12 The SIF values $(K_{\mathrm{II}}/(q_0\sqrt{\pi b}))$ for an elliptical crack under q_0 with different d values $(h = 0.5a)$

$\theta(°)$	$\alpha a = -5$				$\alpha a = 5$			
	$d = 2.0b$	$1.6b$	$1.2b$	$1.0b$	$d = 1.0b$	$1.2b$	$1.6b$	$2.0b$
0.00	0.91933	0.93139	1.06046	1.20684	0.65555	0.77628	0.88388	0.89507
11.25	0.90471	0.91582	1.03276	1.16537	0.66455	0.77445	0.87234	0.88265
22.50	0.88083	0.89002	0.98462	1.10128	0.68886	0.77277	0.85237	0.86092
33.75	0.82110	0.82813	0.89886	0.97011	0.68390	0.73759	0.79729	0.80384
45.00	0.71933	0.72446	0.77590	0.81444	0.62626	0.65675	0.70003	0.70481
56.25	0.59896	0.60261	0.64026	0.66088	0.53502	0.55220	0.58351	0.58691
67.50	0.43793	0.44030	0.46530	0.47631	0.42720	0.40672	0.42720	0.42940
78.75	0.23612	0.23756	0.25357	0.25818	0.22903	0.21637	0.22903	0.23036
90.00	0.00100	0.00171	0.01048	0.01131	–0.00299	–0.00934	–0.00299	–0.00236

(3) Variations of the SIF values with the thickness h of the FGM interlayer

Figures 7.37 and 7.38 show the mode Ⅱ SIF values for different h values where $d = 1.2b$, $\alpha a = -5$ or 5, and Table 7.13 lists the corresponding SIF values. It can be found that the mode Ⅱ SIF values for $0° \leqslant \theta < 90°$ and the absolute ones for $90° < \theta \leqslant 180°$ increase as h increases for $\alpha a = -5$. An opposite behavior for the SIF occurs for $\alpha a = 5$. The SIF values for $\alpha a = -5$ and 5 seem to become steady when $h/a > 2.5$. In Fig. 7.38, the SIF values $(\alpha a = 5)$ for the crack tip at $\theta = 0°, 11.25°, 22.5°$ approach the same value as the FGM thickness increases. This is because the SIF values decrease more rapidly when the crack tip is closer to the graded interlayer for $\alpha > 0$.

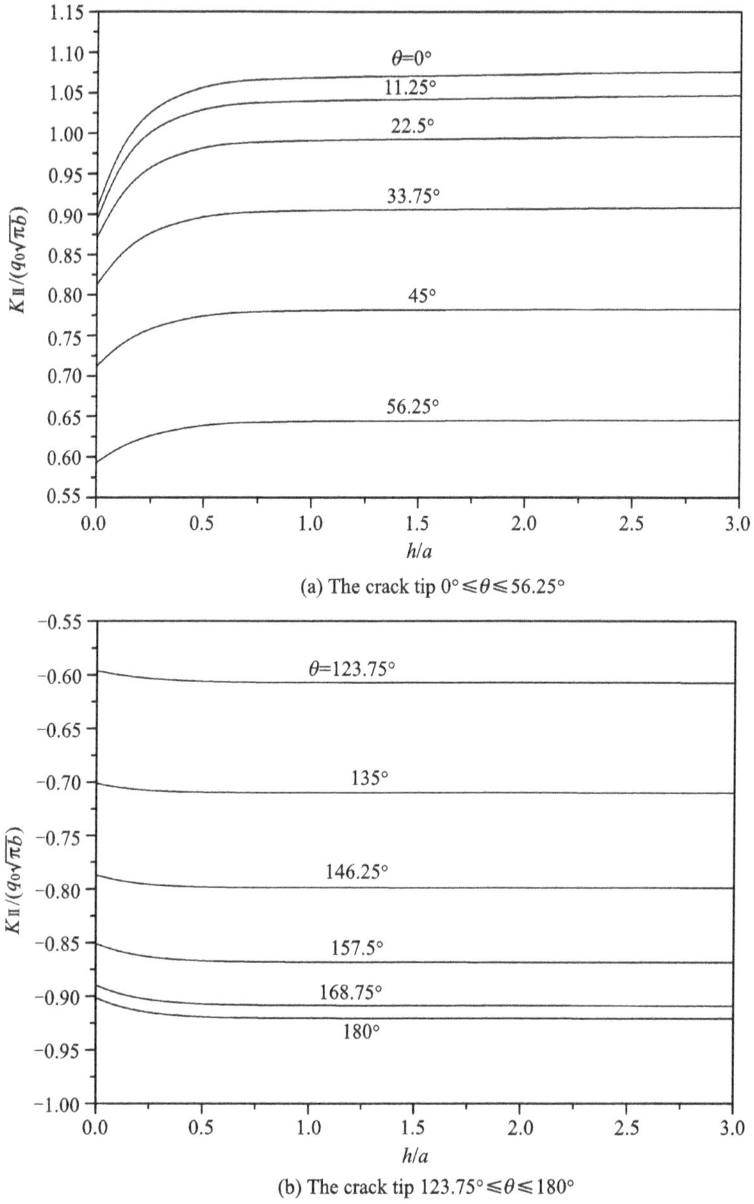

Fig. 7.37 Variations of $K_{II}/(q_0\sqrt{\pi b})$ with h for different θ values $(\alpha a = -5,\ d = 1.2b)$

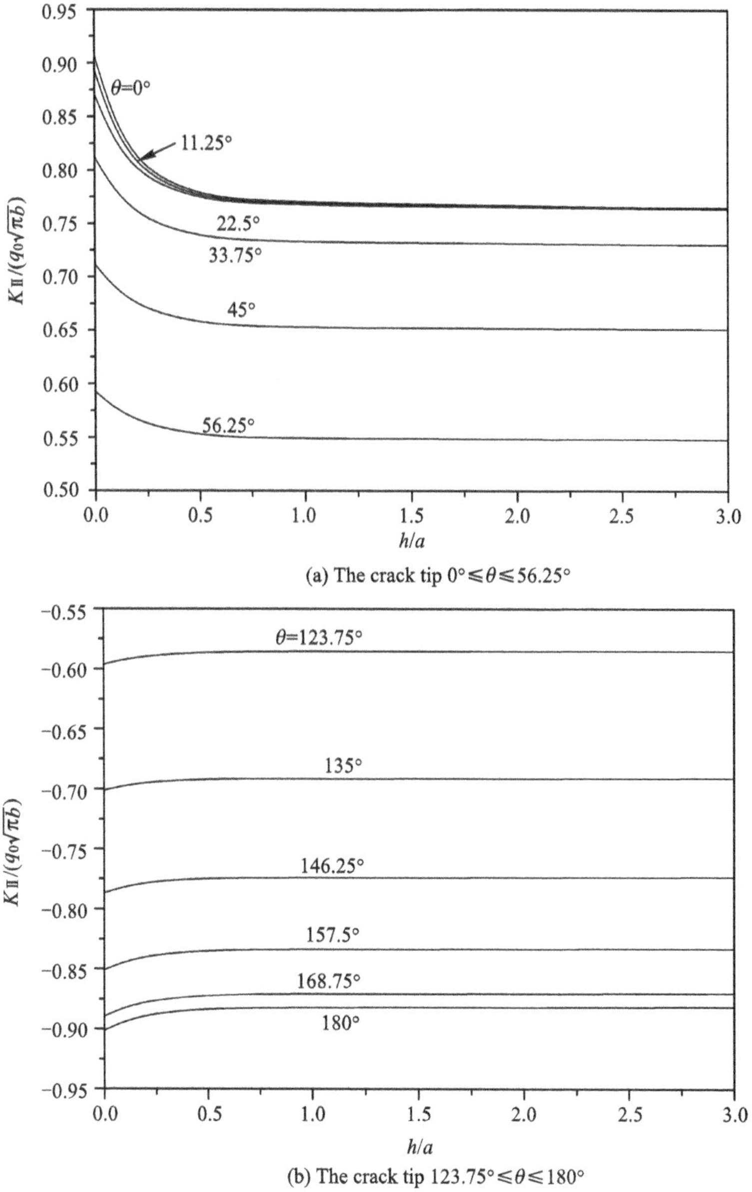

Fig. 7.38 Variations of $K_{II}/(q_0\sqrt{\pi b})$ with h for different θ values ($\alpha a = 5$, $d = 1.2b$)

Table 7.13 The SIF values $(K_{II}/(q_0\sqrt{\pi b}))$ for an elliptical crack under q_0 with different h values $(d = 1.2b)$

$h/(2b)$	$\alpha a = -5$				$\alpha a = 5$			
	$\theta = 0.0°$	11.25°	22.50°	33.75°	$\theta = 0.0°$	11.25°	22.50°	33.75°
0.00	0.90687	0.89339	0.87062	0.81224	0.90687	0.89339	0.87062	0.81224
0.10	0.97994	0.95947	0.92262	0.84948	0.83989	0.8327	0.8226	0.77764
0.25	1.03322	1.00797	0.96284	0.88062	0.79691	0.7934	0.78963	0.75182
0.50	1.06046	1.03276	0.98462	0.89886	0.77628	0.77445	0.77277	0.73759
0.75	1.06669	1.03834	0.9897	0.90331	0.77147	0.77005	0.76875	0.73406
1.00	1.06811	1.03961	0.99086	0.90436	0.77029	0.76898	0.76776	0.73318
1.25	1.06984	1.04116	0.99207	0.90521	0.76888	0.76769	0.76674	0.73245
1.50	1.07042	1.04167	0.99247	0.90549	0.76841	0.76727	0.76641	0.73221
2.00	1.07202	1.04311	0.99354	0.90619	0.76713	0.76611	0.76553	0.73163
2.50	1.07438	1.04524	0.99513	0.90721	0.76527	0.76443	0.76426	0.7308
3.00	1.07527	1.04632	0.99611	0.90798	0.7652	0.76432	0.76403	0.73042

2. Results of the mode III SIF values

(1) Variations of the SIF values with the nonhomogeneity parameter α

Figure 7.39 shows the results for different α values where $h = 0.5a$ and $d = 1.2b$ and

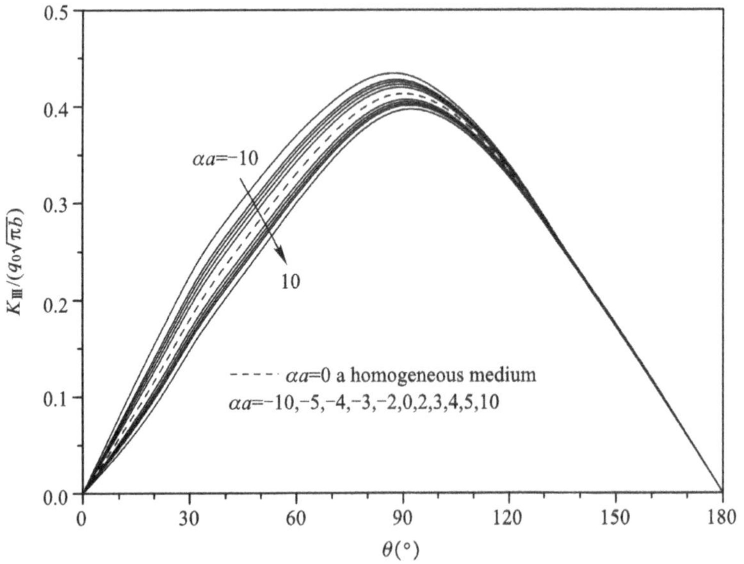

Fig. 7.39 Variations of $K_{III}/(q_0\sqrt{\pi b})$ with θ for different α values $(d = 1.2b, h = 0.5a)$

Table 7.14 The SIF values $(K_{\mathrm{III}}/(q_0\sqrt{\pi b}))$ for an elliptical crack under q_0 with different α values $(h = 0.5a, d = 1.2b)$

$\theta(°)$	$\alpha a = -10$	–5	–4	–3	–2	0	2	3	4	5	10
0.00	0.00000	0.0000	0.00000	0.00000	0.00000	0.00000	0.00000	0.00000	0.00000	0.00000	0.00000
11.25	0.09064	0.08077	0.07796	0.07484	0.07146	0.06443	0.05816	0.05560	0.05346	0.05169	0.04616
22.50	0.17189	0.15633	0.15181	0.14675	0.14124	0.12968	0.11935	0.11516	0.11167	0.10879	0.10007
33.75	0.25272	0.23626	0.23127	0.22561	0.21934	0.20600	0.19395	0.18908	0.18505	0.18177	0.17226
45.00	0.30476	0.28978	0.28500	0.27947	0.27325	0.25979	0.24747	0.24250	0.23841	0.23513	0.22605
56.25	0.35686	0.34369	0.33926	0.33406	0.32812	0.31507	0.30301	0.29814	0.29417	0.29102	0.28270
67.50	0.39842	0.38773	0.38406	0.37971	0.37471	0.36363	0.35331	0.34914	0.34573	0.34304	0.33611
78.75	0.43016	0.42162	0.41866	0.41514	0.4111	0.40210	0.39368	0.39025	0.38746	0.38525	0.37962
90.00	0.43836	0.43215	0.43003	0.42751	0.42462	0.41819	0.41212	0.40963	0.40758	0.40594	0.40170

Table 7.14 lists the corresponding SIF values. The mode Ⅲ SIF shows a similar variation trend to that associated with the mode Ⅱ SIF. The largest variations in the mode Ⅲ SIF are, however, observed within the crack front $30^\circ < \theta < 90^\circ$. The parameter α exerts a weak influence on the mode Ⅲ SIF values at other locations of the crack tip; due to symmetry, the mode Ⅲ SIF values at the crack tip of $\theta = 0^\circ$ and 180° are equal to zero.

(2) Variations of the SIF values with the crack distance d to the FGM interlayer

In Fig. 7.40 the mode Ⅲ SIF values for different d values are presented; here $h = 0.5a$, $\alpha a = -5$ or 5 and Table 7.15 lists the corresponding SIF values. In this case, variations of the mode Ⅲ SIF values are similar to those for the mode Ⅱ SIF values given above; the largest variations of the mode Ⅲ SIF values, however, occur at the front of the elliptical crack near the FGM interlayer.

(3) Variations of the SIF values with the thickness h of the FGM interlayer

Figure 7.41 shows the mode Ⅲ SIF values for different h values where $d = 1.2b$, $\alpha a = -5$ or 5 and Table 7.16 lists the corresponding SIF values. In the figures, we present the trend of the mode Ⅲ SIF values at the crack-tip points for symmetric pairs with respect to the y'-axis. For a pair of crack-tip points, the mode Ⅲ SIF values near the FGM interlayer are more significantly affected by h. Similar to the mode Ⅱ SIF values, the mode Ⅲ SIF values increase for $\alpha a = -5$ and decrease for $\alpha a = 5$ as h increases. It is found that there are no evident changes when $h/a > 2.5$.

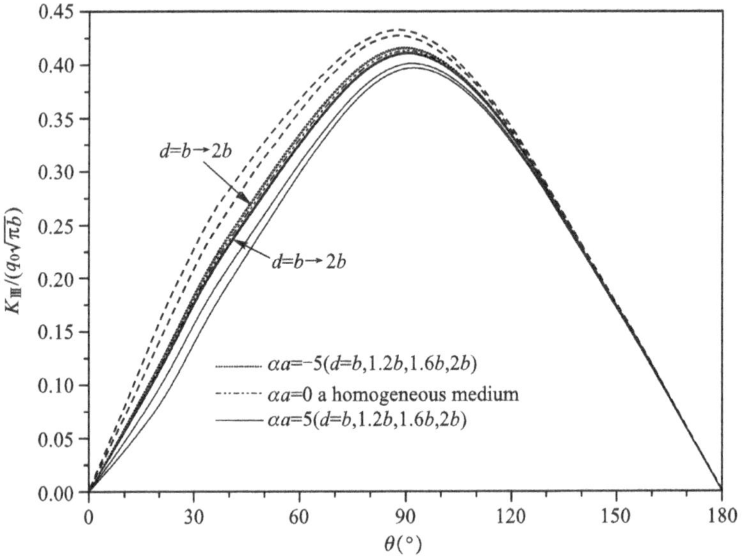

Fig. 7.40 Variations of $K_{\mathrm{III}}/(q_0\sqrt{\pi b})$ with d for different α and θ values $(h = 0.5a)$

Table 7.15 The SIF values $(K_{\mathrm{III}}/(q_0\sqrt{\pi b}))$ for an elliptical crack under q_0 with different d values $(h = 0.5a)$

$\theta(°)$	$\alpha a = -5$				$\alpha a = 5$			
	d=2.0b	1.6b	1.2b	1.0b	d=1.0b	1.2b	1.6b	2.0b
0.00	0.00000	0.00000	0.00000	0.00000	0.00000	0.00000	0.00000	0.00000
11.25	0.06567	0.06702	0.08077	0.08980	0.04261	0.05169	0.06207	0.06327
22.50	0.13191	0.13410	0.15633	0.17903	0.09016	0.10879	0.12564	0.12761
33.75	0.20895	0.21142	0.23626	0.25693	0.16413	0.18177	0.20099	0.20323
45.00	0.26320	0.26560	0.28978	0.30556	0.22159	0.23513	0.25437	0.25657
56.25	0.31877	0.32097	0.34369	0.35534	0.28092	0.29102	0.30952	0.31156
67.50	0.36695	0.36876	0.38773	0.39643	0.33563	0.34304	0.35879	0.36048
78.75	0.40487	0.40632	0.42162	0.42826	0.37965	0.38525	0.39811	0.39947
90.00	0.42015	0.42122	0.43215	0.43733	0.40171	0.40594	0.41534	0.41635

Table 7.16 The SIF values $(K_{\mathrm{III}}/(q_0\sqrt{\pi b}))$ for an elliptical crack under q_0 with different h values $(d = 1.2b)$

$h/2b$	$\alpha a = -5$				$\alpha a = 5$			
	$\theta = 11.25°$	22.50°	33.75°	45.00°	11.25°	22.50°	33.75°	45.00°
0.00	0.06443	0.12968	0.20600	0.25979	0.06443	0.12968	0.2060	0.25979
0.10	0.07196	0.14161	0.21884	0.27179	0.05756	0.11881	0.19421	0.24868
0.25	0.07775	0.15111	0.22978	0.28276	0.05332	0.11177	0.18585	0.23996
0.50	0.08077	0.15633	0.23626	0.28978	0.05169	0.10879	0.18177	0.23513
0.75	0.08144	0.15754	0.23785	0.29160	0.05138	0.10820	0.18087	0.23395
1.00	0.08160	0.15782	0.23823	0.29204	0.05131	0.10806	0.18064	0.23365
1.25	0.08177	0.15809	0.23852	0.29232	0.05118	0.10786	0.18042	0.23344
1.50	0.08183	0.15818	0.23862	0.29241	0.05114	0.10779	0.18035	0.23337
2.00	0.08198	0.15842	0.23886	0.29261	0.05102	0.10760	0.18017	0.23321
2.50	0.08222	0.15878	0.23921	0.29290	0.05084	0.10734	0.17990	0.23298
3.00	0.08227	0.15889	0.23954	0.29302	0.05063	0.10723	0.17978	0.23268

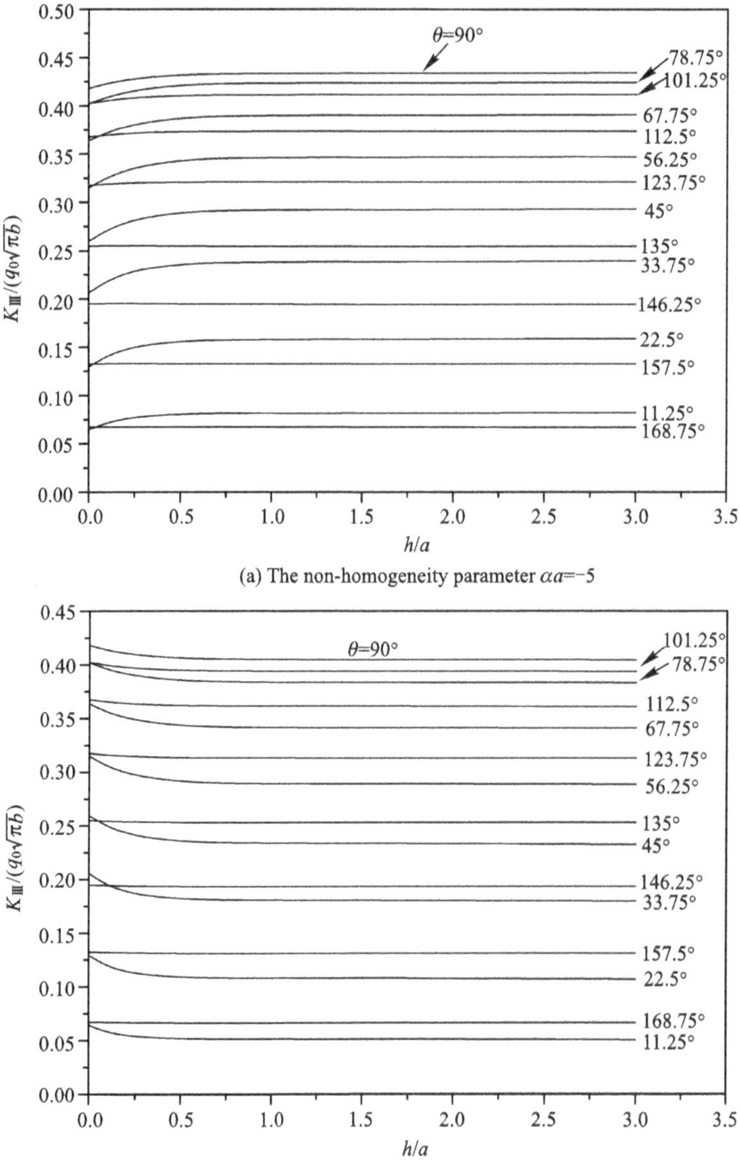

(a) The non-homogeneity parameter $\alpha a=-5$

(b) The non-homogeneity parameter $\alpha a=5$

Fig. 7.41 Variations of $K_{\mathrm{III}}/(q_0\sqrt{\pi b})$ with h for different α and θ values ($\alpha a = 5$, $d = 1.2b$)

7.5 The growth of an elliptical crack perpendicular to the FGM interlayer

7.5.1 Crack growth under a remotely inclined tensile loading

In the following, we will discuss the growth of an elliptical crack shown in Figs. 7.30 and 7.42. In Fig. 7.42, the B point, where the minor axis of the elliptical crack intersects the crack border in Figs. 7.30 and 7.42, is the closest location to the FGM interlayer. The direction of the applied tension σ is always assumed to be parallel to the $O'x'z'$- plane. In Section 7.4, we obtained the SIF values for a crack under uniform compressive and shear loadings. Therefore, we can calculate the SIF values of the crack under a remotely inclined tensile loading σ by using Eqs. (6.10), obtaining the strain energy density factor at the crack tip by using Eq. (6.8) and, finally, determining the directions and critical loadings of an elliptical crack in the FGM.

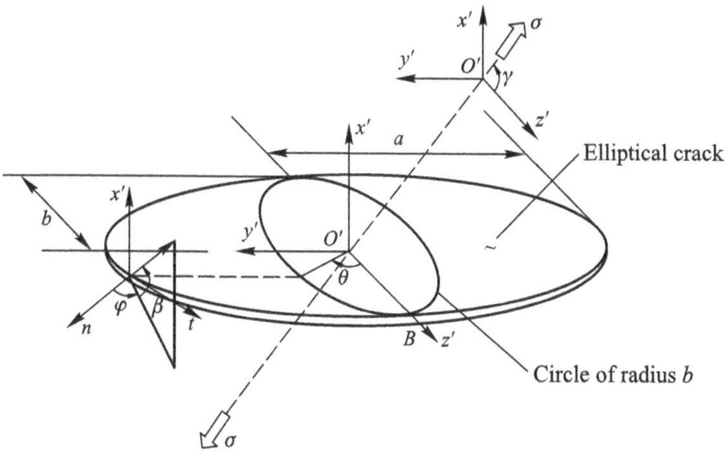

Fig. 7.42 An elliptical crack subjected to a remotely inclined loading

(1) The minimum strain energy density factor at the crack tip

Table 7.17 presents the variations of the spherical angles β_0 and φ_0, corresponding to the minimum strain energy factors $S(\beta_0, \varphi_0)$ (or $S_{\min}$) for various values of θ around the crack periphery, with the material gradient parameter α, where $h = 1.2b$, $d = 0.5a$ and $\gamma = 60°$. As pointed out by Sih and Cha (1974), the values of φ_0 at any crack-tip points are related to the crack geometry whilst the corresponding values of β_0 in the φ_0-plane depend on the crack geometry, the load position and the Poisson's ratio. In Table 7.17, it can be found that the values of β_0 also vary with the parameter α. For the crack tip near the FGM interlayer, the absolute value of β_0 increases as the α value decreases; this is because the SIFs evidently increase as α decreases. However, for the crack tip at a distance away from the FGM interlayer, the β_0 values only change slightly.

Figure 7.43 illustrates the variations of the $S_{\min}$ values with θ around the crack periphery and the parameter α where $h = 1.2b$, $d = 0.5a$ and $\gamma = 60^\circ$. For a homogeneous solid, the maximum $S_{\min}$ value around the whole crack tip appears at the intersection of the minor axis with the crack border. It should be noted that the exact solution of the $S_{\min}$ value, shown in Fig. 7.43, corresponds to those of the SIF values obtained from Eqs. (7.2) and (7.3). The parameter α evidently influences the $S_{\min}$ value at the crack tip near the FGM interlayer. For any given θ value, the $S_{\min}$ value increases as the α value decreases. Figure 7.44 shows the variation of the $S_{\min}$ value with the loading angle γ at the crack tip $\theta = 0^\circ$, which is the closest to the FGM interlayer. It can be observed that for all the loading angles γ (except $\gamma = 0^\circ$) the $S_{\min}$ value at $\theta = 0^\circ$ increases as the α value decreases.

Table 7.17 Values of φ_0 and β_0 for different α values under remote tension loading ($\gamma = 60^\circ$, $d = 1.2b$, $h = 0.5a$)

$\theta(^\circ)$	$\varphi_0(^\circ)$	$\beta_0(^\circ)$				
		$\alpha a = -10$	–5	0	5	10
0.00	0.0000	–45.6277	–44.2690	–42.3123	–41.2497	–41.1606
11.25	–16.0141	–45.0391	–43.7324	–41.8677	–40.8978	–40.8517
22.50	–27.9382	–43.5390	–42.4019	–40.7460	–39.9519	–39.9911
45.00	–36.8699	–37.7375	–37.0193	–35.8388	–35.3243	–35.4144
67.50	–27.9383	–30.3647	–29.8801	–28.9403	–28.4864	–28.5171
90.00	0.0000	–1.6049	–1.1366	–0.07954	–1.0680	–1.35521
135.00	36.8699	35.9082	35.8162	35.9159	36.47158	36.8606
180.00	0.0000	42.3773	42.2875	42.3557	42.87941	43.2730

(2) The critical loadings to initiate crack growth

Figure 7.45 presents the variations of the critical tensile stresses σ_{cr} with the loading angle γ and the parameters α where $h = 1.2b$, $d = 0.5a$. In Fig. 7.43, it can be observed that the maximum $S_{\min}$ value around the whole crack tip is located at the crack front of the intersection of the minor axes with the crack border, this is, $\theta = 0^\circ$ or 180°. For $\alpha > 0$, the position of the maximum $S_{\min}$ value is at $\theta = 180^\circ$ and for $\alpha < 0$ the position is at $\theta = 0^\circ$. So, the σ_{cr} values of the elliptical crack in the FGM can be calculated by setting $S_{\min} = S_{cr}$ at $\theta = 0^\circ$ for $\alpha < 0$ and at $\theta = 180^\circ$ for $\alpha > 0$, respectively. It can be observed that the σ_{cr} value decreases monotonically as the loading angle γ increases. It also decreases as the FGM parameter α decreases under a given γ value.

Figure 7.46 shows the variations of the σ_{cr} value with the crack distance d to the FGM interlayer. Evidently, for a given loading angle γ, the σ_{cr} value decreases for $\alpha a = 5$ and increases for $\alpha a = -5$ as d increases. In the general case, as d increases, the σ_{cr} value decreases for $\alpha > 0$ and increases for $\alpha < 0$.

The variations of the σ_{cr} value with the thickness h of the FGM interlayer are given in Fig. 7.47 for different loading angles γ, when $\alpha a = 5$ and –5. The σ_{cr} value for $h/a = 0$ represents the result in a homogeneous solid. It can be observed that the σ_{cr} value

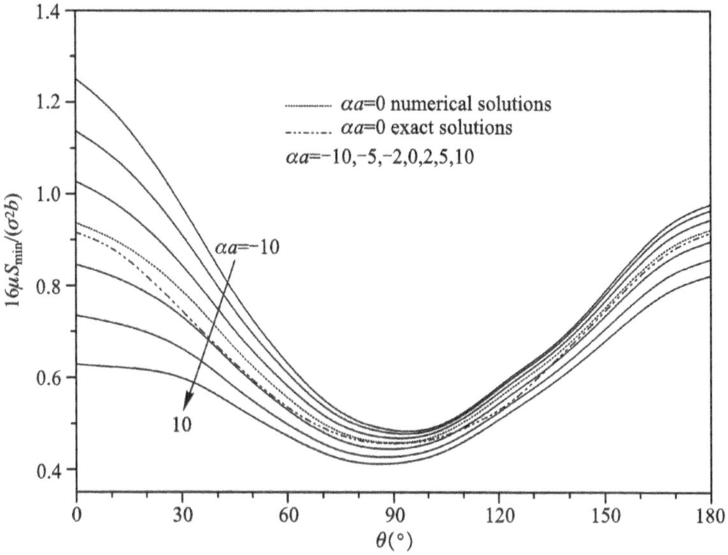

Fig. 7.43 Variations of $S_{\min}$ with θ for different α values ($\gamma = 60°$, $d = 1.2b$, $h = 0.5a$)

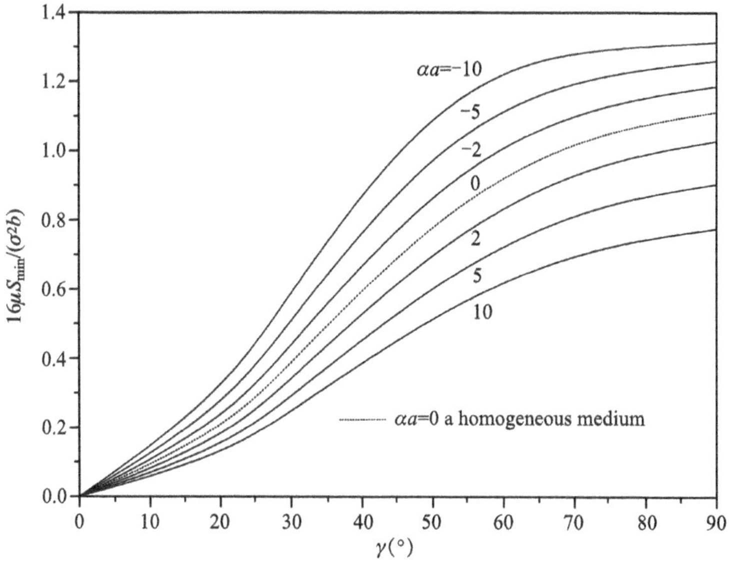

Fig. 7.44 Variations of $S_{\min}$ with γ for different α values ($\theta = 0°$, $d = 1.2b$, $h = 0.5a$)

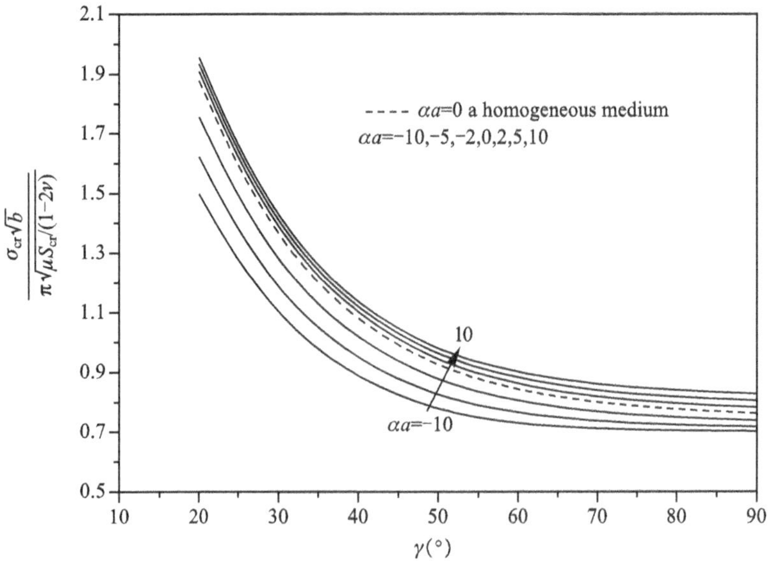

Fig. 7.45 Variations of σ_{cr} with γ for different α angles ($d = 1.2b$, $h = 0.5a$)

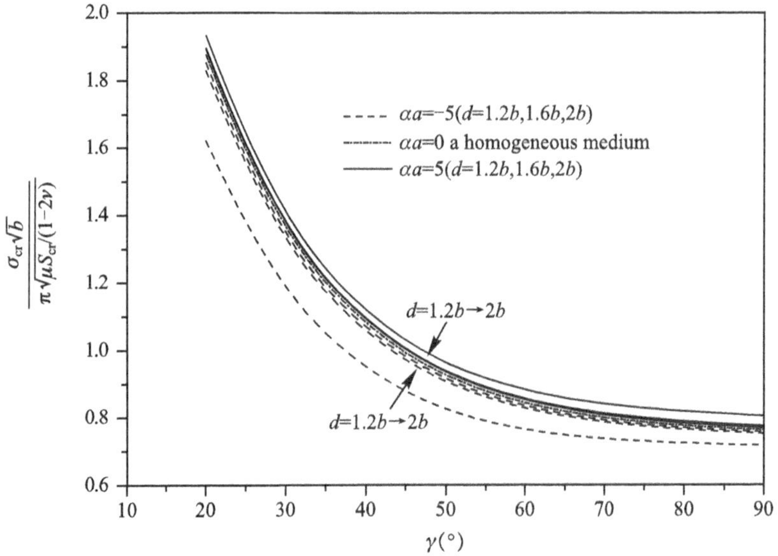

Fig. 7.46 Variations of σ_{cr} with γ for different α and d values ($h = 0.5a$)

increases as h increases for $\alpha a = 5$ whilst the σ_{cr} value decreases for $\alpha a = -5$. Such a variation phenomenon of the σ_{cr} value is evident for $h \leqslant a$ and relatively weak for $h > a$. In the general case, the σ_{cr} value decreases as h increases for $\alpha < 0$ whilst it increases for $\alpha > 0$. The effect of the FGM thickness on the σ_{cr} values is evident for a small loading angle.

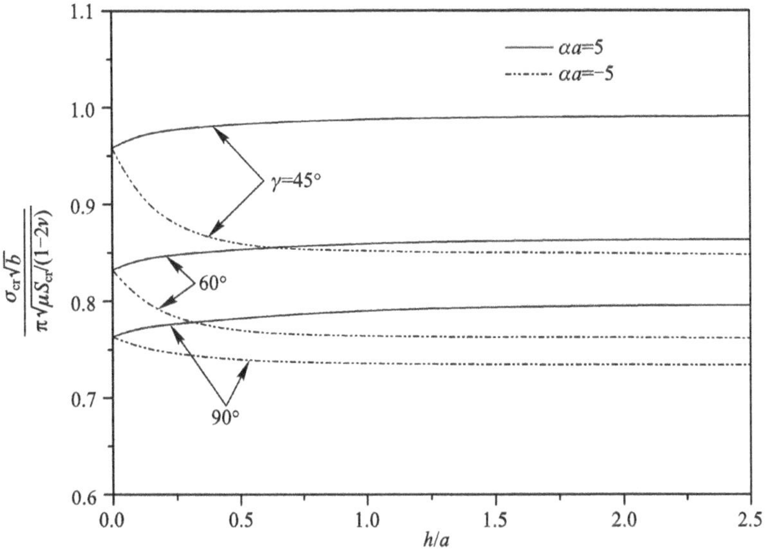

Fig. 7.47 Variations of σ_{cr} with h for different α and γ values ($d = 1.2b$)

7.5.2 Crack growth under a remotely inclined compressive loading

When the direction of the applied loading σ shown in Fig. 7.42 is reversed, the cracked solid with a FGM interlayer is in a state of compression. The surface of the crack initiation is different from the one associated with the tensile loading. For such a compressive loading, we need to replace σ in Eqs. (6.11) with $-\sigma$. In the ensuing, crack growth under compression will be discussed.

(1) The minimum strain energy density factor at the crack tip

The φ_0 value in compression is the same as the one in tension shown in Table 7.18. The angle β_0 in the φ_0-plane around the crack border is tabulated in Table 7.18; it can be observed that the β_0 value varies with the parameter α. Such a variation is evident for the crack tip near the FGM interlayer. The β_0 value decreases as the α value decreases. This can be related to the increase of the absolute K_{I} value. However, for the crack tip far away from the FGM interlayer, the β_0 value changes only slightly.

Figure 7.48 illustrates the variations of the S_{min} value along the crack tip with the parameter α where $\gamma = 60°$, $d = 1.2b$ and $h = 0.5a$. For a homogeneous solid, the maximum

S_{min} value along the crack border is located at the crack front $\theta = 90°$, that is, the intersection of the major axis with the crack border. In Fig. 7.49, the S_{min} value decreases as the parameter α increases for any given crack tip point. The variations are most evident at the crack tip near the FGM interlayer. Figure 7.49 also illustrates the variation of the S_{min} value with the loading angle γ at the crack front $\theta = 0°$, that is, the intersection of the minor axis with the crack border. The curve of the S_{min} value versus the loading angle γ is bell-shaped. The maximum S_{min} value is located at $\gamma \approx 40°$.

Table 7.18 Values of φ_0 and β_0 for different α values under remote compressive loading ($\gamma = 60°$, $d = 1.2b$, $h = 0.5a$)

$\theta(°)$	$\varphi_0(°)$	$\beta_0(°)$				
		$\alpha a = -10$	–5	0	5	10
0.00	0.0000	125.3404	127.1767	129.8379	131.2889	131.4108
11.25	–16.0141	124.9632	126.6473	129.0662	130.3303	130.3906
22.50	–27.9382	124.1535	125.4746	127.4140	128.3514	128.3067
45.00	–36.8699	126.4262	127.2004	128.4899	129.0622	128.9683
67.50	–27.9383	139.2565	139.8790	141.0874	141.6696	141.6275
90.00	0.0000	181.6056	181.1368	180.0000	181.0682	181.3556
135.00	36.8699	–128.4299	–128.5254	–128.4094	–127.8042	–127.3849
180.00	0.00000	–129.7491	–129.8717	–129.7786	–129.0649	–128.5292

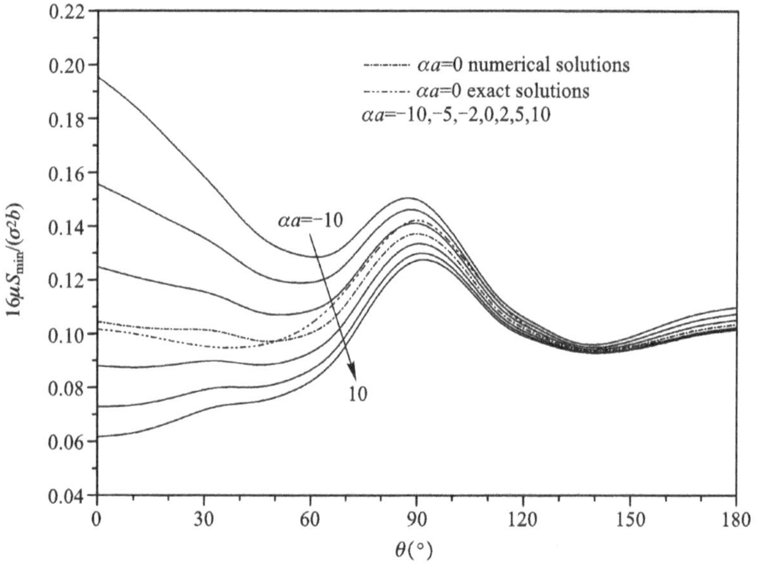

Fig. 7.48 Variations of S_{min} with θ for different α values ($d = 1.2b$, $h = 0.5a$, $\gamma = 60°$)

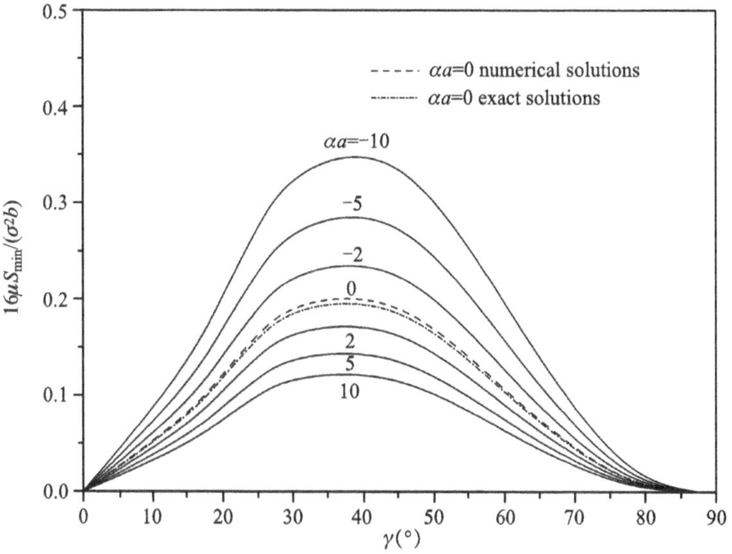

Fig. 7.49 Variations of S_{min} with γ for different α values ($d = 1.2b$, $h = 0.5a$, $\theta = 0°$)

(2) The critical loadings for crack growth

In order to calculate the critical compressive loads σ_{cr} in different cases, we need to find out the maximum S_{min} in the range of $0° \leqslant \theta \leqslant 180°$ and then set $S_{min} = S_{cr}$. Figure 7.50 shows the variations of σ_{cr} with the loading angle γ for different gradient parameters α, where $h = 0.5a$ and $d = 1.2b$. For a homogeneous solid of infinite extent, the minimum σ_{cr} value appears at the loading angle $\gamma \approx 40°$. The influence of α on the σ_{cr} is not evident for $\alpha > 0$. This is because the maximum S_{min} value for $\alpha > 0$ appears at the crack tip, when $\theta = 90°$ or $180°$. The influence of the FGM on the S_{min} value at the crack tip at a distance from the FGM interlayer is relatively weak. Because the maximum S_{min} value for $\alpha < 0$ appears at the crack tip $\theta = 0°$ or $90°$, the influence of α is evident for $\alpha < 0$.

When the loading angle γ is close to either $0°$ or $90°$, the σ_{cr} value approaches a large value. For a given γ value, the σ_{cr} value decreases as the parameter α decreases. This phenomenon is evident for the smaller loading angles in the range of $0° \leqslant \gamma \leqslant 90°$. When the θ value increases for $\gamma \geqslant 45°$, the compressive stress p_0 increases and the shear stress q_0 decreases. When $\gamma = 90°$, the elliptical crack is in compression and this compression keeps the crack stable.

Figure 7.51 illustrates the variations of σ_{cr} with γ for different crack distances to the FGM interlayer. In this figure, it can be observed that σ_{cr} decreases for $\alpha a = 5$ and increases for $\alpha a = -5$ as the crack distance to the interlayer increases. In the general case similar phenomena can be found for $\alpha < 0$ and $\alpha > 0$.

Figure 7.52 presents the variations of the σ_{cr} value with the FGM thickness h where $\gamma = 30°$, $45°$, $60°$ for $\alpha a = -5$ and 5. The case $h/a = 0$ corresponds to an elliptical crack in a homogeneous solid of infinite extent. It can be observed that σ_{cr} decreases for

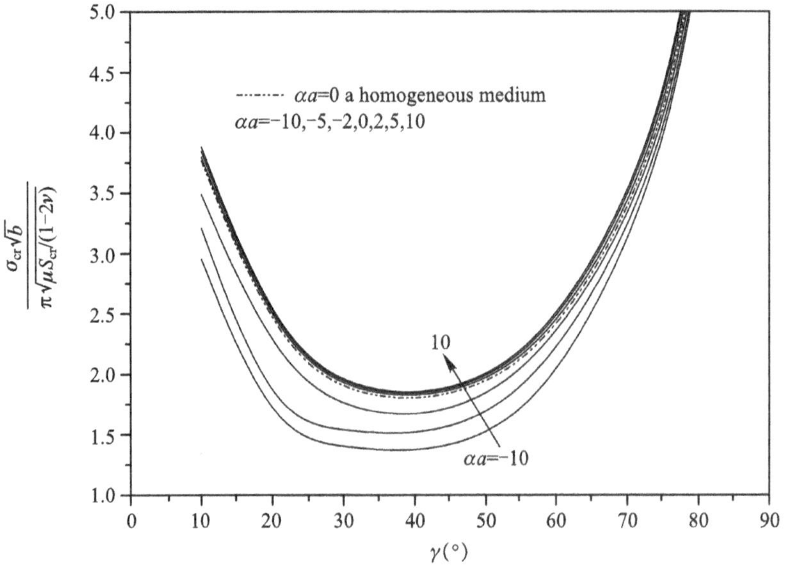

Fig. 7.50 Variations of σ_{cr} with γ for different α values ($h = 0.5a$, $d = 1.2b$)

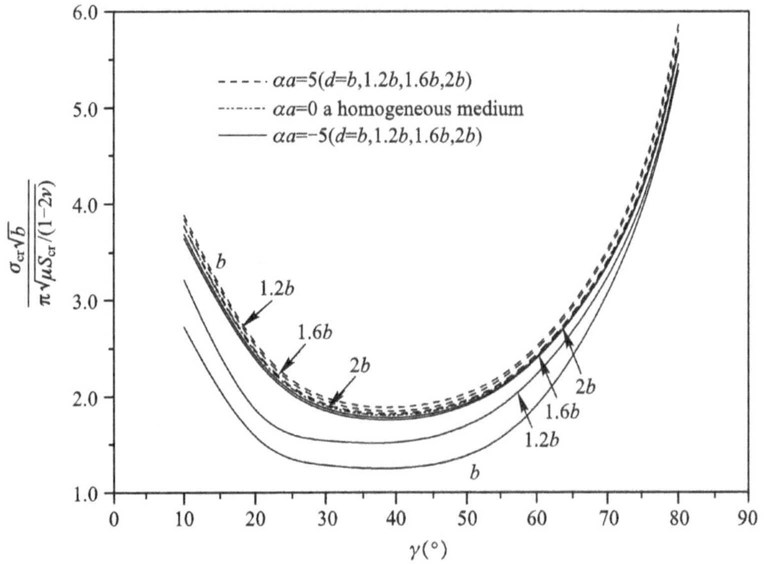

Fig. 7.51 Variations of σ_{cr} with γ for different α and d values ($h = 0.5a$)

$\alpha a = -5$ and increases for $\alpha a = 5$ as h/a increases. Such a variation in σ_{cr} is evident for $h \leqslant 0.5a$ and relatively weak for $h > 0.5a$. The influence of h/a is more evident for a smaller loading angle γ.

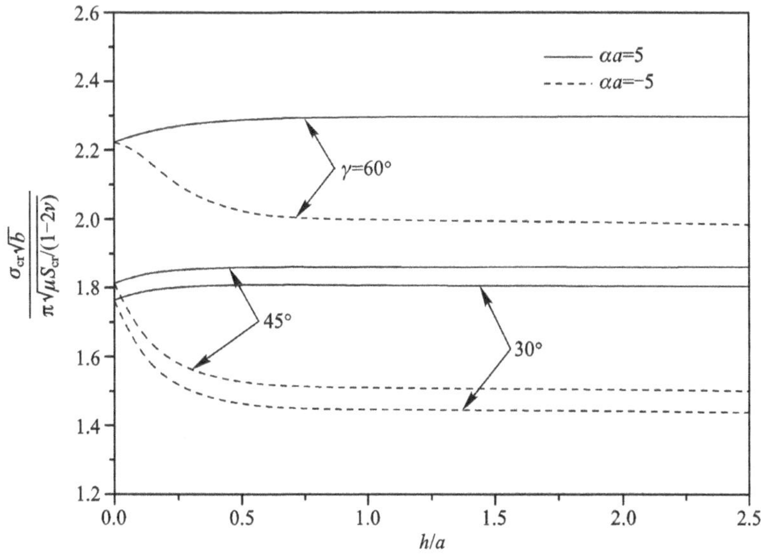

Fig. 7.52 Variations of σ_{cr} with h for different α and γ values $(d = 1.2b)$

7.6 Summary

In this chapter, an elliptical crack in a FGM system has been analyzed. The crack surfaces are parallel to or perpendicular to a graded interlayer bonded to two homogeneous media. The proposed numerical method is used to obtain the SIFs of the crack subjected to uniform compressive and shear tractions. The growth of the elliptical crack under remotely tensile or compressive loadings was analyzed using the SIFs and the principles of fracture mechanics.

In the FGM system, an elliptical crack subjected to a uniform compressive loading has mixed mode I and II deformations and the elliptical crack subject to a uniform shear loading has mixed mode II and III deformations. The SIF values are highly affected by the thickness of the FGM interlayer, the non-homogeneity parameter and the crack distance to the FGM interlayer. It was found that the directions and critical loadings of the crack growth in the FGM system show obvious variations in comparison with the ones of the crack in a homogeneous medium. As in the SIF variations of a crack in a FGM system, the critical loadings for crack growth are highly affected by the above influential factors.

By comparing the results for penny-shaped and elliptical cracks in FGMs, it can be found that both of the cracks have different distributions of the SIF values and the directions and critical loadings of the crack growth at the crack tip. This is because the two cracks have different crack tip geometries.

The research results presented in this chapter have been published in Yue et al. (2004) and Xiao et al. (2005). Many researchers have used and cited the above results (e.g., Baudendistel, 2005; Feng and Su, 2007; Matbuly, 2008; Chang and Wang, 2011; Shanmugavel et al., 2012).

References

Baudendistel CM. Effect of a Graded Layer on the Plastic Dissipation During Mixed-mode Fatigue Crack Growth on Plastically Mismatched Interfaces. Wright State University, 2005.

Chang DM, Wang BL. Thermal shock resistance of brittle ceramic materials with embedded elliptical crack. *Philosophical Magazine Letter*, 2011, 91(10): 648-655.

Feng WJ, Su RKL. Dynamic fracture behaviors of cracks in a functionally graded magneto-electro elastic plate. *European Journal of Mechanics A-Solids*, 2007, 26: 363-379.

Green AE, Sneddon IN. The distribution of the stress in the neighbourhood of a flat elliptical crack in an elastic solid. *Proceedings of the Cambridge Philosophical Society*, 1950, 46: 159-163.

Hachi BK, Rechak S, Belkacemi Y, Maurice G. Modelling of elliptical cracks in infinite and in a pressurized cylinder by a hybrid weight function approach. *International Journal of Pressure Vessels and Piping*, 2005, 127: 165-172.

Irwin GR. Crack-extension force for a part-through crack in a plate. *ASME Journal of Applied Mechanics*, 1962, 29: 651-654.

Kassir MK, Sih GC. Three Dimensional Crack Problems: Mechanics of Fracture. Leyden: Noorhoff International Publishing, 1975.

Lin XB, Smith RA. Fatigue growth prediction of internal surface cracks in pressure vessels. *Journal of Pressure Vessel Technology*, 1998, 120: 17-23.

Matbuly MS. Analysis of mode III crack perpendicular to the interface between two dissimilar strips. *Acta Mechanica Sinica*, 2008, 24(4): 433-438.

Mechab B, Serier B, Bouiadjra BB, Kaddouri K, Feaugas X. Linear and non-linear analyses for semi-elliptical surface cracks in pipes under bending. *International Journal of Pressure Vessels and Piping*, 2011, 88(1): 57-63.

Noda NA, Ohzono R, Chen MC. Analysis of an elliptical crack parallel to a bimaterial interface under tension. *Mechanics of Materials*, 2003, 11: 1059-1076.

Sih GC, Cha CK. A fracture criterion for three-dimensional crack problems. *Engineering Fracture Mechanics*, 1974, 6: 699–723.

Shaha RC, Kobayashia AS. Stress intensity factor for an elliptical crack under arbitrary normal loading. *Engineering Fracture Mechanics*, 1971, 3(1): 71-96.

Shanmugavel P, Bhasker GB, Chadrasekaran M, Mani PS, Srinivasan SP. An overview of fracture analysis in functionally graded in functionally graded materials. *European Journal of Scientific Research*, 2012, 68(3): 412-439.

Standford R J. Principles of Fracture Mechanics. New Jersey: Prentice Hall Inc., 2003.

Tada H, Paris OC, Irwin GR. The Stress Analysis of Cracks Handbook. 3rd ed. London and Bury St Edmunds: Professional Engineering Publishing Limited, 2000.

Xiao HT, Yue ZQ, Tham LG. Analysis of an elliptical crack parallel to graded interfacial zone of bonded bimaterials. *Mechanics of Materials*, 2005, 37: 785-799.

Yue ZQ, Xiao HT, Tham GL. Elliptical crack normal to functionally graded interface of bonded solids. *Theoretical and Applied Fracture Mechanics*, 2004, 42: 227-248.

Zhang Ch, Cui M, Wang J, Gao XW, Sladek J, Sladek V. 3D crack analysis in functionally graded materials. *Engineering Fracture Mechanics*, 2011, 78, 3: 585-604.

Zheng XJ, Glinka G, Dubey R. Calculation of stress intensity factors for semi-elliptical cracks in a thick-wall cylinder. *International Journal of Pressure Vessels and Piping*, 1995, 62: 249-258.

Chapter 8

Yue's Solution-based Dual Boundary Element Method

8.1 Introduction

In Chapters 6 and 7, the multi-region method of Yue's solution-based BEM was applied to analyze penny-shaped and elliptical cracks in FGMs. It was found that the proposed method has the following drawbacks:

- The introduction of artificial boundaries is not unique and thus cannot be implemented into an automatic procedure.
- The method generates a larger system of algebraic equations than strictly required.
- If the artificial boundaries are located in the FGMs, the unknown quantities of linear equations increase greatly because of the variations in the material properties.

In the context of the direct BEM, Watson (1986) first presented the normal derivative of displacement boundary integral equation (BIE) for the development of Hermite cubic elements where the number of unknowns is larger than the one of equations. A literature review shows that the dual integral representation has been proposed for the analysis of degenerate boundaries such as cracks, cusps, corners, and holes of any irregular shape (e. g., Gray et al., 1990).

Dual boundary element formulation is based on a pair of BIEs, namely, the displacement and traction BIEs. The method is single-region based, thus it can model solids with multiple interacting cracks or damage. Two single-region BEMs have been proposed for the study of cracked media and, in the following section, we will discuss some of these developments in analyzing crack problems.

One form of a single-region BEM is that the displacement BIE is collocated on one side of the crack surface and on the uncracked boundary while the traction BIE is collocated on another side of the crack surface. Hong and Chen (1988) presented dual integral equations for solving general linear elasticity problems of finite or infinite domains with degenerate boundaries. Portela et al. (1992) developed an effective numerical implementation of two-dimensional dual integral equations and concluded that the accuracy and efficiency of the implementation make this formulation ideal for the study of crack growth problems under mixed-mode conditions. Mi and Aliabadi (1992) described an effective numerical implementation of the three-dimensional dual BIEs for linear elastic crack problems and demonstrated the accuracy of the proposed method by solving a number of problems, including edge and embedded cracks. Chen and Hong (1999) reviewed the current status of the formulation of dual BEMs with an emphasis on hypersingular integrals and divergent series and Wilde (2000) published the book "*A dual boundary element formulation for three-dimensional fracture analysis*". In these dual boundary element analyses, the displacements on either side of the crack surface are collocated as unknown variables, which may be unnecessary for the calculation of SIFs. Therefore, for an ideal single-region BEM, only one side of the crack surface needs to be discretized.

Another single-region BEM formulation can be achieved by applying a displacement BIE to only the uncracked boundary while the traction BIE is applied to one side of the crack surface. Pan and Amadei (1996) and Pan (1997) proposed such a single-region BEM formulation for two-dimensional anisotropic cracked media. Qin et al. (1997) proposed a similar formulation for three-dimensional isotropic cracked media and this was further developed by Pan and Yuan (2000) for three-dimensional cracks in anisotropic solids. Cisilino (1999) and dell'Erba (2000) analyzed, respectively, linear and nonlinear crack growth and thermoelastic fracture mechanics by using the single-region BEM. The dual boundary element formulation for a transversely isotropic bi-material developed by Yue et al. (2007) belongs to this single-region BEM category. In these dual boundary element analyses, the relative crack opening displacement is treated as an unknown quantity on the crack surface; the number of unknown quantities to be obtained obviously decreases, which simplifies the analysis of crack problems. The traction integral equation for a crack in an infinite space resembles the displacement discontinuity or dislocation method. This approach was used for investigating some crack problems (see, e.g., Gray et al., 1990; Xu et al., 1997).

Although dual BEMs have been widely applied, only a few of the crack problems in nonhomogeneous media have been investigated. In this chapter, we will present a numerical implementation of the Yue's solution based dual BIEs and analyze several numerical examples to validate the solutions of the square-shaped crack in an infinite space.

8.2 Yue's solution-based dual boundary integral equations

8.2.1 The displacement boundary integral equation

In Chapter 4, the displacement BIE based on Yue' solution was reported. By collocating the source point on the uncracked boundary shown in Fig. 8.1, this integral equation (4.13) can be written as

$$\begin{aligned} c_{ij}(P_S)u_j(P_S) &+ \int_{S+\Gamma^++\Gamma^-} t^Y_{ij}(P_S,Q)u_j(Q)\mathrm{d}S(Q) \\ &= \int_{S+\Gamma^++\Gamma^-} u^Y_{ij}(P_S,Q)t_j(Q)\mathrm{d}S(Q), \quad (i,j=x,y,z) \end{aligned} \tag{8.1}$$

where P_S and Q are the source and field points, respectively, $t^Y_{ij}(P_S,Q)$ and $u^Y_{ij}(P_S,Q)$ are the tractions and displacements of Yue's solution, respectively, $t_j(Q)$ and $u_j(Q)$ are the tractions and displacements of the field point Q on the boundaries, S is the uncracked boundary of the cracked body, Γ^+ and Γ^- are two crack surfaces, $c_{ij}(P_S)$ is a coefficient that is dependent on the local boundary geometry at the source point P_S.

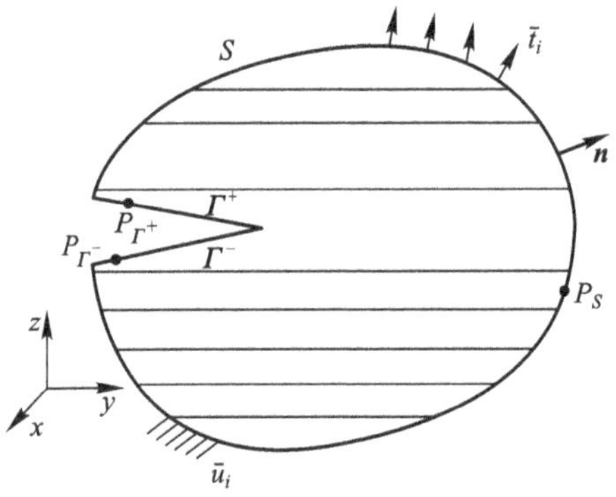

Fig. 8.1 Geometry of the three-dimensional crack in a multilayered solid. S is the uncracked boundary, Γ^+ and Γ^- are the upper and lower crack surfaces, respectively

Assume that the points Q_{Γ^-} and Q_{Γ^+} on the two crack surfaces are completely coincident before loading so that there are opposite outward normal directions on the two points. Thus, the following relationships of kernel functions exist for the two points Q_{Γ^-} and Q_{Γ^+}

$$t_{ij}^Y(P_S, Q_{\Gamma^+}) = -t_{ij}^Y(P_S, Q_{\Gamma^-}) \tag{8.2a}$$

$$u_{ij}^Y(P_S, Q_{\Gamma^+}) = u_{ij}^Y(P_S, Q_{\Gamma^-}) \tag{8.2b}$$

Assume that there is a balanced relationship of tractions as follows

$$t_j(Q_{\Gamma^+}) = -t_j(Q_{\Gamma^-}) \tag{8.3}$$

The relative crack opening displacement can be described as

$$\Delta u_j(Q_{\Gamma^+}) = u_j(Q_{\Gamma^+}) - u_j(Q_{\Gamma^-}) \tag{8.4}$$

By using the above relationships Eqs. (8.2) ~ (8.4), the two integrals in Eq. (8.1) can be written as

$$\int_{S+\Gamma^+ + \Gamma^-} t_{ij}^Y(P_S, Q)u_j(Q)\mathrm{d}S(Q) = \int_S t_{ij}^Y(P_S, Q)u_j(Q)\mathrm{d}S(Q) + \int_{\Gamma^+} t_{ij}^Y(P_S, Q)\Delta u_j(Q)\mathrm{d}S(Q) \tag{8.5a}$$

$$\int_{S+\Gamma^+ + \Gamma^-} u_{ij}^Y(P_S, Q)t_j(Q)\mathrm{d}S(Q) = \int_S u_{ij}^Y(P_S, Q)t_j(Q)\mathrm{d}S(Q) \tag{8.5b}$$

Thus, Eq. (8.1) can be rewritten as

$$c_{ij}(P_S)u_j(P_S) + \int_S t_{ij}^Y(P_S,Q)u_j(Q)\mathrm{d}S(Q) + \int_{\Gamma^+} t_{ij}^Y(P_S,Q)\Delta u_j(Q)\mathrm{d}S(Q)$$
$$= \int_S u_{ij}^Y(P_S,Q)t_j(Q)\mathrm{d}S(Q), \quad (i,j=x,y,z) \tag{8.6}$$

where the integral domain contains the uncracked boundary S and the crack surface Γ^+.

The integral equation (8.6) is a general form of the displacement BIE based on Yue's solution. This formulation has been extensively used in solving crack problems for different fundamental solutions (see, e.g., Cisilino, 1999; Pan and Yuan, 2000).

8.2.2 The traction boundary integral equation

The stress BIE can be obtained by taking the limiting form of the internal stress equation (4.62). In this analysis, let an internal point p go to the boundary point P in the same way as was done before with the displacement BIE. If P is assumed to be on a smooth boundary, u_i is differentiable and its derivatives are Hölder continuous, the limiting process gives the stress BIE as

$$\frac{1}{2}\sigma_{ij}(P) + \int_S T_{ijk}^Y(P,Q)u_k(Q_S)\mathrm{d}S(Q) = \int_S U_{ijk}^Y(P,Q)t_k(Q)\mathrm{d}S(Q), \quad (i,j,k=x,y,z) \tag{8.7}$$

where T_{ijk}^Y and U_{ijk}^Y are the new kernel functions obtained by using the numerical difference of the derivatives of Yue's tractions and displacements as described in Eqs. (4.63a) and (4.63b).

From Eq. (8.7), the traction BIE can be obtained using Eq. (2.4b) as follows

$$\frac{1}{2}t_j(P) + n_i(P)\int_S T_{ijk}^Y(P,Q)u_k(Q)\mathrm{d}S(Q) = n_i(P)\int_S U_{ijk}^Y(P,Q)t_k(Q)\mathrm{d}S(Q) \tag{8.8}$$

where $n_i(P)$ denotes the component of the outward unit normal to the boundary at the source point P.

It is worth noting that the integral equation (8.8) is valid if the following conditions hold (Widle, 2000):

- the local geometry at the source point is smooth,
- the displacement derivative field is Hölder continuous, and
- the traction field is Hölder continuous.

If the crack surface possesses a discontinuous tangential plane either at a certain point or along certain lines, the numerical execution of the traction BIE needs to be carried out carefully. The most general strategy to satisfy the three conditions on the crack surface involves the use of discontinuous elements where the source points are moved to the interior of the discontinuous elements.

By collocating the stress BIE on the source point P_{Γ^+} lying on the crack surface Γ^+ shown in Fig. 8.1, Eq. (8.7) can be written as

$$\frac{1}{2}\sigma_{ij}(P_{\Gamma^+})+\frac{1}{2}\sigma_{ij}(P_{\Gamma^-})+\int_{S+\Gamma^++\Gamma^-} T^Y_{ijk}(P_{\Gamma^+},Q)u_k(Q)\mathrm{d}S(Q)$$
$$=\int_{S+\Gamma^++\Gamma^-} U^Y_{ijk}(P_{\Gamma^+},Q)t_k(Q)\mathrm{d}S(Q) \tag{8.9}$$

Multiplying Eq. (8.9) by the outward unit normal $n_i(P_{\Gamma^+})$ and noting that $n_i(P_{\Gamma^+}) = -n_i(P_{\Gamma^-})$, the traction BIE on the crack surface results in

$$\frac{1}{2}t_j(P_{\Gamma^+})-\frac{1}{2}t_j(P_{\Gamma^-})+n_i(P_{\Gamma^+})\int_{S+\Gamma^++\Gamma^-} T^Y_{ijk}(P_{\Gamma^+},Q)u_k(Q)\mathrm{d}S(Q)$$
$$=n_i(P_{\Gamma^+})\int_{S+\Gamma^++\Gamma^-} U^Y_{ijk}(P_{\Gamma^+},Q)t_k(Q)\mathrm{d}S(Q) \tag{8.10}$$

As described in Section 8.2.1, the points Q_{Γ^-} and Q_{Γ^+} on the two crack surfaces are completely coincident and there are opposite outward normal directions on these two points. Thus, the following relationships of the kernel functions of the points exist on two crack surfaces

$$T^Y_{ijk}(P_S,Q_{\Gamma^+})=-T^Y_{ijk}(P_S,Q_{\Gamma^-}) \tag{8.11a}$$

$$U^Y_{ijk}(P_S,Q_{\Gamma^+}=U^Y_{ijk}(P_S,Q_{\Gamma^-}) \tag{8.11b}$$

By applying Eqs. (8.3), (8.4) and (8.11), the two integrals in Eq. (8.10) can be written as

$$\int_{S+\Gamma^++\Gamma^-} T^Y_{ijk}(P_{\Gamma^+},Q)u_k(Q)\mathrm{d}S(Q)=\int_S T^Y_{ijk}(P_{\Gamma^+},Q)u_k(Q)\mathrm{d}S(Q)+$$
$$\int_{\Gamma^+} T^Y_{ijk}(P_{\Gamma^+},Q)\Delta u_k(Q)\mathrm{d}S(Q) \tag{8.12a}$$

$$\int_{S+\Gamma^++\Gamma^-} U^Y_{ijk}(P_{\Gamma^+},Q)t_k(Q)\mathrm{d}S(Q)=\int_S U^Y_{ijk}(P_{\Gamma^+},Q)t_k(Q)\mathrm{d}S(Q) \tag{8.12b}$$

Using Eqs. (8.3) and (8.12), Eq. (8.10) can be further rewritten as

$$t_j(P_{\Gamma^+})+n_i(P_{\Gamma^+})\int_S T^Y_{ijk}(P_{\Gamma^+},Q)u_k(Q)\mathrm{d}S(Q)+n_i(P_{\Gamma^+})\int_{\Gamma^+} T^Y_{ijk}(P_{\Gamma^+},Q)\Delta u_k(Q)\mathrm{d}S(Q)$$
$$=n_i(P_{\Gamma^+})\int_S U^Y_{ijk}(P_{\Gamma^+},Q)t_k(Q)\mathrm{d}S(Q),\quad (i,j,k=x,y,z) \tag{8.13}$$

where the integral domain contains the uncracked boundary S and the crack surface Γ^+. Eq. (8.13) is independent of Eq. (8.6) and is the Yue's solution-based traction BIE.

8.2.3 The dual boundary integral equations for crack problems

Equations (8.6) and (8.13) give explicit expressions for the dual BIEs based on Yue's solution. These two integral equations do not contain the integrations on the interfaces of multilayered media because Yue's solution strictly satisfies the interface conditions. Collocating Eq. (8.6) on the uncracked boundary S and Eq. (8.13) on Γ^+ constitutes the dual BIEs for crack problems. The implementation of the two integral equations is similar to that described in Chapter 4 for the conventional BEM. A detailed description of the implementation will be given in the next section. In these equations, the unknown quantities are the tractions and displacements on the uncracked boundary and the discontinuous displacements on the crack surfaces.

After obtaining these unknown quantities using Eqs. (8.6) and (8.13), we can calculate the displacements $u_j(Q_{\Gamma^-})$ on the crack surface Γ^- by applying the displacement BIE (8.6) on the crack surface Γ^-. Although this will require extra integrations, there is no system of equations to be solved because all the tractions and displacements are already known. Hence, the displacements $u_j(Q_{\Gamma^+})$ on the crack surface Γ^+ can be easily obtained from the following

$$u_j(Q_{\Gamma^+}) = \Delta u_j(Q_{\Gamma^+}) + u_j(Q_{\Gamma^-}) \tag{8.14}$$

When the dual BIEs are applied for the study of the crack problems in a multilayered medium of infinite extent, the displacement BIE is not required; therefore, the traction BIE degenerates into the basic equation of the displacement discontinuity method (Crouch and Starfield, 1983). In this case, Eq. (8.13) can be simplified to

$$t_j(P_{\Gamma^+}) + n_i(P_{\Gamma^+}) \int_{\Gamma^+} T^Y_{ijk}(P, Q_{\Gamma^+}) \Delta u_k(Q_{\Gamma^+}) \mathrm{d}S(Q_{\Gamma^+}) = 0, \quad (i, j, k = x, y, z) \tag{8.15}$$

8.3 Numerical implementation

8.3.1 Boundary discretization

In the context of the dual BIEs, we will apply four- to eight-node isoparametric elements to discretize the uncracked boundary. Thus, the numerical methods and the corresponding computer codes described in Chapter 4 can be further utilized to develop the numerical method of the dual BIEs, i.e., the dual BEM.

As shown in Fig. 8.2, three types of nine-noded quadrilateral curved elements are employed to discretize the crack surfaces. The coordinates at any point in each element can be related to its element nodal coordinates as follows:

$$x = \sum_{\alpha=1}^{9} N_\alpha x^\alpha \tag{8.16a}$$

$$y = \sum_{\alpha=1}^{9} N_\alpha y^\alpha \tag{8.16b}$$

$$z = \sum_{\alpha=1}^{9} N_\alpha z^\alpha \tag{8.16c}$$

where $N_\alpha(\alpha = 1 \sim 9)$ are the shape functions.

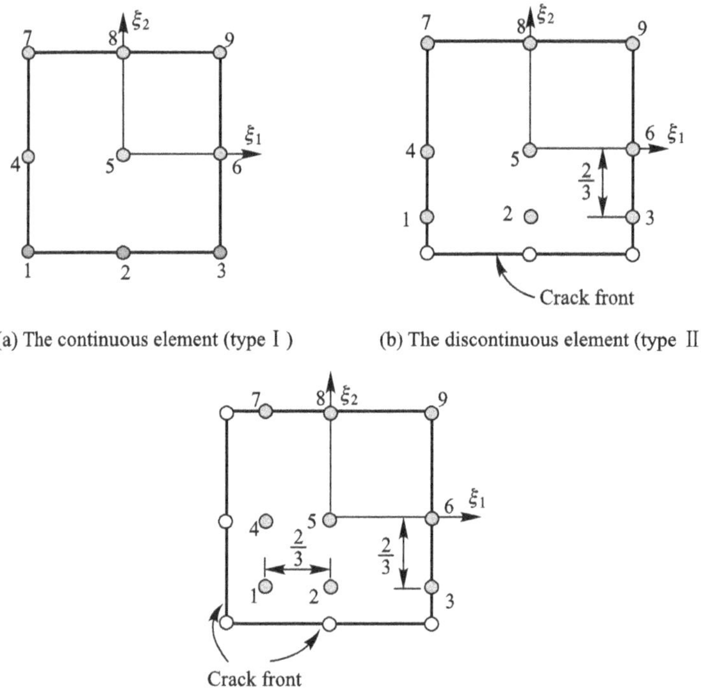

(a) The continuous element (type I)

(b) The discontinuous element (type II)

(c) The discontinuous element (type III)

○ Geometry node ◎ Collocation node

Fig. 8.2 The continuous and discontinuous elements

The shape functions of the type I element, shown in Fig. 8.2a, take the form

$$\begin{aligned}
N_1 &= 0.25\xi_1\xi_2(\xi_1-1)(\xi_2-1),\\
N_2 &= 0.5\xi_2(1-\xi_1^2)(\xi_2-1),\\
N_3 &= 0.25\xi_1\xi_2(\xi_1+1)(\xi_2-1),\\
N_4 &= 0.5\xi_1(1-\xi_2^2)(\xi_1-1),\\
N_5 &= (1-\xi_1^2)(1-\xi_2^2),\\
N_6 &= 0.5\xi_1(1-\xi_2^2)(\xi_1+1),\\
N_7 &= 0.25\xi_1\xi_2(\xi_1-1)(\xi_2+1),\\
N_8 &= 0.5\xi_2(1-\xi_1^2)(\xi_2+1),\\
N_9 &= 0.25\xi_1\xi_2(1+\xi_1)(1+\xi_2)
\end{aligned} \tag{8.17}$$

The type I element is continuous and isoparametric, and displaced on the crack surface away from the crack tip. The relative crack opening displacements (CODs) on the crack surfaces can be expressed as

$$\Delta u_i = \sum_{\alpha=1}^{9} N_\alpha \Delta u_i^\alpha \tag{8.18}$$

where $N_\alpha(\alpha = 1 \sim 9)$ are the shape functions and Δu_i^α are the CODs at node α.

As mentioned in Chapter 6, the singularity and angular distribution of the crack-tip stress field for the nonhomogeneous material are identical to those in the homogeneous material as long as the material properties are continuous. Hence, the displacement and stress fields at the crack tip of a crack in FGMs can be analyzed by using the same analytical methods as in a homogeneous medium. Considering the displacement and stress fields, two types of discontinuous elements, shown in Fig. 8.2b and 8.2c, are employed, numbered type II and type III elements, respectively.

As shown in Fig. 8.2b, the nodes 1, 2 and 3 on the type II element are at a position that is 1/3 from the side $\xi_2 = -1$. Thus, the shape functions of the type II element are as follows

$$\begin{aligned}
N_1 &= 0.45\xi_1\xi_2(\xi_1 - 1)(\xi_2 - 1), \\
N_2 &= 0.9\xi_2(1 - \xi_1^2)(\xi_2 - 1), \\
N_3 &= 0.45\xi_1\xi_2(\xi_1 + 1)(\xi_2 - 1), \\
N_4 &= 0.75\xi_1(1 - \xi_2)(2/3 + \xi_2)(\xi_1 - 1), \\
N_5 &= 1.5(\xi_1^2 - 1)(\xi_2 - 1)(2/3 + \xi_2), \\
N_6 &= 0.75\xi_1(1 - \xi_2)(2/3 + \xi_2)(\xi_1 + 1), \\
N_7 &= 0.3\xi_1\xi_2(\xi_1 - 1)(2/3 + \xi_2), \\
N_8 &= 0.6\xi_2(1 - \xi_1^2)(2/3 + \xi_2), \\
N_9 &= 0.3\xi_1\xi_2(\xi_1 + 1)(2/3 + \xi_2)
\end{aligned} \tag{8.19}$$

In order to capture the specific characteristics of CODs near a crack front, the CODs of the type II element on the crack surface can be expressed as

$$\Delta u_i = \sum_{\alpha=1}^{9} \sqrt{1 + \xi_2} N_\alpha \Delta u_i^\alpha \tag{8.20}$$

where $N_\alpha(\alpha = 1 \sim 9)$ are the shape functions and Δu_i^α are the CODs at node α. The coefficient $\sqrt{1 + \xi_2}$ is introduced to capture the crack behavior at the crack front. This type of element is positioned at the smooth crack front and the side $\xi_2 = -1$ of the element is located at the crack front.

Figure 8.2c shows that the nodes 1, 2 and 3 of the type III element are positioned at a distance 1/3 from the side $\xi_2 = -1$ and the nodes 1, 4 and 7 are at a position that is 1/3 from the side $\xi_1 = -1$. The shape functions of the type III element are given as

$$
\begin{aligned}
N_1 &= 0.81\xi_1\xi_2(\xi_1-1)(\xi_2-1),\\
N_2 &= 1.35\xi_2(1-\xi_1)(2/3+\xi_1)(\xi_2-1),\\
N_3 &= 0.54\xi_1\xi_2(2/3+\xi_1)(\xi_2-1),\\
N_4 &= 1.35\xi_1(1-\xi_2)(2/3+\xi_2)(\xi_1-1),\\
N_5 &= 2.25(1-\xi_1)(2/3+\xi_1)(1-\xi_2)(2/3+\xi_2),\\
N_6 &= 0.9\xi_1(1-\xi_2)(2/3+\xi_2)(2/3+\xi_1),\\
N_7 &= 0.54\xi_1\xi_2(\xi_1-1)(2/3+\xi_2),\\
N_8 &= 0.9\xi_2(2/3+\xi_2)(1-\xi_1)(2/3+\xi_1),\\
N_9 &= 0.36\xi_1\xi_2(2/3+\xi_1)(2/3+\xi_2)
\end{aligned}
\tag{8.21}
$$

The CODs of the type Ⅲ element on the crack surfaces can be expressed as

$$
\Delta u_i = \sum_{\alpha=1}^{9} \sqrt{1+\xi_1}\sqrt{1+\xi_2}N_\alpha \Delta u_i^\alpha \tag{8.22}
$$

where $N_\alpha(\alpha = 1 \sim 9)$ are the shape functions and Δu_i^α are CODs at node α. The coefficient $\sqrt{1+\xi_1}\sqrt{1+\xi_2}$ is introduced to capture the crack behavior at the crack front. This type of element is positioned at the non-smooth crack front and the sides $\xi_1 = -1$ and $\xi_2 = -1$ of the element are located at the crack front.

8.3.2 The discretized boundary integral equation

Having defined the displacements, tractions and CODs within an element in terms of the element nodal values, we can now treat the latter as the discrete variables of the problem. Once these quantities are determined, the displacements, tractions and CODs on the boundaries are known everywhere. Having discretized the uncracked boundary$S$ into $ne1$ elements and the crack surface Γ^+ into $ne2$ elements, we can now rewrite the displacement BIE (8.6) in terms of these parameters, which will be determined using the shape functions; thus

$$
\begin{aligned}
& c_{ij}(P_S)u_j(P_S) + \sum_{e=1}^{ne1}\left\{\sum_{\alpha=1}^{m} u_j^\alpha(Q^\alpha)\int_{S_e} t_{ij}^Y[P_S, Q(\xi_1,\xi_2)]N_\alpha(\xi_1,\xi_2)J(\xi_1,\xi_2)\mathrm{d}\xi_1\mathrm{d}\xi_2\right\} +\\
& \sum_{e=1}^{ne2}\left\{\sum_{\alpha=1}^{9} \Delta u_j^\alpha(Q^\alpha)\int_{\Gamma_e^+} t_{ij}^Y[P_S, Q(\xi_1,\xi_2)]g(\xi_1,\xi_2)N_\alpha(\xi_1,\xi_2)J(\xi_1,\xi_2)\mathrm{d}\xi_1\mathrm{d}\xi_2\right\}\\
& = \sum_{e=1}^{ne1}\left\{\sum_{\alpha=1}^{m} t_j^\alpha(Q^\alpha)\int_{S_e} u_{ij}^Y[P_S, Q(\xi_1,\xi_2)]N_\alpha(\xi_1,\xi_2)J(\xi_1,\xi_2)\mathrm{d}\xi_1\mathrm{d}\xi_2\right\},\\
& (i,j = x,y,z)
\end{aligned}
\tag{8.23}
$$

where the source point P_S is collocated on the uncracked boundary S of the cracked body, Q^α and Q are the nodal points of an element and the points within the element, respectively, $J(\xi_1,\xi_2)$ is the Jacobian determinant, and $g(\xi_1,\xi_2)$ is defined as

- for the type I element, $g(\xi_1,\xi_2)=1$;
- for the type II element, $g(\xi_1,\xi_2)=\sqrt{1+\xi_2}$; and
- for the type III element, $g(\xi_1,\xi_2)=\sqrt{1+\xi_1}\sqrt{1+\xi_2}$.

We can now rewrite the traction BIE (8.13) using the above arrangements as

$$\begin{aligned}
&t_j(P_{\Gamma^+})+n_i(P_{\Gamma^+})\sum_{e=1}^{ne1}\left\{\sum_{\alpha=1}^{m}u_k^\alpha(Q^\alpha)\int_{S_e}T_{ijk}^Y[P_{\Gamma^+},Q(\xi_1,\xi_2)]N_\alpha(\xi_1,\xi_2)J(\xi_1,\xi_2)\mathrm{d}\xi_1\mathrm{d}\xi_2\right\}+\\
&n_i(P_{\Gamma^+})\sum_{e=1}^{ne2}\left\{\sum_{\alpha=1}^{9}\Delta u_k^\alpha(Q^\alpha)\int_{\Gamma_e^+}T_{ijk}^Y[P_{\Gamma^+},Q(\xi_1,\xi_2)]g(\xi_1,\xi_2)N_\alpha(\xi_1,\xi_2)J(\xi_1,\xi_2)\mathrm{d}\xi_1\mathrm{d}\xi_2\right\}\\
&\quad=n_i(P_{\Gamma^+})\sum_{e=1}^{ne1}\left\{\sum_{\alpha=1}^{m}t_k^\alpha(Q^\alpha)\int_{S_e}U_{ijk}^Y[P_{\Gamma^+},Q(\xi_1,\xi_2)]N_\alpha(\xi_1,\xi_2)J(\xi_1,\xi_2)\mathrm{d}\xi_1\mathrm{d}\xi_2\right\},\\
&\quad(i,j,k=x,y,z)
\end{aligned}\tag{8.24}$$

where the source point P_{Γ^+} is collocated on the crack surface Γ^+, Q^α and Q are the nodal points of an element and the points within the element, respectively, and $J(\xi_1,\xi_2)$ is the Jacobian determinant.

8.4 Numerical integrations

8.4.1 Numerical integrations for the displacement BIE

According to the nature of the kernel functions and the relative position of the source point with respect to the element on which the integration is carried out, the two integrals in Eq. (8.23) are regular or non-regular and need to be evaluated carefully.

(1) The integral on the right-hand side of the displacement BIE

There is the following integral on the right-hand of Eq. (8.23)

$$\int_{S_e}u_{ij}^Y[P_S,Q(\xi_1,\xi_2)]N_\alpha(\xi_1,\xi_2)J(\xi_1,\xi_2)\mathrm{d}\xi_1\mathrm{d}\xi_2 \tag{8.25}$$

When the source point P_S is positioned on the element S_e and $P_S=Q^\alpha$, the integral (8.25) is of the order $o(r^{-1})$ and weakly singular; it can be evaluated using a variable transformation, i.e., the triangle to square transformation as used in Chapter 4.

When the source point P_S is positioned on the element S_e and $P_S\neq Q^\alpha$ or the source point P_S is not positioned on the element S_e, the integral (8.25) is regular and can be evaluated using the Gaussian quadrature rules.

(2) The first integral on the left-hand side of the displacement BIE

The following integral occurs on the left-hand side of Eq. (8.23)

$$\int_{S_e} t_{ij}^Y[P_S, Q(\xi_1, \xi_2)] N_\alpha(\xi_1, \xi_2) J(\xi_1, \xi_2) \mathrm{d}\xi_1 \mathrm{d}\xi_2 \tag{8.26}$$

When the source point P_S is positioned on the element S_e and $P_S = Q^\alpha$, the singularity of the kernel in the integral (8.26) is $o(r^{-2})$ and the integral (8.26) is strongly singular. This singular integral, together with the coefficient $c_{ij}(P_S)$, can be evaluated by using an indirect method, i.e., the rigid-body-motion constraint. Substituting unit rigid-body displacements and the tractions equal to zero into the displacement BIE (8.23), we obtain the following expression, which is the same as the one given in Chapter 4,

$$\begin{aligned} c_{ij}(P_S) + & \sum_{\substack{e=1 \\ P \in S_e \text{ and } P = Q^\alpha}}^{ne1} \int_{S_e} t_{ij}^Y[P_S, Q(\xi_1, \xi_2)] N_\alpha(\xi_1, \xi_2) J(\xi_1, \xi_2) \mathrm{d}\xi_1 \mathrm{d}\xi_2 \\ = & -\sum_{e=1}^{ne1} \sum_{\substack{\alpha=1 \\ P \neq Q^\alpha}}^{m} \int_{S_e} t_{ij}^Y[P_S, Q(\xi_1, \xi_2)] N_\alpha(\xi_1, \xi_2) J(\xi_1, \xi_2) \mathrm{d}\xi_1 \mathrm{d}\xi_2 \end{aligned} \tag{8.27}$$

When the source point P_S is positioned on the element S_e and $P_S \neq Q^\alpha$ or the source point P_S is not positioned on the element S_e, the integral (8.26) is regular and can be evaluated using the Gaussian quadrature rule.

(3) The second integral on the left-hand side of the displacement BIE

The following integral occurs on the left-hand side of Eq. (8.23)

$$\int_{\Gamma_e^+} t_{ij}^Y[P_S, Q(\xi_1, \xi_2)] g(\xi_1, \xi_2) N_\alpha(\xi_1, \xi_2) J(\xi_1, \xi_2) \mathrm{d}\xi_1 \mathrm{d}\xi_2 \tag{8.28}$$

In Eq. (8.23), the source point P_S is always collocated on the uncracked boundary. Hence, the integral (8.28) on the crack surface is regular and can be evaluated using the Gaussian quadrature rule.

8.4.2 Numerical integrations for the traction BIE

In the traction BIE (8.24), there are three integrals to be evaluated; two types of integrals will be considered in the following.

(1) The regular integrals

In Eq. (8.24), the source point P_{Γ^+} is collocated on the crack surface Γ^+. The first integral on the left-hand side of Eq. (8.24) and the integral on the right-hand side are carried out for the uncracked boundary and are regular. When the source point P_{Γ^+} is positioned on the element Γ_e^+ and $P_{\Gamma^+} \neq Q^\alpha$ or P_{Γ^+} is not positioned on Γ_e^+, the second integral on the left-hand side of Eq. (8.24) is also regular. Hence, these integrals can be evaluated using the Gaussian quadrature rule.

(2) The hypersingular integral

The following integral appears on the left-hand side of Eq. (8.24):

$$\int_{\Gamma_e^+} T_{ijk}^Y[P_{\Gamma^+}, Q(\xi_1,\xi_2)]g(\xi_1,\xi_2)N_\alpha(\xi_1,\xi_2)J(\xi_1,\xi_2)\mathrm{d}\xi_1\mathrm{d}\xi_2 \tag{8.29}$$

When the source point P_{Γ^+} is positioned on the element Γ_e^+ and $P_{\Gamma^+} = Q^\alpha$, the singularity of the kernel in the integral (8.29) is $o(r^{-3})$ and the integral (8.29) is hypersingular. Guiggiani and Gigante (1990) and Guiggiani et al. (1992) developed a technique to evaluate the hypersingular integral. In this technique, the singular part of the fundamental solution in the kernel is subtracted out and integrated analytically or semi-analytically exactly, leaving the remaining integrand well posed. The technique is based on a Taylor series expansion of the kernel function, shape functions and the Jacobian determinant. Because of the complexity of the expressions in Yue's solution, it is difficult for the above technique to be used to evaluate the hypersingular integral in Eq. (8.29). In the following, we utilize the method proposed by Pan and Yuan (2000) to evaluate this hypersingular integral.

The integral (8.29) can be written in a local coordinate system as

$$\int_{-1}^{1}\int_{-1}^{1} T_{ijk}^Y[P_{\Gamma^+}(\xi_1^c,\xi_2^c), Q(\xi_1,\xi_2)]g(\xi_1,\xi_2)N_\alpha(\xi_1,\xi_2)J(\xi_1,\xi_2)\mathrm{d}\xi_1\mathrm{d}\xi_2 \tag{8.30}$$

where (ξ_1^c,ξ_2^c) and (ξ_1,ξ_2) are, respectively, the local coordinates of the source point P_{Γ^+} and the field point Q.

The polar coordinate system is introduced at the source point (ξ_1^c,ξ_2^c)

$$\xi_1 = \xi_1^c + r\cos\theta \tag{8.31a}$$

$$\xi_2 = \xi_2^c + r\sin\theta \tag{8.31b}$$

The element is then divided into several triangle domains. Equation (8.30) can be rewritten as

$$\sum_M \int_{\theta_1}^{\theta_2}\int_0^{R(\theta)} T_{ijk}^Y[P_{\Gamma^+}(\xi_1^c,\xi_2^c), Q(r,\theta)]g(r,\theta)N_\alpha(r,\theta)J(r,\theta)r\mathrm{d}r\mathrm{d}\theta \tag{8.32}$$

where the summation over M is for all the integrals on the element. Without loss of generality, we consider the following three cases shown in Figs. 8.3 to 8.5. When the source point P_{Γ^+} is located at the other nodes, similar analyses can be employed.

(i) The collocation point is located at the corner node of the element

For this case, the element is subdivided into two triangles and the summation on M in Eq. (8.32) is from the triangle 1 to the triangle 2. In Fig. 8.3, the collocation point is located at node 1 (as shown in Fig. 8.2a) and $(\xi_1^c,\xi_2^c) = (-1,-1)$. Equation (8.31) can be rewritten as

$$\xi_1 = -1 + r\cos\theta \tag{8.33a}$$

$$\xi_2 = -1 + r\sin\theta \tag{8.33b}$$

For the triangle 1, $\theta \in [0, \pi/4]$ and $R(\theta) = 2/\cos\theta$; for the triangle 2, $\theta \in [\pi/4, \pi/2]$ and $R(\theta) = 2/\cos(\theta - \pi/2)$.

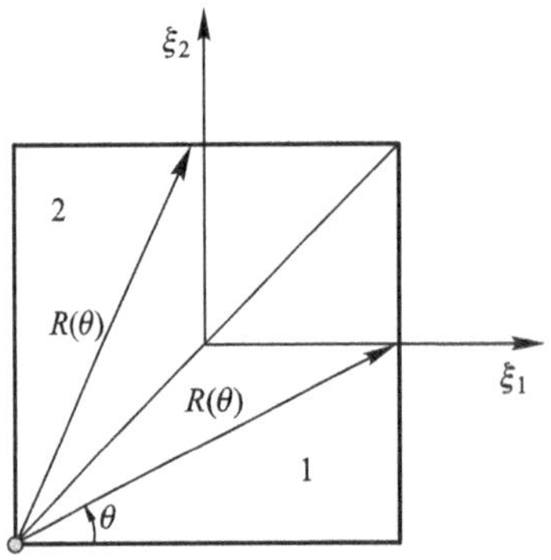

Fig. 8.3 The triangle domains divided in an element when the source point is a corner node

(ii) The collocation point is located on the element side

In this case, the element needs to be divided into three triangles and the summation on M in Eq. (8.32) is from the triangle 1 to the triangle 3. In Fig. 8.4, the collocation point is located at node 1 (as shown in Fig. 8.2b) and $(\xi_1^c, \xi_2^c) = (-1, -2/3)$. Equation (8.31) can be rewritten as

$$\xi_1 = -1 + r\cos\theta \tag{8.34a}$$

$$\xi_2 = -2/3 + r\sin\theta \tag{8.34b}$$

For the different triangles, we have the following integral parameters

- For the triangle 1, $\theta \in [-\pi/2, -\text{atan}(1/6)]$ and $R(\theta) = 1/(3\cos(\theta + \pi/2))$;
- For the triangle 2, $\theta \in [-\text{atan}(1/6), \text{atan}(5/6)]$ and $R(\theta) = 2/\cos\theta$;
- For the triangle 3, $\theta \in [\text{atan}(5/6), \pi/2]$ and $R(\theta) = 5/(3\cos(\theta - \pi/2))$.

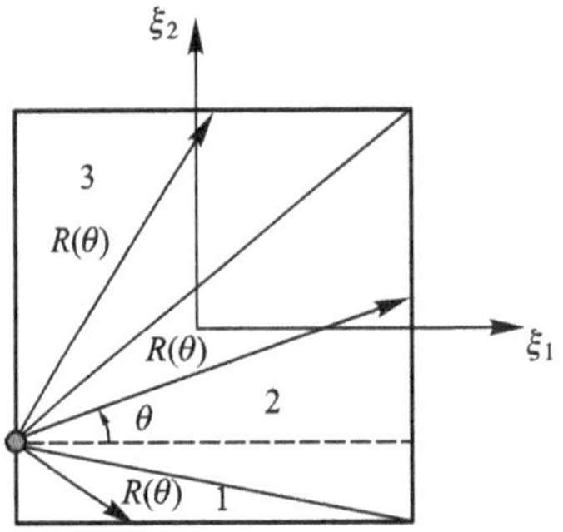

Fig. 8.4 The triangle domains divided in an element when the source point is a side node

(iii) The collocation point is located at an internal node

For this case, the element needs to be divided into four triangles and the summation on M in Eq. (8.32) is from the triangle 1 to the triangle 4. In Fig. 8.5, the collocation point is located at node 1 (as shown in Fig. 8.2c) and $(\xi_1^c, \xi_2^c) = (-2/3, -2/3)$. Eq. (8.31) can be rewritten as

$$\xi_1 = -2/3 + r\cos\theta \tag{8.35a}$$

$$\xi_2 = -2/3 + r\sin\theta \tag{8.35b}$$

For the different triangles, we have the following integral parameters

- For the triangle 1, $\theta \in [-\text{atan}(1/5), \pi/4]$ and $R(\theta) = 5/(3\cos\theta)$;
- For the triangle 2, $\theta \in [\pi/4, \pi/2 + \text{atan}(1/5)]$ and $R(\theta) = 5/(3\cos(\theta - \pi/2))$;
- For the triangle 3, $\theta \in [\pi/2 + \text{atan}(1/5), 5\pi/4]$ and $R(\theta) = 1/(3\cos(\theta - \pi))$;
- For the triangle 4, $\theta \in [5\pi/4, 2\pi - \text{atan}(1/5)]$ and $R(\theta) = 1/(3\cos(\theta - 3\pi/2))$.

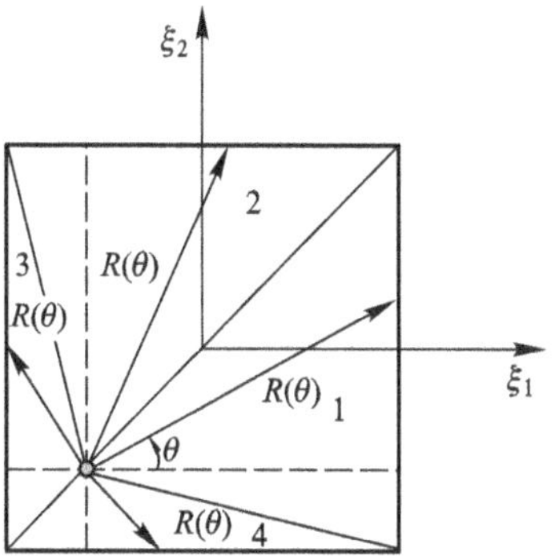

Fig. 8.5 The triangle domains divided in an element when the source point is an internal point

After a coordinate transform, the inner and outer integrals are given with respect to the polar coordinates r and θ and there is a polar coordinate r in the integrand of the integral (8.32). It can be observed that the integrand in Eq. (8.32) has a singularity $o(r^{-2})$. The numerical quadrature given in Kutt (1975a, b) can be utilized to evaluate the inner finite-part integral with respect to r. On the other hand, the outer integral with respect to θ is regular and can be calculated using the regular Gaussian quadrature.

In Appendix 4, the finite-part integral is introduced and the corresponding formulae for evaluating this type of integral are presented. For a given Gaussian point θ_s, the inner integral in Eq. (8.32) can be approximated by Kutt's N-point equispace quadrature as follows

$$\int_0^R \frac{f(r)}{r^2}\mathrm{d}r \approx \frac{1}{R}\sum_{l=1}^{N}(w_l + c_l \ln R) f\left(\frac{l-1}{N}R\right) \tag{8.36}$$

where w_l are the weights and c_l the coefficients given in Kutt (1975a, b), and the integrand is given by

$$f(r) = T_{ijk}^{Y}[P_{\Gamma^+}(\xi_1^c, \xi_2^c), Q(r, \theta_s)]g(r, \theta_s)N_\alpha(r, \theta_s)J(r, \theta_s)r^3 \tag{8.37}$$

Note that, when deriving the N-point equispace quadrature (8.36), it is assumed that the integrand $f(r) \in C^0[0, R]$ and $f(r) \in C^2$ in the neighborhood of $r = 0$. If $l = 1$ in Eq. (8.36), then $r = 0$ in Eq. (8.37). Because T_{ijk}^{Y} is of the singularity $o(r^{-3})$, the integrand in Eq. (8.30) does not exist. For the case $l = 1$, take $r = 10^{-8}$. In the following numerical examples, Kutt's 20-point equispace quadrature is used in the finite-part integral with respect to r, and 20 Gaussian points are used for the regular outer integral with respect to θ.

If the crack surface is flat, continuous elements can be used to discretize the interior crack surface, with discontinuous elements for the crack front only. However, if the crack surface is curved, then discontinuous elements are needed for the whole crack surface in order to satisfy the continuity requirement for $f(r)$. Pan and Yuan (2000) and dell'Erba (2000) introduced different kinds of discontinuous elements.

8.5 Linear equation systems for the discretized dual BIEs

For the discretized dual BIEs (8.23) and (8.24), the linear equation systems need to be established to solve the unknown quantities of displacements, tractions and CODs. In the following, we will discuss the establishment of the linear equation systems.

In Eq. (8.23), we introduce the following expressions

$$G1_{ij}^{e\alpha} = \int_{S_e} t_{ij}^{Y}[P_S, Q(\xi_1, \xi_2)]N_\alpha(\xi_1, \xi_2)J(\xi_1, \xi_2)\mathrm{d}\xi_1\mathrm{d}\xi_2 \tag{8.38a}$$

$$C1_{ij}^{e\alpha} = \int_{\Gamma_e^+} t_{ij}^{Y}[P_S, Q(\xi_1, \xi_2)]g(\xi_1, \xi_2)N_\alpha(\xi_1, \xi_2)J(\xi_1, \xi_2)\mathrm{d}\xi_1\mathrm{d}\xi_2 \tag{8.38b}$$

$$H1_{ij}^{e\alpha} = \int_{S_e} u_{ij}^{Y}[P_S, Q(\xi_1, \xi_2)]N_\alpha(\xi_1, \xi_2)J(\xi_1, \xi_2)\mathrm{d}\xi_1\mathrm{d}\xi_2 \tag{8.38c}$$

In Eq. (8.38a), the superscripts and subscripts of $G1_{ij}^{e\alpha}$ have the following meanings

Element number
↓
$G1_{ij}^{e\alpha}$ ← Collocation point
← The j direction of point Q (8.39)
↑
The i direction of point P_S

In Eqs. (8.38b) and (8.38c), the superscripts and subscripts of $C1_{ij}^{e\alpha}$ and $H1_{ij}^{e\alpha}$ have the meanings similar to those in Eq. (8.39).

Assume that there are $N1$ and $N2$ nodes on the uncracked boundary and the crack surface, respectively. Using Eqs. (8.38), we can rewrite Eq. (8.23) as

$$c_{ij}^P u_j^P + \sum_{e=1}^{ne1} \sum_{\alpha=1}^{m} u_j^\alpha G1_{ij}^{e\alpha} + \sum_{e=1}^{ne2} \sum_{\alpha=1}^{9} \Delta u_j^\alpha C1_{ij}^{e\alpha} = \sum_{e=1}^{ne1} \sum_{\alpha=1}^{m} t_j^\alpha H1_{ij}^{e\alpha}, \quad (P = 1,2,\cdots,N1) \tag{8.40}$$

where P corresponds to the source point P_S.

In Eq. (8.40), we introduce the following notations

$$\boldsymbol{C}^P = \begin{bmatrix} c_{xx}^P & c_{xy}^P & c_{xz}^P \\ c_{yx}^P & c_{yy}^P & c_{yz}^P \\ c_{zx}^P & c_{zy}^P & c_{zz}^P \end{bmatrix}, \overline{\boldsymbol{G}}1^{P,n} = \begin{bmatrix} G1_{xx}^{Pn} & G1_{xy}^{Pn} & G1_{xz}^{Pn} \\ G1_{yx}^{Pn} & G1_{yy}^{Pn} & G1_{yz}^{Pn} \\ G1_{zx}^{Pn} & G1_{zy}^{Pn} & G1_{zz}^{Pn} \end{bmatrix}, \quad (n = 1,2,\cdots,N1),$$

$$\boldsymbol{H}1^{P,n} = \begin{bmatrix} H1_{xx}^{Pn} & H1_{xy}^{Pn} & H1_{xz}^{Pn} \\ H1_{yx}^{Pn} & H1_{yy}^{Pn} & H1_{yz}^{Pn} \\ H1_{zx}^{Pn} & H1_{zy}^{Pn} & H1_{zz}^{Pn} \end{bmatrix}, \quad (n = 1,2,\cdots,N1),$$

$$\boldsymbol{C}1^{P,n} = \begin{bmatrix} C1_{xx}^{Pn} & C1_{xy}^{Pn} & C1_{xz}^{Pn} \\ C1_{yx}^{Pn} & C1_{yy}^{Pn} & C1_{yz}^{Pn} \\ C1_{zx}^{Pn} & C1_{zy}^{Pn} & C1_{zz}^{Pn} \end{bmatrix}, \quad (n = 1,2,\cdots,N2),$$

$$\boldsymbol{u}^P = (u_x^P \quad u_y^P \quad u_z^P)^{\mathrm{T}}, \quad \Delta\boldsymbol{u}^P = (\Delta u_x^P \quad \Delta u_y^P \quad \Delta u_z^P)^{\mathrm{T}}, \quad \boldsymbol{t}^P = (t_x^P \quad t_y^P \quad t_z^P)^{\mathrm{T}}$$

where the superscript T stands for the transpose of a matrix.

In the above expressions, the elements of $\overline{\boldsymbol{G}}1^{P,n}$, $\boldsymbol{H}1^{P,n}$ and $\boldsymbol{C}1^{P,n}$ are, respectively, the summations of $G1_{ij}^{e\alpha}$, $H1_{ij}^{e\alpha}$ and $C1_{ij}^{e\alpha}$ from different elements. Thus, Eq. (8.40) can be further rewritten as

$$\boldsymbol{C}^P\boldsymbol{u}^P + [\overline{\boldsymbol{G}}1^{P,1} \quad \overline{\boldsymbol{G}}1^{P,2} \quad \cdots \quad \overline{\boldsymbol{G}}1^{P,N1}] \begin{Bmatrix} \boldsymbol{u}^1 \\ \boldsymbol{u}^2 \\ \vdots \\ \boldsymbol{u}^{N1} \end{Bmatrix} +$$

$$[\boldsymbol{C}1^{P,1} \quad \boldsymbol{C}1^{P,2} \quad \cdots \quad \boldsymbol{C}1^{P,N2}] \begin{Bmatrix} \Delta\boldsymbol{u}^1 \\ \Delta\boldsymbol{u}^2 \\ \vdots \\ \Delta\boldsymbol{u}^{N2} \end{Bmatrix}$$

$$= [\boldsymbol{H}1^{P,1} \quad \boldsymbol{H}1^{P,2} \quad \cdots \quad \boldsymbol{H}1^{P,N1}] \begin{Bmatrix} \boldsymbol{t}^1 \\ \boldsymbol{t}^2 \\ \vdots \\ \boldsymbol{t}^{N1} \end{Bmatrix}, \quad (P = 1,2,\cdots,N1) \tag{8.41}$$

On the left-hand side of Eq. (8.41), the source point P of the first term always has the same point among the $N1$ nodes on the uncracked boundary. Rearranging the coefficients of displacements on the left-hand side of Eq. (8.41), we obtain

$$\begin{bmatrix} G1^{P,1} & G1^{P,2} & \cdots & G1^{P,N1} \end{bmatrix} \begin{Bmatrix} \boldsymbol{u}^1 \\ \boldsymbol{u}^2 \\ \vdots \\ \boldsymbol{u}^{N1} \end{Bmatrix} + \begin{bmatrix} C1^{P,1} & C1^{P,2} & \cdots & C1^{P,N2} \end{bmatrix} \begin{Bmatrix} \Delta\boldsymbol{u}^1 \\ \Delta\boldsymbol{u}^2 \\ \vdots \\ \Delta\boldsymbol{u}^{N2} \end{Bmatrix}$$

$$= \begin{bmatrix} \boldsymbol{H}1^{P,1} & \boldsymbol{H}1^{P,2} & \cdots & \boldsymbol{H}1^{P,N1} \end{bmatrix} \begin{Bmatrix} \boldsymbol{t}^1 \\ \boldsymbol{t}^2 \\ \vdots \\ \boldsymbol{t}^{N1} \end{Bmatrix}, \quad (P = 1,2,\cdots,N1) \tag{8.42}$$

where

$$G1^{P,n} = \begin{cases} C^P + \overline{G}1^{P,n}, & P = n \\ \overline{G}1^{P,n}, & P \neq n \end{cases} \quad \text{and } n = 1,2,\cdots,N1$$

For each node on the uncracked boundary, we can establish a set of displacement BIEs (8.42). Thus, the simultaneous linear equations for all the nodes on the uncracked boundary can be obtained as follows

$$\begin{bmatrix} G1^{1,1} & G1^{1,2} & \cdots & G1^{1,N1} \\ G1^{2,1} & G1^{2,2} & \cdots & G1^{2,N1} \\ \vdots & \vdots & & \vdots \\ G1^{N1,1} & G1^{N1,2} & \cdots & G1^{N1,N1} \end{bmatrix} \begin{Bmatrix} \boldsymbol{u}^1 \\ \boldsymbol{u}^2 \\ \vdots \\ \boldsymbol{u}^{N1} \end{Bmatrix} + \begin{bmatrix} C1^{1,1} & C1^{1,2} & \cdots & C1^{1,N2} \\ C1^{2,1} & C1^{2,2} & \cdots & C1^{2,N2} \\ \vdots & \vdots & & \vdots \\ C1^{N1,1} & C1^{N1,2} & \cdots & C1^{N1,N2} \end{bmatrix} \begin{Bmatrix} \Delta\boldsymbol{u}^1 \\ \Delta\boldsymbol{u}^2 \\ \vdots \\ \Delta\boldsymbol{u}^{N2} \end{Bmatrix}$$

$$= \begin{bmatrix} \boldsymbol{H}1^{1,1} & \boldsymbol{H}1^{1,2} & \cdots & \boldsymbol{H}1^{1,N1} \\ \boldsymbol{H}1^{2,1} & \boldsymbol{H}1^{2,2} & \cdots & \boldsymbol{H}1^{2,N1} \\ \vdots & \vdots & & \vdots \\ \boldsymbol{H}1^{N1,1} & \boldsymbol{H}1^{N1,2} & \cdots & \boldsymbol{H}1^{N1,N1} \end{bmatrix} \begin{Bmatrix} \boldsymbol{t}^1 \\ \boldsymbol{t}^2 \\ \vdots \\ \boldsymbol{t}^{N1} \end{Bmatrix} \tag{8.43}$$

There are the arrays of displacements and tractions in Eq. (8.43). Multiplying the coefficient matrices by the known quantities, we can obtain the array $(\boldsymbol{F}^d)_{N1\times 1}$. We let $(\boldsymbol{UT})_{N1\times 1}$, $(\boldsymbol{A}^d)_{N1\times N1}$, $(\Delta\boldsymbol{U})_{N2\times 1}$, and $(\boldsymbol{B}^d)_{N1\times N2}$ stand for the unknown quantities (displacements and tractions), the corresponding coefficient matrix, the unknown quantities of the CODs, and the corresponding coefficient matrix, respectively. Then, Eq. (8.43) can be rewritten as

$$\boldsymbol{A}^d\boldsymbol{UT} + \boldsymbol{B}^d\Delta\boldsymbol{U} = \boldsymbol{F}^d \tag{8.44}$$

where the superscript d stands for the displacement BIE.

In Eq. (8.24), we introduce the following expressions

$$G2^{e\alpha}_{jk} = n_i \int_{S_e} T^Y_{ijk}[P_{\Gamma^+}, Q(\xi_1,\xi_2)] N_\alpha(\xi_1,\xi_2) J(\xi_1,\xi_2)\mathrm{d}\xi_1\mathrm{d}\xi_2 \tag{8.45a}$$

$$C2_{jk}^{e\alpha} = n_i \int_{\Gamma_e^+} T_{ijk}^Y [P_{\Gamma^+}, Q(\xi_1,\xi_2)] g(\xi_1,\xi_2) N_\alpha(\xi_1,\xi_2) J(\xi_1,\xi_2) \mathrm{d}\xi_1 \mathrm{d}\xi_2 \tag{8.45b}$$

$$H2_{jk}^{e\alpha} = n_i \int_{S_e} U_{ijk}^Y [P_{\Gamma^+}, Q(\xi_1,\xi_2)] N_\alpha(\xi_1,\xi_2) J(\xi_1,\xi_2) \mathrm{d}\xi_1 \mathrm{d}\xi_2 \tag{8.45c}$$

In Eq. (8.45), the superscripts and subscripts of $G1_{jk}^{e\alpha}$, $C2_{jk}^{e\alpha}$ and $H2_{jk}^{e\alpha}$ have the meanings similar to those given in Eq. (8.39). Thus, Eq. (8.24) can be rewritten as

$$t_j^P + \sum_{e=1}^{ne1} \sum_{\alpha=1}^{m} u_k^\alpha G2_{jk}^{e\alpha} + \sum_{e=1}^{ne2} \sum_{\alpha=1}^{9} \Delta u_k^\alpha C2_{jk}^{e\alpha} = \sum_{e=1}^{ne1} \sum_{\alpha=1}^{m} t_k^\alpha H2_{jk}^{e\alpha}, \quad (P = 1, 2, \cdots, N2) \tag{8.46}$$

where P corresponds to the source point P_{Γ^+}.

In Eq. (8.46), we introduce the following notations

$$\boldsymbol{G2}^{P,n} = \begin{bmatrix} G2_{xx}^{Pn} & G2_{xy}^{Pn} & G2_{xz}^{Pn} \\ G2_{yx}^{Pn} & G2_{yy}^{Pn} & G2_{yz}^{Pn} \\ G2_{zx}^{Pn} & G2_{zy}^{Pn} & G2_{zz}^{Pn} \end{bmatrix}, \quad (n = 1, 2, \cdots, N1),$$

$$\boldsymbol{H2}^{P,n} = \begin{bmatrix} H2_{xx}^{Pn} & H2_{xy}^{Pn} & H2_{xz}^{Pn} \\ H2_{yx}^{Pn} & H2_{yy}^{Pn} & H2_{yz}^{Pn} \\ H2_{zx}^{Pn} & H2_{zy}^{Pn} & H2_{zz}^{Pn} \end{bmatrix}, \quad (n = 1, 2, \cdots, N1),$$

$$\boldsymbol{C2}^{P,n} = \begin{bmatrix} C2_{xx}^{Pn} & C2_{xy}^{Pn} & C2_{xz}^{Pn} \\ C2_{yx}^{Pn} & C2_{yy}^{Pn} & C2_{yz}^{Pn} \\ C2_{zx}^{Pn} & C2_{zy}^{Pn} & C2_{zz}^{Pn} \end{bmatrix}, \quad (n = 1, 2, \cdots, N2)$$

In the above expressions, the elements of $\boldsymbol{G2}^{P,n}$, $\boldsymbol{H2}^{P,n}$ and $\boldsymbol{C2}^{P,n}$ are, respectively, the summations of $G2_{ij}^{e\alpha}$, $H2_{ij}^{e\alpha}$ and $C2_{ij}^{e\alpha}$ from different elements. Thus, Eq. (8.46) can be rewritten as

$$\boldsymbol{t}^P + [\boldsymbol{G2}^{P,1} \quad \boldsymbol{G2}^{P,2} \quad \cdots \quad \boldsymbol{G2}^{P,N1}] \begin{Bmatrix} \boldsymbol{u}^1 \\ \boldsymbol{u}^2 \\ \vdots \\ \boldsymbol{u}^{N1} \end{Bmatrix} +$$

$$[\boldsymbol{C2}^{P,1} \quad \boldsymbol{C2}^{P,2} \quad \cdots \quad \boldsymbol{C2}^{P,N2}] \begin{Bmatrix} \Delta\boldsymbol{u}^1 \\ \Delta\boldsymbol{u}^2 \\ \vdots \\ \Delta\boldsymbol{u}^{N2} \end{Bmatrix}$$

$$= [\boldsymbol{H2}^{P,1} \quad \boldsymbol{H2}^{P,2} \quad \cdots \quad \boldsymbol{H2}^{P,N1}] \begin{Bmatrix} \boldsymbol{t}^1 \\ \boldsymbol{t}^2 \\ \vdots \\ \boldsymbol{t}^{N1} \end{Bmatrix}, \quad (P = 1, 2, \cdots, N2) \tag{8.47}$$

For each node on the crack surfaces, we can establish a set of traction BIEs (8.47). Thus, the simultaneous linear equations of all the nodes on the crack surface can be obtained as follows

$$\begin{Bmatrix} t^1 \\ t^2 \\ \vdots \\ t^{N2} \end{Bmatrix} + \begin{bmatrix} G2^{1,1} & G2^{1,2} & \cdots & G2^{1,N1} \\ G2^{2,1} & G2^{2,2} & \cdots & G2^{2,N1} \\ \vdots & \vdots & & \vdots \\ G2^{N2,1} & G2^{N2,2} & \cdots & G2^{N2,N1} \end{bmatrix} \begin{Bmatrix} u^1 \\ u^2 \\ \vdots \\ u^{N1} \end{Bmatrix} + \begin{bmatrix} C2^{1,1} & C2^{1,2} & \cdots & C2^{1,N2} \\ C2^{2,1} & C2^{2,2} & \cdots & C2^{2,N2} \\ \vdots & \vdots & & \vdots \\ C2^{N2,1} & C2^{N2,2} & \cdots & C2^{N2,N2} \end{bmatrix} \begin{Bmatrix} \Delta u^1 \\ \Delta u^2 \\ \vdots \\ \Delta u^{N2} \end{Bmatrix}$$
$$= \begin{bmatrix} H2^{1,1} & H2^{1,2} & \cdots & H2^{1,N1} \\ H2^{2,1} & H2^{2,2} & \cdots & H2^{2,N1} \\ \vdots & \vdots & & \vdots \\ H2^{N2,1} & H2^{N2,2} & \cdots & H2^{N2,N1} \end{bmatrix} \begin{Bmatrix} t^1 \\ t^2 \\ \vdots \\ t^{N1} \end{Bmatrix} \tag{8.48}$$

The first item on the left-hand side of Eq. (8.48) is a known array consisting of the prescribed tractions on the crack surface. The second item on the left-hand side of Eq. (8.48) has an array consisting of the displacements, some of which are prescribed and others are unknown. Multiplying the known displacements with the corresponding coefficients, we can obtain the second known array. The item on the right-hand side of Eq. (8.48) has an array consisting of the tractions, some of which are prescribed and others are unknown. Multiplying the known tractions with the corresponding coefficients, we can obtain the third known array. According to the above descriptions, the new notations can be introduced as follows

- $(\boldsymbol{F}^t)_{N2\times 1}$ denotes the summation of the three known arrays,
- $(\boldsymbol{UT})_{N1\times 1}$ denotes the unknown quantities of displacements and tractions and $(\boldsymbol{A}^t)_{N2\times N1}$ denotes the corresponding coefficient matrix,
- $(\Delta \boldsymbol{U})_{N2\times 1}$ denotes the array of the CODs on the left-hand side of Eq. (8.48) and $(\boldsymbol{B}^t)_{N2\times N2}$ denotes the corresponding coefficient matrix.

Thus, Eq. (8.48) can be rewritten as

$$\boldsymbol{A}^t\boldsymbol{UT} + \boldsymbol{B}^t\Delta\boldsymbol{U} = \boldsymbol{F}^t \tag{8.49}$$

where the superscript t stands for the traction BIE.

Putting Eqs. (8.44) and (8.49) together, we have the simultaneous linear equations

$$\begin{bmatrix} \boldsymbol{A}^d & \boldsymbol{B}^d \\ \boldsymbol{A}^t & \boldsymbol{B}^t \end{bmatrix} \begin{Bmatrix} \boldsymbol{UT} \\ \Delta\boldsymbol{U} \end{Bmatrix} = \begin{Bmatrix} \boldsymbol{F}^d \\ \boldsymbol{F}^t \end{Bmatrix} \tag{8.50a}$$

or

$$\boldsymbol{AX} = \boldsymbol{F} \tag{8.50b}$$

where $\boldsymbol{A}$ is a coefficient matrix, $\boldsymbol{X}$ is the array containing the boundary unknowns u_i, t_i and Δu_i, and $\boldsymbol{F}$ is the array for known components. All boundary nodal values of u_i, t_i and Δu_i become known after solving Eq. (8.50).

For a crack in a multilayered medium of infinite extent, only the traction BIE (8.48) is needed and this can be simplified to

$$\begin{bmatrix} C2^{1,1} & C2^{1,2} & \cdots & C2^{1,N2} \\ C2^{2,1} & C2^{2,2} & \cdots & C2^{2,N2} \\ \vdots & \vdots & & \vdots \\ C2^{N2,1} & C2^{N2,2} & \cdots & C2^{N2,N2} \end{bmatrix} \begin{Bmatrix} \Delta u^1 \\ \Delta u^2 \\ \vdots \\ \Delta u^{N2} \end{Bmatrix} = -\begin{Bmatrix} t^1 \\ t^2 \\ \vdots \\ t^{N2} \end{Bmatrix} \tag{8.51}$$

When the tractions on the crack surfaces are prescribed and the CODs are unknown, Eq. (8.51) can be used to obtain the CODs on the crack surfaces.

Based on the numerical method presented above, computer programs have been written in FORTRAN to calculate the displacements, tractions and CODs of a multilayered dissimilar elastic solid containing cracks. The program incorporates all the functions proposed in this chapter and Chapter 4.

8.6 Numerical verifications

8.6.1 Calculation of stress intensity factors

Let (x', y', z') be a local Cartesian coordinate system attached to the crack front, shown in Fig. 8.6. The x'-axis is tangential to the crack front, the z'-axis is normal to the crack surface and the y'-axis is thus formed by the interaction of the plane normal to the crack front and the plane tangential to the crack plane. The CODs can be defined as

$$\Delta u_i(x', y', z') = u_i^+(x', y', z') - u_i^-(x', y', z'), \quad (i = x', y', z') \tag{8.52}$$

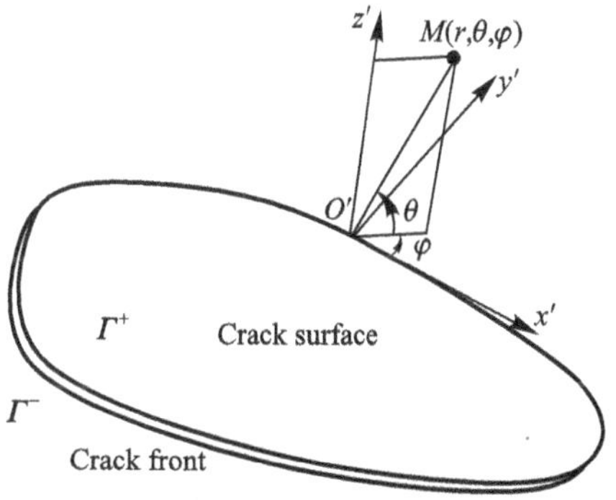

Fig. 8.6 Local coordinates at the crack front

where the superscripts + and – correspond to the crack surfaces, which are located at $z' = 0^+$ and $z' = 0^-$, respectively.

As described in Eq. (2.11), the SIFs can be calculated by using the displacements on the crack surface. For a crack in a FGM, stresses near the crack tip have a square-root singularity and the dimensionless angular functions are the same as those for homogeneous materials. The SIFs depend on the material gradients, external load and geometry. However, material gradients do not affect the order of the singularity and the angular functions. Thus, the SIFs of a crack in FGMs can be calculated by using the CODs as follows

$$K_{\mathrm{I}} = \frac{E}{4(1-\nu^2)}\sqrt{\frac{\pi}{2r}}\Delta u_{z'}\left(r, \theta = \pm\pi, \varphi = -\frac{\pi}{2}\right) \tag{8.53a}$$

$$K_{\mathrm{II}} = \frac{E}{4(1-\nu^2)}\sqrt{\frac{\pi}{2r}}\Delta u_{y'}\left(r, \theta = \pm\pi, \varphi = -\frac{\pi}{2}\right) \tag{8.53b}$$

$$K_{\mathrm{III}} = \frac{E}{4(1+\nu)}\sqrt{\frac{\pi}{2r}}\Delta u_{x'}\left(r, \theta = \pm\pi, \varphi = -\frac{\pi}{2}\right) \tag{8.53c}$$

where (r, θ, φ) are the spherical coordinates located at the crack front.

The CODs near the crack front can be used to obtain highly accurate SIF values. For this reason, the codes of calculating the CODs at any point of the element are written by using the CODs at nodes, the shape functions N_α and the function $g(\xi_1, \xi_2)$. In the FORTRAN code, the CODs of the points $\xi_1 = -1 + 10^{-5}$ or $\xi_2 = -1 + 10^{-5}$ along the side of the discontinuous element perpendicular to the crack front are presented for calculating the SIF values. It is obvious that more accurate SIF values can be obtained by using the CODs of these points.

8.6.2 The effect of different meshes and the coefficient *D* on the SIF values

In order to verify the accuracy of the proposed method and the corresponding program, two aspects of the analyses are carried out in the following.

Let us first consider the effect of different meshes on the SIF values. A square crack (with a side length $2c$) in a homogeneous region of infinite extent is studied and the crack surfaces are subjected to a uniform compressive stress p. Four element meshes, given in Table 8.1, are employed; the meshes 1 and 4 are shown in Fig. 8.7. Figure 8.8 illustrates the variations of the SIF values for the four different meshes. In this figure, it can be seen that the maximum value occurs at the middle of the square side and decreases to zero at the corners. The maximum SIF values ($K_{\mathrm{I}}/(p\sqrt{\pi c})$) predicted from the present DBEM are 0.7529, 0.7469, 0.7388 and 0.7564, respectively for meshes 1 to 4. These values compare well with the values of 0.74 given in Weaver (1977), 0.76 given in Murakami (1987), and the value 0.7626 given in Pan and Yuan (2000). In the following, mesh 4 is used to study the effect of the coefficient *D* on the SIF values.

Table 8.1 Meshes of the crack surface

Mesh no.	Element number				Node number
	Type I	Type II	Type III	Total number	
1	25	20	4	49	225
2	36	24	4	64	289
3	49	28	4	81	361
4	64	32	4	100	441

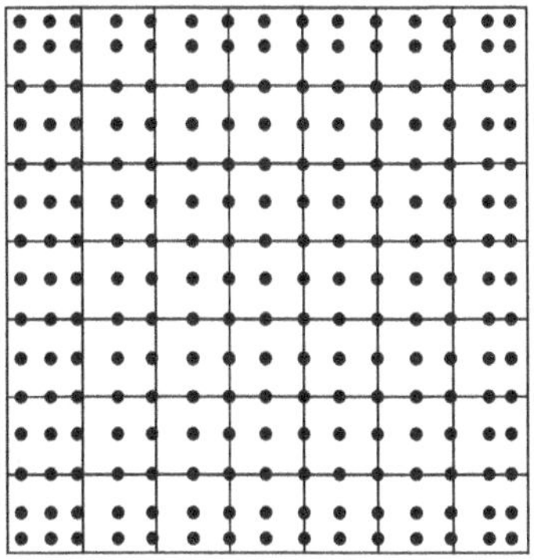

(a) Mesh 1: 49 elements and 225 nodes

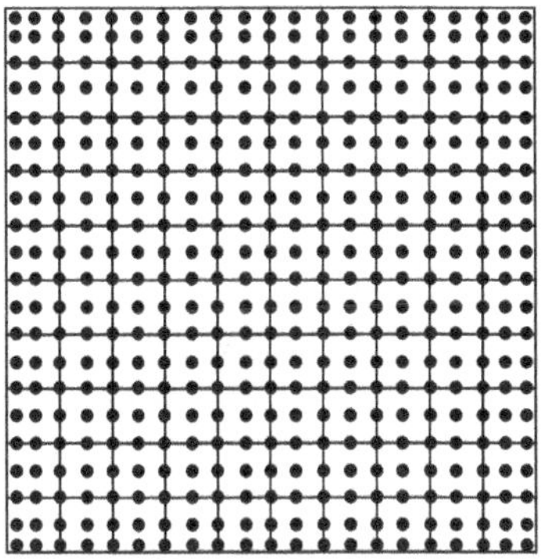

(b) Mesh 4: 100 elements and 441 nodes

Fig. 8.7 Meshes of the crack surface for a square crack

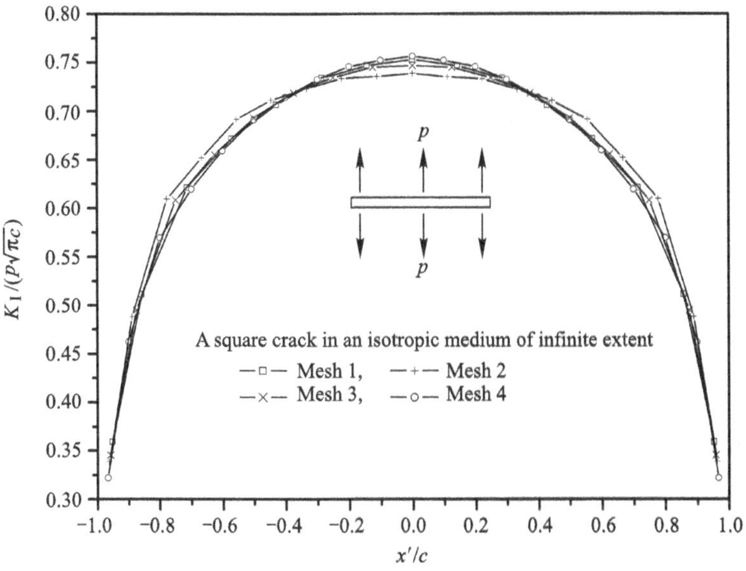

Fig. 8.8 Variations of the SIF values for four element meshes

In Eq. (4.64), the coefficient D is introduced to calculate the kernels T^{Y}_{ijk} and U^{Y}_{ijk}. In order to obtain the best value, five D values are analyzed and compared: $10^{-4}r$, $10^{-5}r$, $10^{-6}r$, $10^{-7}r$ and $10^{-8}r$, where r is the distance between the source and field points.

For the five D values, the maximum SIF values $(K_{\mathrm{I}}/(p\sqrt{\pi c}))$ at the middle of the square side of the crack are $0.7503, 0.7503, 0.7605, 0.7748$ and 0.6661, respectively. These results indirectly verify the best D value given in Tonon et al. (2001), that is, $D = 10^{-6}r$, which can be used to obtain highly accurate kernel values.

8.7 Summary

This chapter developed the dual boundary element method for the analysis of crack problems in the FGMs. The method is based on Yue's solution and uses a pair of BIEs, namely, the displacement and traction BIEs. The former is collocated exclusively on the uncracked boundary, and the latter is collocated only on one side of the crack surface. The displacements and/or tractions are used as unknown variables on the uncracked boundary and the relative crack opening displacement (i.e., displacement discontinuity) is treated as an unknown quantity on the crack surface. The $4 \sim 8$ noded isoparametric elements are used to discretize the uncracked boundary, and 9-noded continuous and discontinuous elements are used to discretize the crack surface. Discontinuous elements are introduced to capture the crack-tip behavior. The coordinate transform and the rigid-body motion method are used, respectively, to calculate the weakly and strongly singular integrals in the displacement BIE. Kutt's (1975a, b) numerical method is used to calculate the hypersingular integral in the traction BIE.

For a square-shaped crack in an infinite homogeneous domain, the SIFs obtained using the present formulation are in very good agreement with existing numerical results. It is shown that the proposed method can be used to analyze different cracks subjected to complex loads in graded materials and to model non-homogeneous solids with multiple interacting cracks. The above results have been published in Xiao and Yue (2008, 2011). In the next chapter, we will introduce some developments in the fracture mechanics analysis of FGMs using the proposed method.

Appendix 4 Finite-part integrals and Kutt's numerical quadrature

A4.1 Introduction

Broadly speaking, the adjective "singular" is applied to an integral equation if its kernel is in a class of functions that makes the usual Riemann or Lebesque definition impossible. The integral of a function $f(x)$ with weight $\ln|x-s|$ is weakly singular and has to be interpreted in the sense of Riemann. The integral of a function $f(x)$ with weight $1/(x-s)$ is Cauchy-type singular and can not be interpreted in the sense of Riemann and has to be interpreted in the Cauchy principal value sense.

The integral of a function $f(x)$ with the weight $1/(x-s)^{\lambda}$ ($\lambda \geqslant 2$) is Hadamard-type singular and can not be interpreted in Riemann and Cauchy principal value senses. It was Hadamard (1923) who first suggested that a practically useful definition can be made by considering some limit process. He called the resulting notion a finite-part integral. The finite-part integral of a function $f(x) \in c^2(a,b)$ with weight $1/(x-s)^2$ is defined as

$$f.\,p.\int_a^b \frac{f(x)}{(x-s)^2}\mathrm{d}x = \lim_{\varepsilon\to 0}\left\{\int_a^{s-\varepsilon}\frac{f(x)}{(x-s)^2}\mathrm{d}x + \int_{s+\varepsilon}^b \frac{f(x)}{(x-s)^2}\mathrm{d}x - \frac{2f(s)}{\varepsilon}\right\} \qquad \text{(A4.1)}$$

where $s \in (a,b)$ and $f.\,p.$ denotes the Hadamard finite-part integral.

The integral defined in Eq. (A4.1) can be interpreted and calculated directly in some specific instances. Considering $f(x) = 1$ in Eq. (A4.1), we have

$$\begin{aligned} f.\,p.\int_a^b \frac{1}{(x-s)^2}\mathrm{d}x &= \lim_{\varepsilon\to 0}\left\{\int_a^{s-\varepsilon}\frac{1}{(x-s)^2}\mathrm{d}x + \int_{s+\varepsilon}^b \frac{1}{(x-s)^2}\mathrm{d}x - \frac{2}{\varepsilon}\right\} \\ &= \lim_{\varepsilon\to 0}\left\{\left(\frac{1}{\varepsilon} - \frac{1}{s-a}\right) + \left(\frac{1}{s-b} + \frac{1}{\varepsilon}\right) - \frac{2}{\varepsilon}\right\} \\ &= -\left(\frac{1}{b-s} + \frac{1}{s-a}\right) \qquad \text{(A4.2)} \end{aligned}$$

Obviously, it is only possible to analytically calculate the integral (A4.1) for some specific functions $f(x)$. In the general case, the integral (A4.1) cannot be calculated analytically and may be calculated by using Kutt's numerical quadrature.

A4.2 Kutt's numerical quadrature

We present coefficients for the numerical computation of

$$\mathrm{I} := f.\,p.\int_s^r \frac{f(x)}{(x-s)^{\lambda}}\mathrm{d}x, \quad \lambda \text{ real and } \lambda \geqslant 1 \qquad \text{(A4.3)}$$

where $f(x)$ is a real function of the real variable x. We require $f(x) \in C$ in an interval containing $[s,r]$ and $f(x) \in C^{\lambda}$ in a neighborhood of s.

Kutt (1975a, b) developed the numerical quadrature of finite-part integrals: By considering the distribution of integral stations, Kutt's numerical quadrature can be classified into equispaced and Gaussian type formulae. We choose Kutt's numerical method with equispaced stations for calculating the integral (A4.3). When λ is an integral, we have the following formula for the integral (A4.3)

$$f.\,p.\int_s^r \frac{f(x)}{(x-s)^\lambda}\mathrm{d}x \approx (r-s)^{1-\lambda}\sum_{l=1}^{N}[w_l + c_l \ln|r-s|/(\lambda-1)!]f[(r-s)x_l+s] \quad \text{(A4.4)}$$

where w_l are the weights and c_l are the coefficients given by Kutt, both of which correspond to the equispaced stations $x_l = (l-1)/N$. For $N = 20$, the weights and coefficients for equispaced stations are expressed as $a_l \times 10^{n_l}$ and presented in Table A4.1.

Table A4.1 The weights and coefficients of Kutt's numerical quadrature

l	$w_l (= a_l \times 10^{n_l})$		$c_l (= a_l \times 10^{n_l})$	
	a_l	n_l	a_l	n_l
1	0.2674245013 1277256847	3	–0.7095479314 2873638229	2
2	–0.1731163597 4921140910	4	0.3800000000 0000000000	3
3	0.8173065888 6572005224	4	–0.1710000000 0000000000	4
4	–0.3116139931 4889658198	5	0.6460000000 0000000000	4
5	0.9391585238 2195858836	5	–0.1938000000 0000000000	5
6	–0.2259631011 9959965565	6	0.4651200000 0000000000	5
7	0.4400530960 0041874958	6	–0.9044000000 0000000000	5
8	–0.7011764845 8919860447	6	0.1439657142 8571428571	6
9	0.9208533719 3542211997	6	–0.1889550000 0000000000	6
10	–0.1000766004 7140010366	7	0.2052844444 4444444444	6
11	0.9007856450 5148905623	6	–0.1847560000 0000000000	6
12	–0.6699342676 1846350388	6	0.1374218181 8181818181	6
13	0.4092681653 0463205332	6	–0.8398000000 0000000000	5
14	–0.2032996907 9805166638	6	0.4174153846 1538461538	5
15	0.8082607885 7219770331	5	–0.1661142857 1428571428	5
16	–0.2510701124 4399961739	5	0.5168000000 0000000000	4
17	0.5869740773 8872411772	4	–0.1211250000 0000000000	4
18	–0.9704242001 1767101655	3	0.2011764705 8823529411	3
19	0.1009020480 0811358669	3	–0.2111111111 1111111111	2
20	–0.4795467029 0640279529	1	0.1052631578 9473684210	1

References

Chen JT, Hong HK. Review of dual boundary element methods with emphasis on hypersingular integrals and divergent series. ASME *Applied Mechanics Reviews*, 1999, 52: 17-32.

Cisilino A. Linear and Nonlinear Crack Growth Using Boundary Elements. Southampton: WIT Press, 1999.

Crouch SL, Starfield AM. Boundary Element Methods in Solid Mechanics: With applications in Rock Mechanics and Geological Engineering. London: George Allen and Unwin Publishers, 1983.

dell'Erba DN. Thermoelastic Fracture Mechanics Using Boundary Elements. Southampton: WIT Press, 2000.

Gray LJ, Martha LF, Ingraffea AR. Hypersingular integrals in boundary element fracture analysis. *International Journal for Numerical Methods in Engineering*, 1990, 29: 1135-1158.

Guiggiani M, Gigante A. A general algorithm for multi-dimensional Cauchy principal value integrals in the boundary element method. *ASME Journal of Applied Mechanics*, 1990, 57: 906-915.

Guiggiani M, Krishnasamy G, Rudolphi TJ, Rizzo FJ. A general algorithm for the numerical solution of hypersingular boundary integral equations. *ASME Journal of Applied Mechanics*, 1992, 59: 604-614.

Hadamard J. Lectures on Cauchy's Problems in Linear Partial Differential Equations. New York: Yale University Press, 1923.

Hong HK, Chen JT. Derivations of integral equations of elasticity. *ASCE Journal of Engineering Mechanics*, 1988, 114: 1028-1044.

Kutt HR. Quadrature Formulae for Finite-part Integrals, Special Report WISK 178. Pretoria: National Research Institute for Mathematical Sciences, 1975a.

Kutt HR. On the Numerical Evaluation of Finite-part Integrals Involving an Algebraic Singularity, Special Report WISK 179. Pretoria: National Research Institute for Mathematical Sciences, 1975b.

Mi Y, Aliabadi MH. Dual boundary element method for three-dimensional fracture mechanics analysis. *Engineering Analysis with Boundary Elements*, 1992, 10: 161-171.

Murakami Y. Stress Intensity Factors Handbook. Oxford: Pergamon Press, 1987.

Pan E. A general boundary element analysis of 2D linear elastic fracture mechanics. *International Journal of Fracture*, 1997, 88: 41-59.

Pan E, Amadei B. A 3-D boundary element formulation of anisotropic elasticity with gravity. *Applied Mathematical Modeling*, 1996, 20: 114-120.

Pan E, Yuan FG. Boundary element analysis of three-dimensional crack in anisotropic solids. *International Journal for Numerical Methods in Engineering*, 2000, 48: 211-237.

Portela A, Aliabadi MH, Rooke DP. Dual boundary element method: efficient implementation for cracked problems.*International Journal for Numerical Methods in Engineering*, 1992, 33: 1269-1287.

Qin TY, Chen WJ, Tang RJ. Three-dimensional crack problem analysis using boundary element method with finite-part integral. *International Journal of Fracture*, 1997, 84: 191-202.

Tonon F, Pan E, Amadei B. Green's functions and boundary element method formulation for 3D anisotropic media. *Computers and Structures*, 2001, 79: 469-482.

Watson JO. Hermitian cubic and singular elements for plane strain. Developments in Boundary Element Methods 4. Banerjee PK and Watson JO. Elsevier Appl. Sci. Publ., Barking, 1986, 1-28.

Weaver J. Three-dimensional crack analysis. *International Journal of Solids and Structures*, 1977, 13: 321-330.

Wilde A. A Dual Boundary Element Formulation for Three-dimensional Fracture Analysis. Southampton: WIT Press, 2000.

Xiao HT, Yue ZQ. Dual boundary element analysis of rectangular-shaped cracks in graded materials (in Chinese). *Chinese Journal of Theoretical and Applied Mechanics*, 2008, 40(6): 840-847.

Xiao HT, Yue ZQ. A three-dimensional displacement discontinuity method for crack problems in layered rocks. *International Journal of Rock Mechanics and Mining Sciences*, 2011, 48: 412-420.

Xu Y, Moran B, Belytschko T. Self-similar crack expansion method for three-dimensional crack analysis. ASME *Journal of Applied Mechanics*, 1997, 64: 729-737

Yue ZQ, Xiao HT, Pan E. Stress intensity factors of square crack inclined to interface of transversely isotropic bi-material. *Engineering Analysis with Boundary Elements*, 2007, 31: 50-56.

Chapter 9

Analysis of Rectangular Cracks in the FGMs

9.1 Introduction

Rectangular cracks are another type of cracks that are studied in fracture mechanics. But, because of their complex geometry, it is difficult to determine exact solutions of rectangular cracks. In some cases, rectangular cracks can be analyzed using numerical methods. Weaver (1977) utilized a system of integral equations that is defined over the crack surface only for the analysis of rectangular cracks in an isotropic medium of infinite extent. Murakami (1987) analyzed rectangular cracks with non-smooth surfaces in an isotropic medium of semi-infinite extent and presented the SIF values for rectangular cracks under the action of remotely uniform stresses. Bains et al. (1992) presented a new procedure for the determination of the weight functions that are used to evaluate the SIFs of a straight-fronted crack in a rectangular bar subjected to various loadings. Wen et al. (1998) developed the indirect BEM coupled with the weight function concept and presented the results of mixed-mode SIFs for a square bar with either a rectangular or a circular crack under a static load. Pan and Yuan (2000) presented a boundary element analysis of rectangular cracks in anisotropic solids. By using a special fundamental solution, Yue et al. (2007) developed the dual BEM to study a rectangular crack in a transversely isotropic bi-material. Kassir (1982), Itou (1983) and Qin and Noda (2003) also analyzed different types of rectangular cracks.

After nearly thirty years of efforts, there have been considerable developments in fracture mechanics for FGMs. However, to the authors' best knowledge, there has been no research into rectangular cracks in FGMs. In this chapter, we apply the proposed dual BEM derived in Chapter 8 for the analysis of rectangular cracks in FGMs under the action of different loads and discuss the effect of the non-homogeneity parameter, the thickness of the FGM interlayer and the relative position of the crack in the FGM system on the fracture properties.

9.2 A square crack in FGMs of infinite extent

9.2.1 General

When analyzing crack problems in FGMs, it is often assumed that the elastic properties of the FGMs are approximated by exponential or power functions. Although this assumption is made for the mathematical formulation, it is certainly of significance in understanding the fracture mechanics of FGMs. In this chapter, a FGM system with two half-spaces and the interlayer bonded with the half-spaces, as shown in Fig. 9.1, will be analyzed. It is assumed that the elastic modulus of the FGM interlayer can be approximated by Eq. (6.1a) and the Poisson's ratio is kept constant, i.e., $\nu = 0.3$. For convenient analyses, two coordinate systems, i.e., the global coordinates $Oxyz$ and the local coordinates $O'x'y'z'$, are established. In the coordinate system $O'x'y'z'$, the coordinate plane $O'x'y'$ is on the crack surface and the z'-axis is perpendicular to the crack surface. Also the z-axis and z'-axis have an angle θ. The sides AB and CD of the crack are parallel to the FGM interlayer.

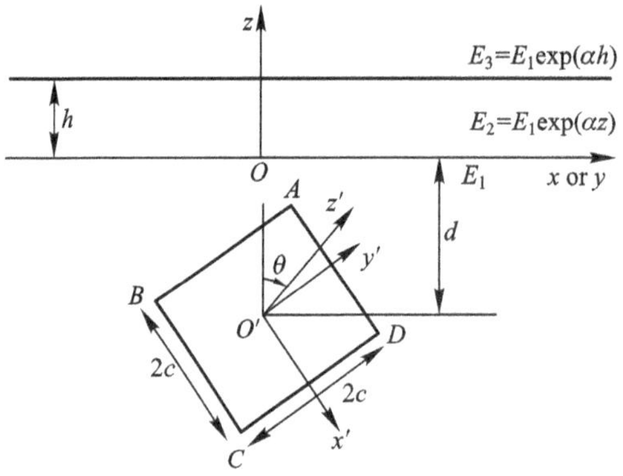

Fig. 9.1 Geometry of a rectangular crack ($2c \times 2c$) in a FGM. The normal direction of the crack has an angle θ with the z direction

The layered discretization technique is applied to approximate the variation in the elastic properties of the medium with depth. In Fig. 9.1, we take $\theta = 0°$, i.e., the square crack surfaces are parallel to the interlayer of the FGM. The crack surfaces are subjected to uniform compressive stresses. To analyze the crack problem, we use the following parameters: $h/c = 0.5$, $\alpha c = 2$ and $d = c$. Mesh 4, shown in Fig. 8.7b, is used for this square crack.

Figure 9.2 illustrates the variations in the mode I and II SIF values with the layer discretization number. We have chosen four discretization numbers, i.e., $n = 5, 10, 15, 20$. For a crack parallel to the FGM interlayer, the mode I and II deformations are coupled together. It can be observed from Fig. 9.2, that as the layer discretization number n increases, the SIF values approach stable values. In the ensuing, we will use a large number of piecewise homogeneous layers, i.e., $n = 20$, for the analysis of a square crack in a FGM.

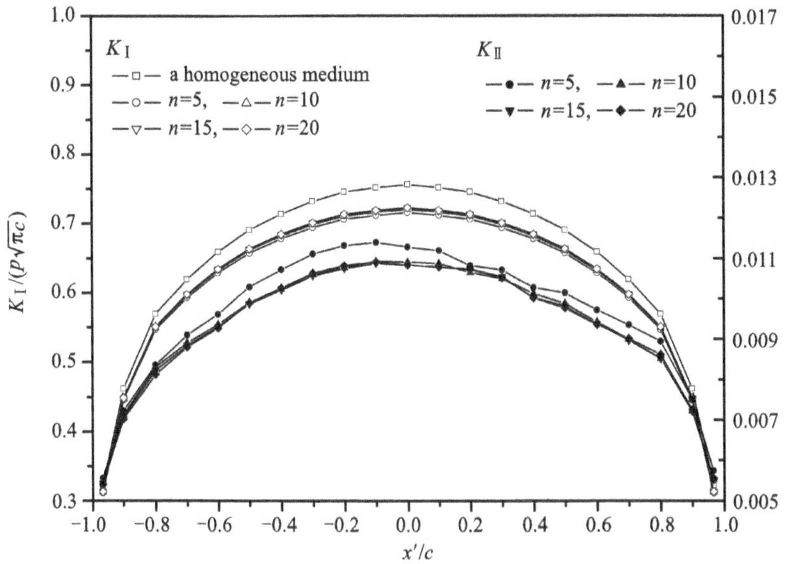

Fig. 9.2 Variations of the SIF values with layer number n of the discretized FGM interlayer ($\alpha c = 2, \theta = 0°$)

9.2.2 A square crack parallel to the FGM interlayer

In Fig. 9.1, the crack is located in material 1 that is bonded to the FGM interlayer; the crack surfaces are subjected to a uniform compressive stress p. We let $\alpha c = 2$ and –2 and take $\theta = 0°$, i.e., the crack surfaces are parallel to the FGM interlayer. Figures 9.3 ~ 9.6 illustrate the variations of the SIF values with the crack distance d to the FGM interlayer. In these figures, K_I and K_{II} are symmetrical to the x'-axis. Table 9.1 presents the mode I and II SIF values along the side of the crack $y' = -c$ ($x' < 0$).

Table 9.1 The SIF values along the crack front $y' = -c$ for a crack parallel to the FGM interlayer

x'/c	$K_I/(p\sqrt{\pi c})$				$K_{II}/(p\sqrt{\pi c})$			
	$\alpha c = 2$		$\alpha c = -2$		$\alpha c = 2$		$\alpha c = -2$	
	d/c=0.5	0.1	0.5	0.1	0.5	0.1	0.5	0.1
–0.96667	0.309459	0.302081	0.346046	0.348807	0.010377	0.01638	–0.01481	–0.02267
–0.5	0.640905	0.601978	0.785909	0.838126	0.022462	0.042103	–0.03346	–0.06849
–0.3	0.673369	0.63274	0.843534	0.901429	0.024838	0.046142	–0.03725	–0.0769
–0.1	0.688848	0.647753	0.871885	0.931778	0.026021	0.048214	–0.03933	–0.08113
0.0	0.692352	0.651432	0.877141	0.937371	0.026068	0.048454	–0.03955	–0.08157

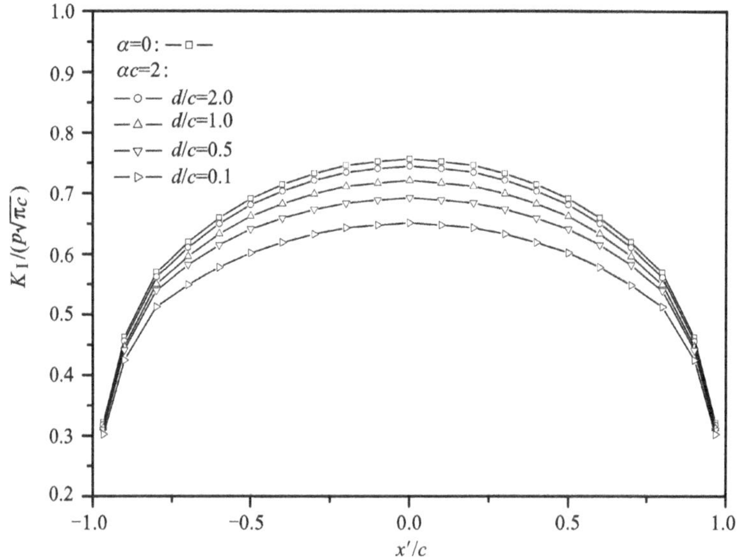

Fig. 9.3 Variations of $K_I/(p\sqrt{\pi c})$ with the crack distance d ($\alpha c = 2$, $\theta = 0°$)

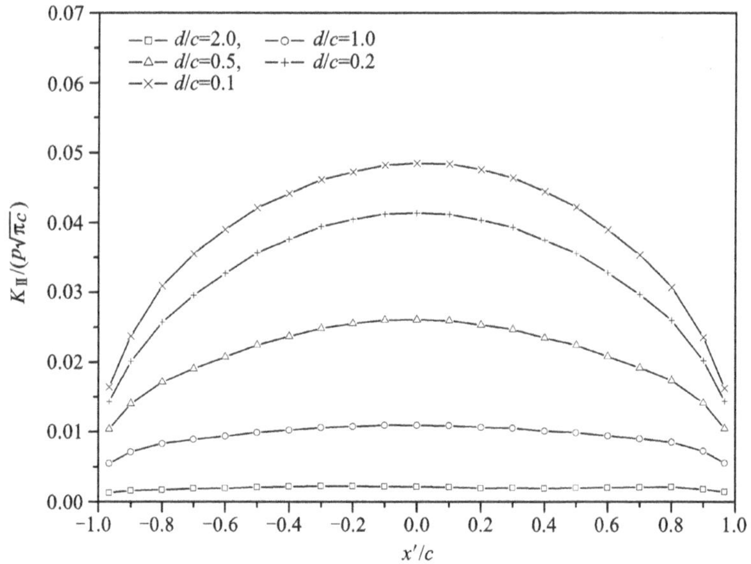

Fig. 9.4 Variations of $K_{II}/(p\sqrt{\pi c})$ with the crack distance d ($\alpha c = 2$, $\theta = 0°$)

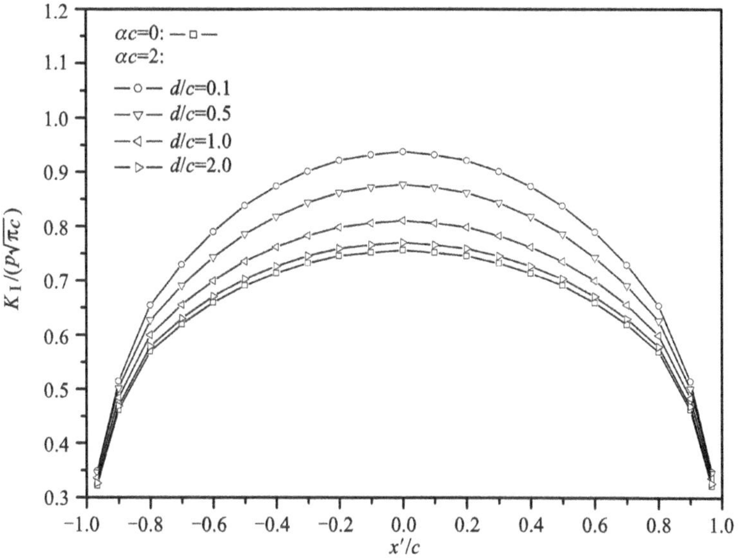

Fig. 9.5 Variations of $K_{\mathrm{I}}/(p\sqrt{\pi c})$ with the crack distance d ($\alpha c = -2$, $\theta = 0°$)

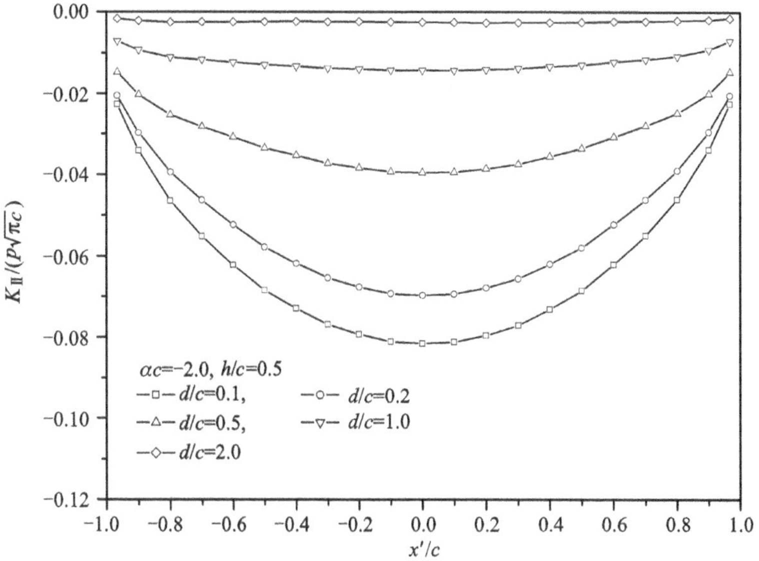

Fig. 9.6 Variations of $K_{\mathrm{II}}/(p\sqrt{\pi c})$ with the crack distance $d(\alpha c = -2$, $\theta = 0°)$

For a crack under the action of a compressive stress in FGM, crack opening displacements exist and crack sliding displacements along the crack surfaces also appear. Hence the mode I and II SIF values are coupled together. For $\alpha c = 2$, the K_{I} values decrease and the K_{II} values increase as the crack distance to the FGM interlayer decreases. For $\alpha c = -2$, the K_{I} values increase and the K_{II} values decrease as the crack distance to the FGM interlayer decreases. For a crack in a more compliant body, $\alpha > 0$, the crack opening is constrained by the stiffer upper half-space and the constraints becomes more pronounced as the distance of the crack to the FGM interlayer decreases. On the other hand, for $\alpha < 0$, the crack is located in a stiffer solid, and the compliant upper body tends to increase the crack opening. For $\alpha < 0$ and $\alpha > 0$, the upper and lower surfaces slide in different directions along the crack surfaces.

9.2.3 A square crack having a 45° angle with the FGM interlayer

We take $\theta = 45°$, i.e., the normal direction of the crack surface has an angle of 45° with the FGM interlayer. Consider that the crack surfaces are subjected to a uniform compressive stress p. In order to plot the SIF values along the crack lines, a line coordinate L is used to measure the crack front lines from AB, BC, CD to DA (Fig. 9.1). The line coordinate L starts at the corner point A of the square crack, i.e., $L/c = 0$. It increases along the lines AB, BC, CD to DA. Correspondingly, L/c increases from $0 \sim 2$, from $2 \sim 4$, from $4 \sim 6$, and from $6 \sim 8$, respectively.

In Fig. 9.1, the sides $x' = \pm c$ of the square crack are parallel to the FGM interlayer and the sides $y' = \pm c$ are at a 45° angle to the FGM interlayer. Considering the geometrical dimensions and positions of the crack in the FGM, it can be seen that there are the same SIF values at the crack front along the sides $y' = \pm c$. Figures 9.7 and 9.8 give the variations in the SIF values with the crack distance to the FGM interlayer, where $\alpha c = 2$ and –2; Table 9.2 presents the SIF values at the crack front $x' = -c$ $(y' < 0)$ and $y' = c(x' < 0)$. Note that the case $d/c = 0.7071$ indicates that the crack side $x' = -c$ contacts the bottom side $z = 0$ of the FGM interlayer.

Table 9.2 The SIF values $(K_{\mathrm{I}}/(p\sqrt{\pi c}))$ along the crack front for crack surfaces at an angle of 45° to the FGM interlayer

y'/c or x'/c	Crack front $x' = -c$				Crack front $y' = -c$			
	$\alpha c = 2$		$\alpha c = -2$		$\alpha c = 2$		$\alpha c = -2$	
	d/c=1.0	0.7071	1.0	0.7071	1.0	0.7071	1.0	0.7071
–0.96667	0.31009	0.29517	0.33904	0.35634	0.31016	0.29758	0.33899	0.35364
–0.5	0.65112	0.5918	0.74988	0.82272	0.66147	0.64135	0.73562	0.76505
–0.3	0.68657	0.62316	0.79969	0.88589	0.70344	0.68064	0.77658	0.80983
–0.1	0.70356	0.63825	0.82389	0.91876	0.72609	0.70471	0.79371	0.82514
0.0	0.70759	0.63865	0.82894	0.92758	0.73209	0.71227	0.79604	0.82545

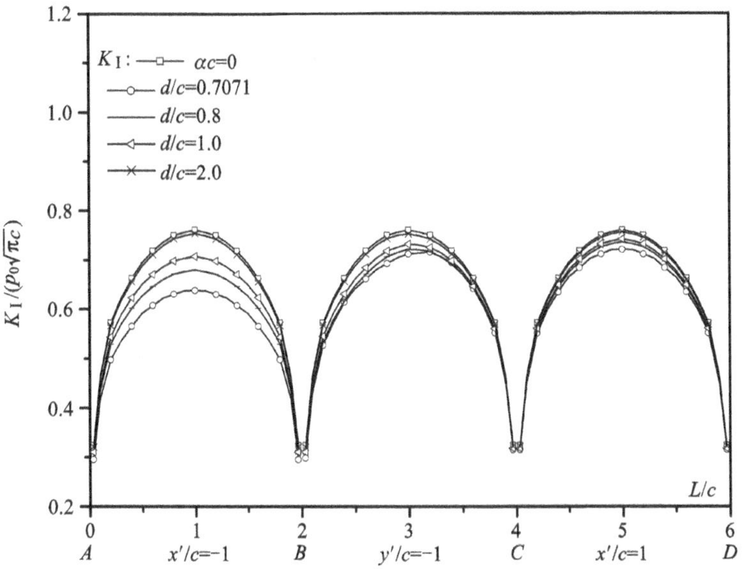

Fig. 9.7 Variations of $K_I/(p\sqrt{\pi c})$ with the crack distance d ($\alpha c = 0, 2, \theta = 45°$)

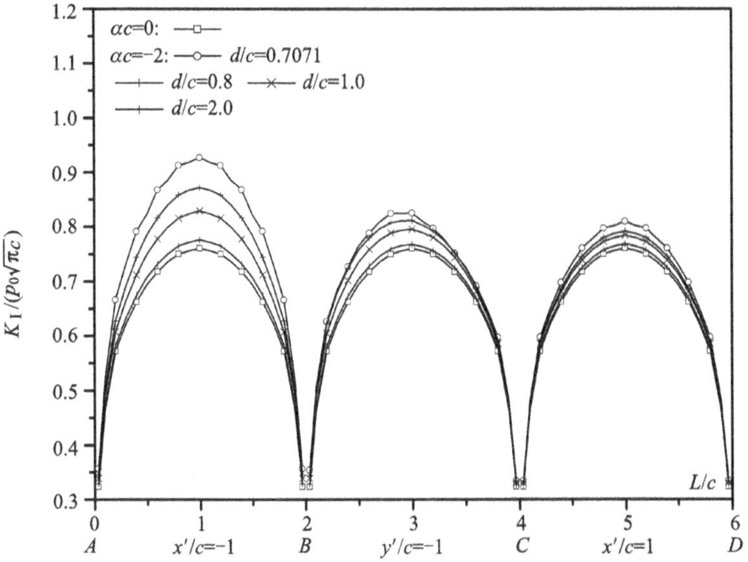

Fig. 9.8 Variations of $K_I/(p\sqrt{\pi c})$ with the crack distance d ($\alpha c = 0, -2, \theta = 45°$)

When $\alpha c = 2$, the stiffer upper half-space, i.e., materials 2 and 3, constrains the crack opening. As the crack approaches the FGM interlayer, the SIF values decrease. The FGM interlayer exerts a more obvious influence on the side $x' = -c$ than on the side $x' = c$. As the crack front moves away from the FGM interlayer, the influence of the upper solid becomes weaker. For $\alpha c = -2$, the compliant upper half-space tends to increase the crack opening and the SIF values increase correspondingly. As was the case for $\alpha c = 2$, the FGM interlayer exerts a more obvious influence on the side $x' = -c$ than on the side $x' = c$. As the crack front moves away from the FGM interlayer, the influence of the upper solid becomes negligible.

For the same reasons, the K_{II} values become small as the crack moves away from the FGM interlayer. For $d/c = 2$, the absolute values of K_{II} are less than 0.01, which approaches the error of the numerical method presented in Chapter 8. As the angle between the crack surfaces and the interlayer becomes larger, part of the crack front moves away from the interlayer and the influence of the interlayer on these sides becomes weaker. However, because a numerical error exists, it is difficult to determine the K_{II} values and therefore we have not presented the K_{II} values.

9.2.4 A square crack perpendicular to the FGM interlayer

Let $\theta = 90°$, i.e., the crack surfaces are perpendicular to the FGM interlayer and subject to a uniform compressive stress p. For this case, there is only the mode I deformation. Figures 9.9 and 9.10 illustrate the variations in the K_{I} values with changes in the crack distance d to the FGM interlayer.

The sides $x' = \pm c$ of the square crack are parallel to the FGM interlayer and the sides $y' = \pm c$ are perpendicular to it; because the geometrical dimensions and loadings of the cracked body are symmetrical to the coordinate plane $O'x'z'$, the crack fronts of the sides

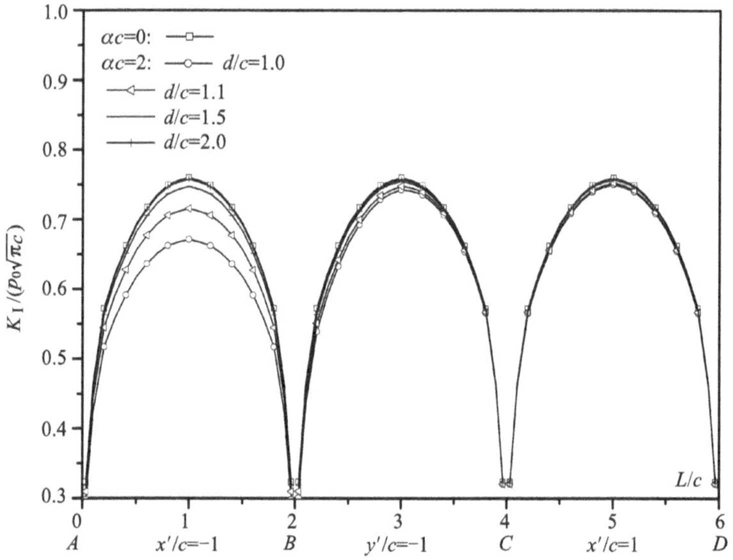

Fig. 9.9 Variations of $K_{\mathrm{I}}/(p\sqrt{\pi c})$ with the crack distance d ($\alpha c = 2$, $\theta = 90°$)

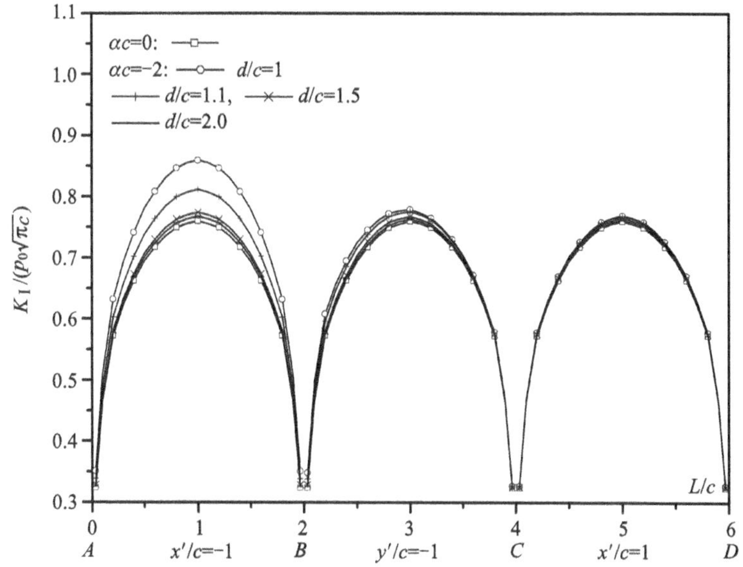

Fig. 9.10 Variations of $K_\mathrm{I}/(p\sqrt{\pi c})$ with the crack distance d ($\alpha c = -2$, $\theta = 90°$)

$y' = \pm c$ have the same SIF values. Hence, only the SIF values of the crack front $y' = -c$ are presented in Figs. 9.9 and 9.10. For $d = c$, the side $x' = -c$ contacts the bottom surface, i.e., $z = 0$, of the interlayer. It can be found that the case $\theta = 90°$ has the same variation tendencies in the SIF values. This is also found in the case where $\theta = 45°$. The FGM interlayer exerts a less obvious influence on the crack front $x' = c$ away from the interlayer.

9.3 A square crack in the FGM interlayer

In Fig. 9.11, a square crack is located in the FGM interlayer and the normal direction of the crack surface $z = 0^-$ is the same as the gradient direction of the FGM. Let $\alpha c = 2$ and $h/c = 0.5$. The five cases, $d/c = 0.05, 0.15, 0.25, 0.35, 0.45$, are analyzed. Mesh 4, shown in Fig. 8.7b, is used for this square crack.

Figures 9.12 and 9.13 show the variations in the K_I and K_II values as the crack distance d to the FGM interlayer changes. The corresponding SIF values are listed in Table 9.3. It can be found that as d increases, the K_I and K_II values increase. From $d = 0$ to $d = h$, the elastic modulus on the crack position becomes large and the constraint on the crack opening becomes weak so that the K_I values increase. At the same time, the relative sliding of the crack surfaces along the x and y directions occurs so that the K_II values are not equal to zero and increase as d increases. In discussing the plane crack with a normal direction along the gradient one of the FGMs, Delale et al. (1988) reached the same conclusions.

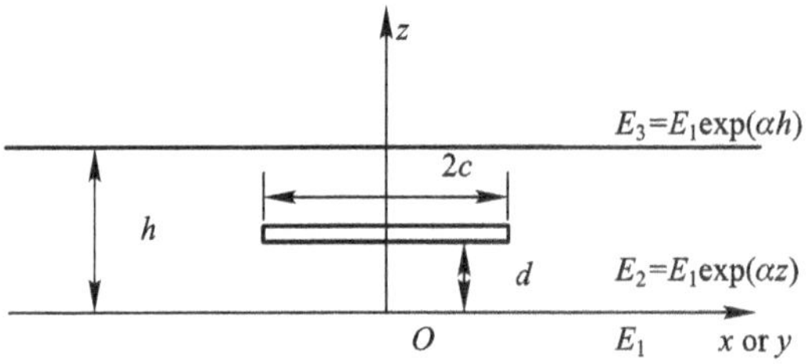

Fig. 9.11 Geometry of a rectangular crack $(2c \times 2c)$ in the FGM interlayer

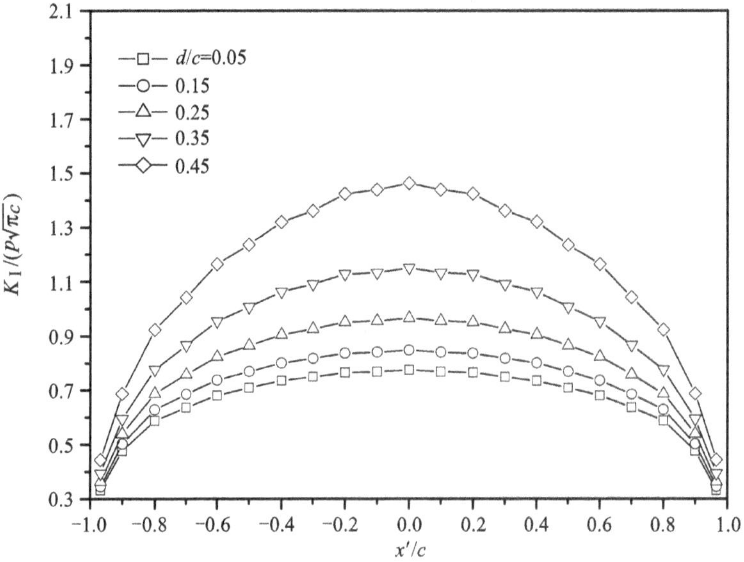

Fig. 9.12 Variations of $K_{\mathrm{I}}/(p\sqrt{\pi c})$ with the crack distance $d(\alpha c = 2,\, h/c = 0.5)$

Table 9.3 The SIF values along the crack front $y' = -c$ for a crack in the FGM interlayer $(\alpha c = 2)$

x'/c	$K_{\mathrm{I}}/(p\sqrt{\pi c})$				$K_{\mathrm{II}}/(p\sqrt{\pi c})$			
	d/c=0.05	0.15	0.35	0.45	d/c=0.05	0.15	0.35	0.45
–0.96667	0.331305	0.34405	0.393429	0.443214	0.025115	0.0291	0.036255	0.040577
–0.5	0.708009	0.76925	1.006743	1.236308	0.071905	0.088505	0.123238	0.143138
–0.3	0.749307	0.817576	1.090587	1.361846	0.077154	0.094836	0.135446	0.160237
–0.1	0.76867	0.840494	1.132501	1.439826	0.077868	0.095742	0.138898	0.166073
0.0	0.774953	0.847567	1.150289	1.463176	0.07797	0.095891	0.138941	0.166147

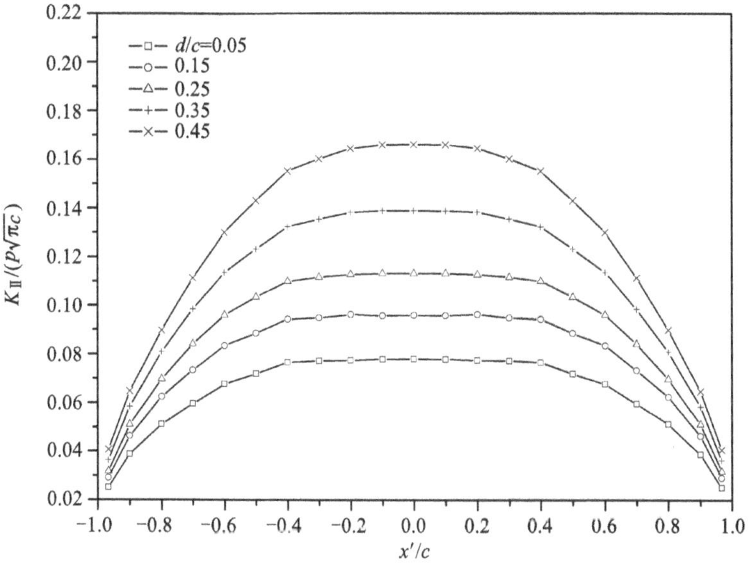

Fig. 9.13 Variations of $K_{II}/(p\sqrt{\pi c})$ with the crack distance $d(\alpha c = 2, h/c = 0.5)$

Referring to the results for the case where $\alpha c = 2$ and $h/c = 0.5$, the variations in the SIF values for the case $\alpha c = -2$ and $h/c = 0.5$ can be obtained easily; in this, the crack position d should be calculated from the plane $z/c = 0.5$, the K_I values are positive and the K_{II} values are negative. The K_I values and the absolute values of K_{II} are the same as for the case $\alpha c = 2$ and $h/c = 0.5$.

9.4 A rectangular crack in FGMs of infinite extent

9.4.1 General

Consider now a rectangular crack in a FGM of infinite extent. The crack has the dimensions of the long side $4c$ and the short side $2c$ and is subjected to a uniform compressive stress p. The crack surface is discretized into a boundary mesh with the 60 type I elements, the 34 type II elements and the 4 type III elements and 435 nodes. The rectangular crack in a homogeneous medium of infinite extent is analyzed to verify the accuracy of the proposed method and the mesh. Numerical analysis shows that $K_I/(p\sqrt{\pi c}) = 0.7934$ for the middle point along the short side of the crack front and $K_I/(p\sqrt{\pi c}) = 0.88307$ for the middle point along the long side of the crack front. When compared with the square crack, it can be found that, as the ratio of the long side to the short side increases, the K_I values increase at the crack front and this variation is more obvious in the middle of the sides. It can be predicted that, as the ratio of the long side to the short increases, the

$K_I/(p\sqrt{\pi c})$ values at the middle point of the long side approaches one that is the K_I value for a plane crack in an infinite plane under the action of a uniform compressive stress on the crack surface. Weaver (1977) also discussed the variations of the SIF values as the ratio of the long side to the short increases and drew the same conclusions.

In the following, we will discuss the variations of the SIF values for rectangular cracks parallel to and perpendicular to the FGM interlayer and subjected to a uniform compressive stress p. Similar to the line coordinate system discussed in Section 9.2.3, a line coordinate system L is established along the crack front from AB, BC, CD to DA. Shown in Fig. 9.14, the origin of the new coordinate system is located at the corner point A of the crack front and the L/c values of other corner points (B, C and D) are 2, 6 and 8, respectively.

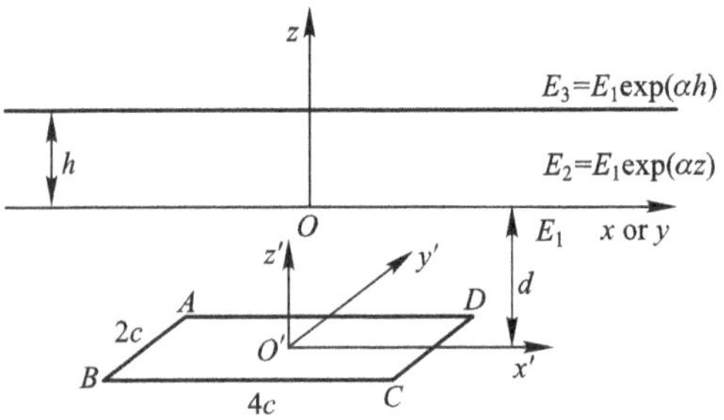

Fig. 9.14 Geometry of a rectangular crack $(2c \times 4c)$ parallel to the FGM interlayer

9.4.2 A rectangular crack parallel to the FGM interlayer

In Fig. 9.14, the rectangular crack is parallel to the FGM interlayer. The mode I and II deformations of this crack are coupled together. Figures 9.15 ~ 9.18 illustrate the variations of the SIF values with the crack distance d to the interlayer, where $\alpha c = 2$ and –2.

For $\alpha c = 2$, as the crack approaches the interlayer, the K_I values decrease and the K_{II} values increase. This is because the stiffer FGM interlayer constrains the crack opening and causes sliding of the crack surfaces. For $\alpha c = -2$, the K_I values are positive and the K_{II} values negative; as the crack approaches the interlayer, the K_I values increase and the K_{II} values decrease. This is because the compliant FGM interlayer increases the crack opening and causes sliding of the crack surfaces. For $\alpha < 0$ and $\alpha > 0$, the two crack surfaces slide along different directions.

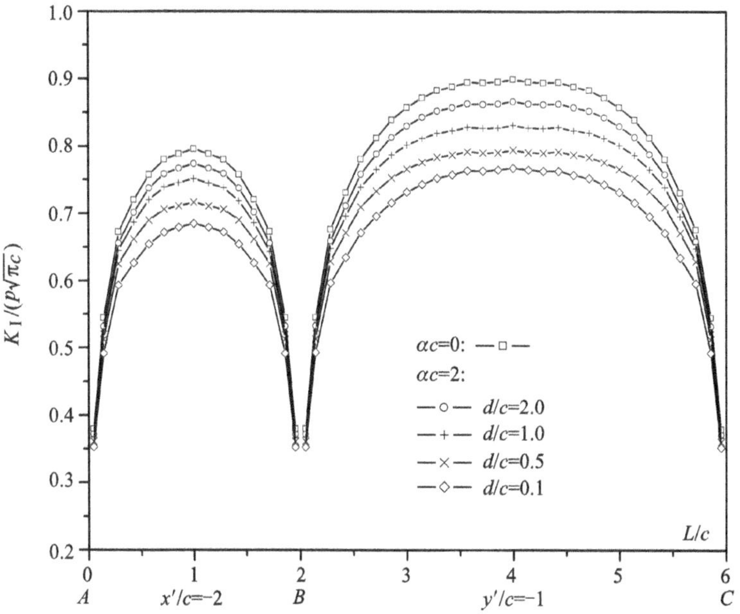

Fig. 9.15 Variations of $K_{\text{I}}/(p\sqrt{\pi c})$ with the crack distance d ($\alpha c = 0$, 2)

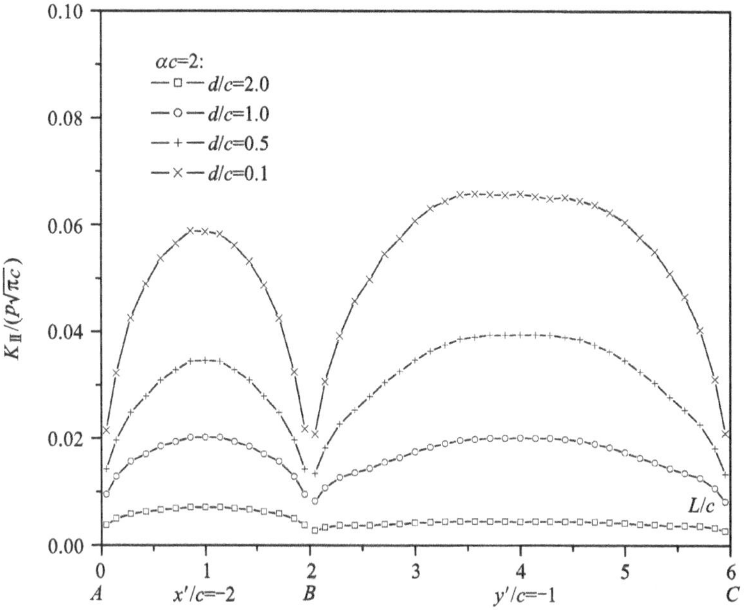

Fig. 9.16 Variations of $K_{\text{II}}/(p\sqrt{\pi c})$ with the crack distance d ($\alpha c = 2$)

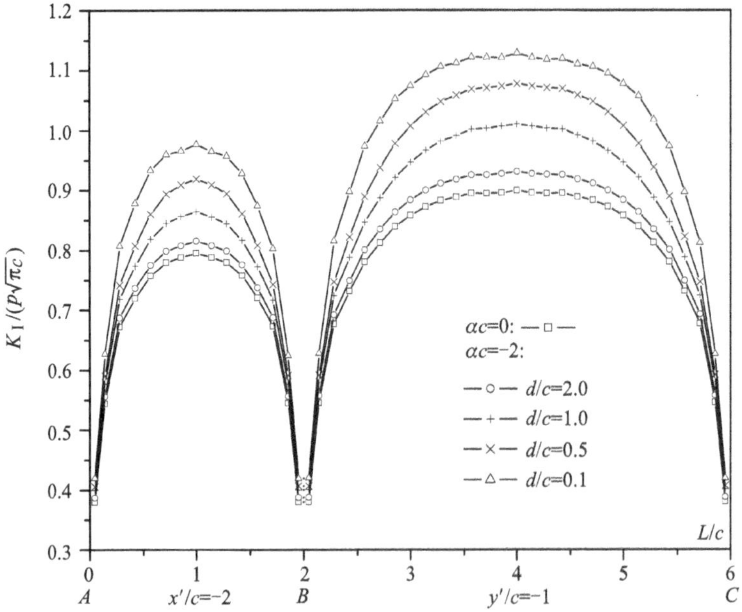

Fig. 9.17 Variations of $K_I/(p\sqrt{\pi c})$ with the crack distance d ($\alpha c = 0, -2$)

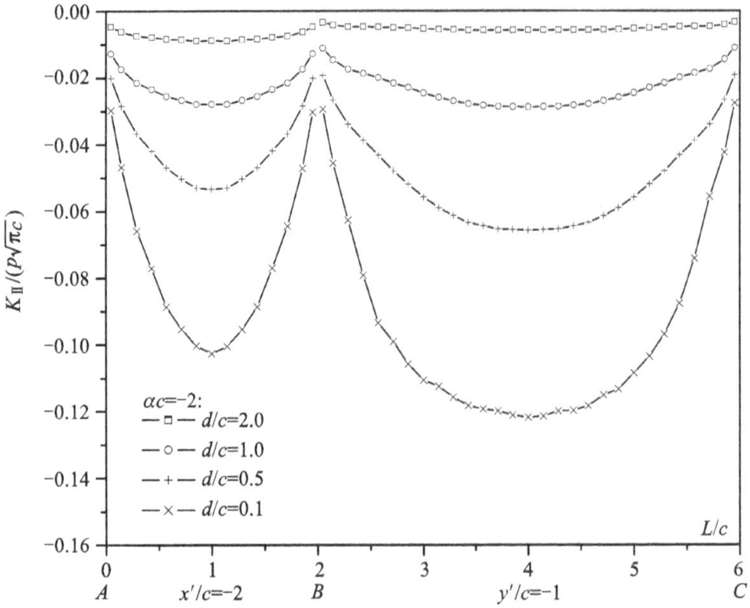

Fig. 9.18 Variations of $K_{II}/(p\sqrt{\pi c})$ with the crack distance d ($\alpha c = 0, -2$)

9.4.3 A rectangular crack with long sides perpendicular to the FGM interlayer

In Fig. 9.19, the long sides of a rectangular crack are perpendicular to the FGM interlayer. There is only mode I deformation for this crack problem. Figures 9.20 and 9.21 illustrate the variations of the SIF values as the crack distance d to the interlayer changes, where $\alpha c = 2$ and –2.

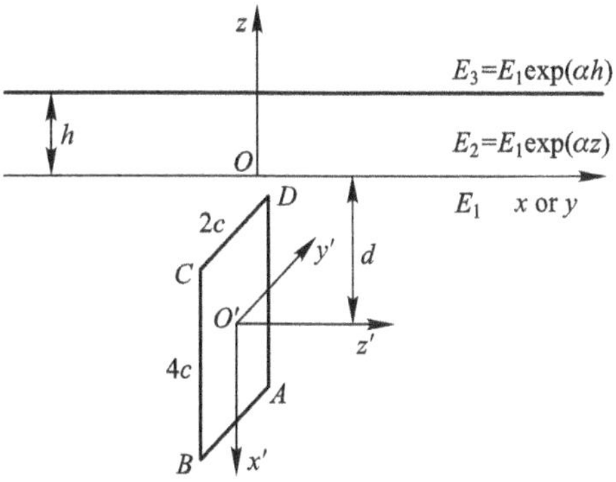

Fig. 9.19 Geometry of a rectangular crack $(4c \times 2c)$ perpendicular to the FGM interlayer

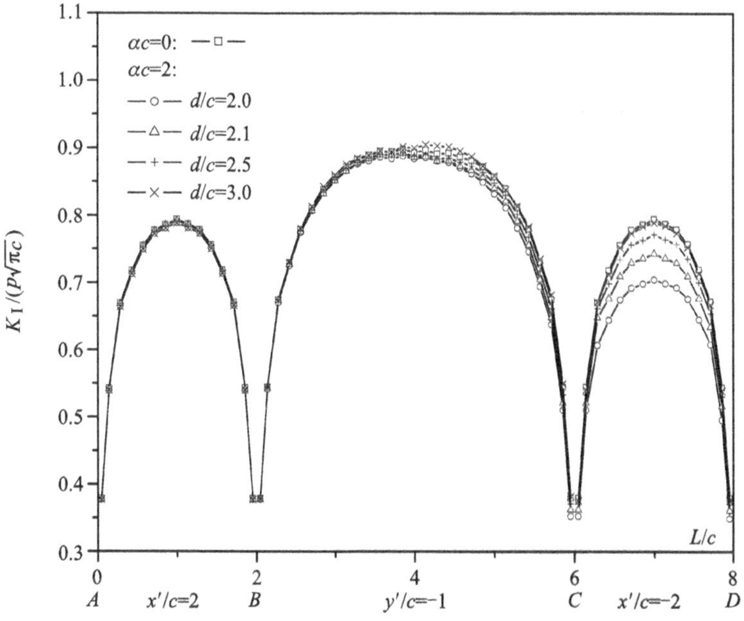

Fig. 9.20 Variations of $K_I/(p\sqrt{\pi c})$ with the crack distance d $(\alpha c = 0, 2)$

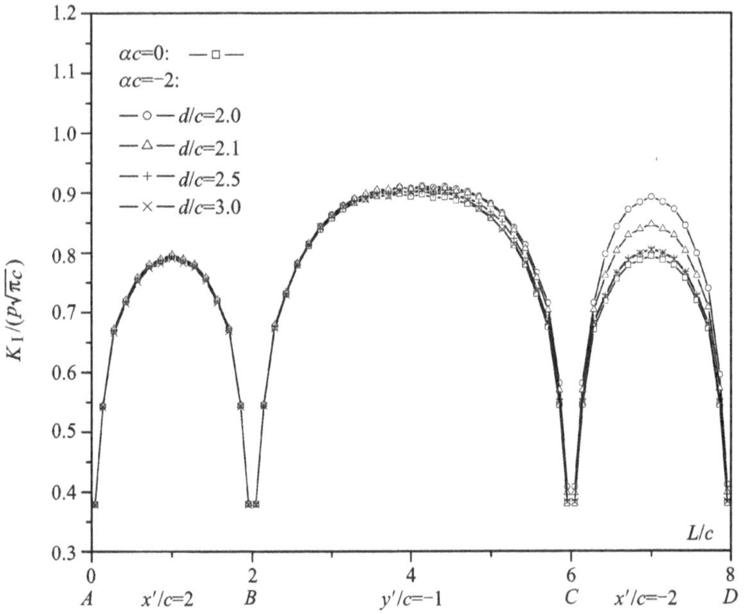

Fig. 9.21 Variations of $K_{\mathrm{I}}/(p\sqrt{\pi c})$ with the crack distance d ($\alpha c = 0, -2$)

For $\alpha c = 2$, the SIF values along the side $x'/c = -2$ are less than those along $x'/c = 2$ because of the constraint imposed by the stiffer FGM interlayer; this constraint is more obvious around the middle of the shorter side. For $\alpha c = -2$, the compliant FGM interlayer increases the crack opening; therefore, the SIF values along the side $x'/c = -2$ are larger than those along $x'/c = 2$. This change is more obvious around the middle of the short side. For $\alpha c = 2$ and –2, along the long side of the crack, the SIF values of the crack front adjacent to the interlayer vary more obviously than those away from it.

9.4.4 A rectangular crack with short sides perpendicular to the FGM interlayer

In Fig. 9.22, the short sides of a rectangular crack are perpendicular to the FGM interlayer. There is mode I deformation for this crack problem. Figures 9.23 and 9.24 illustrate the variations of the SIF values with the crack distance d to the interlayer where $\alpha c = 2$ and –2.

For $\alpha c = 2$, the stiffer FGM interlayer constrains the crack opening and the SIF values along the side $x'/c = -1$ are less than those along $x'/c = 1$; this change is more obvious around the middle of the long side. For $\alpha c = -2$, the more compliant FGM interlayer increases the crack opening and the SIF values along the side $x'/c = -1$ are obviously larger than the ones along $x'/c = 1$; this change is more obvious around the middle of the

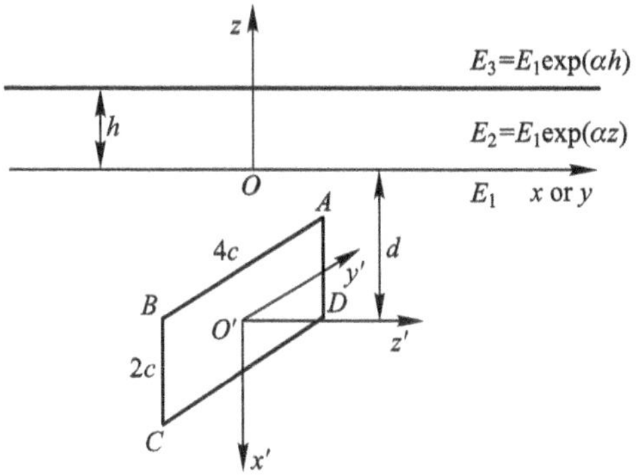

Fig. 9.22 Geometry of a rectangular crack ($2c \times 4c$) perpendicular to the FGM interlayer

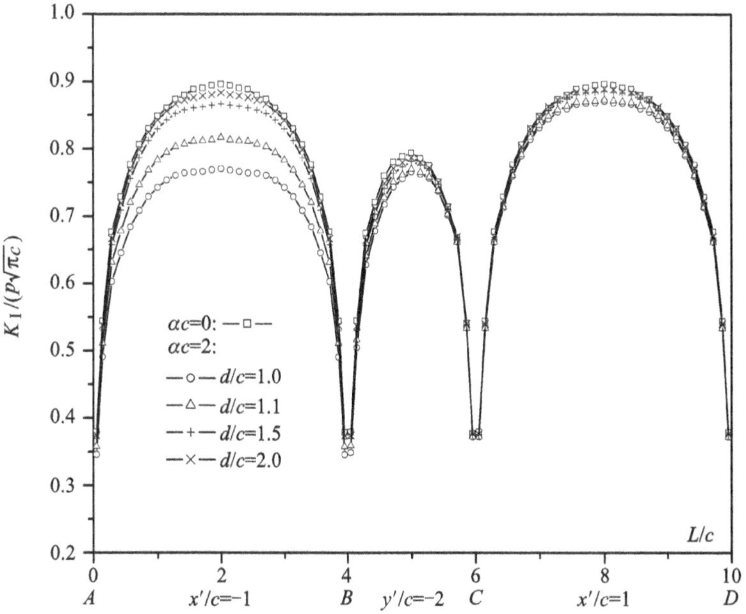

Fig. 9.23 Variations of $K_I/(p\sqrt{\pi c})$ with the crack distance d ($\alpha c = 0, 2$)

long side. For $\alpha c = 2$ and –2, along the short side of the crack, the SIF values of the crack front adjacent to the interlayer vary more obviously than the ones further away from it.

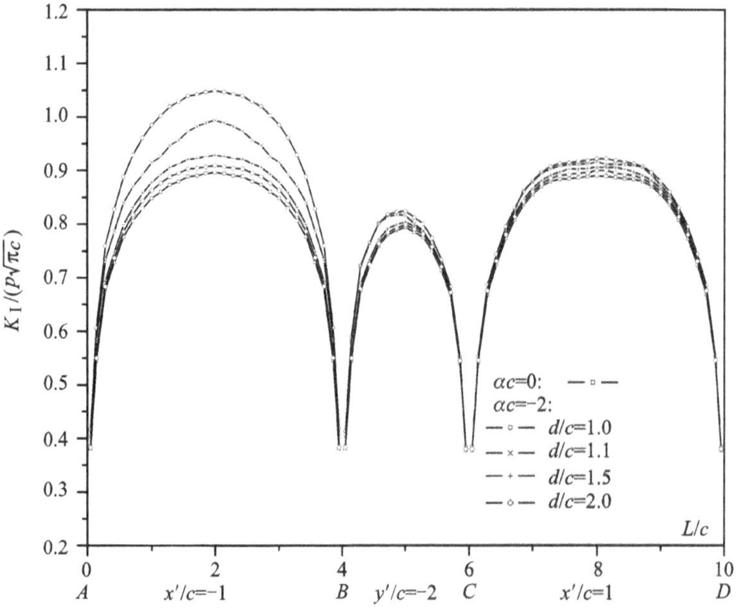

Fig. 9.24 Variations of $K_I/(p\sqrt{\pi c})$ with the crack distance d ($\alpha c = 0, -2$)

9.5 A square crack in a FGM of finite extent

Let us consider a square crack in a cube composed of three bonded materials, shown in Fig. 9.25. The finite cube with a square crack is subjected to a uniform tension p at the top and bottom faces along the direction of the z-axis. Of the three materials, materials 1 and 3 are homogeneous media and material 2 is a graded medium. The elastic modulus of the materials is approximated by Eq. (6.1a). Let $\alpha c = 2$ and –2. The Poisson's ratio of the FGM is kept constant and is equal to 0.3. The square crack is located in material 1 and is parallel to the FGM interlayer. The cube has a height $2H$ and a width W. In particular, $W = 4c$, $W = H$ and $h = 0.5c$. The side length of the square crack is $2c$. The non-crack boundaries are discretized into a mesh system with 250 eight-node isoparametric elements and 752 nodes. The square crack is discretized using the same mesh as that in Fig. 8.7.

The displacement BIE (8.6) is not applied when analyzing a square crack in an infinite domain; however, the displacement BIE (8.6) and traction BIE (8.13) are both applied to analyze the square crack in a finite domain. For the square crack in the homogeneous cube ($d = 0$), the $K_I/(p\sqrt{\pi c})$ value is 0.8112 by using the present method and is 0.81 if the method of Wen et al. (1998) is used and 0.8183 if Pan et al.'s (2000) method is employed.

Figures 9.26 to 9.29 illustrate the variations of the SIF values as the crack distance d to the FGM interlayer changes and Table 9.4 presents the corresponding SIF data. As the crack approaches the FGM interlayer, the K_I values decrease and K_{II} values increase for $\alpha c = 2$ and the K_I values and the absolute K_{II} values increase for $\alpha c = -2$. It can be seen

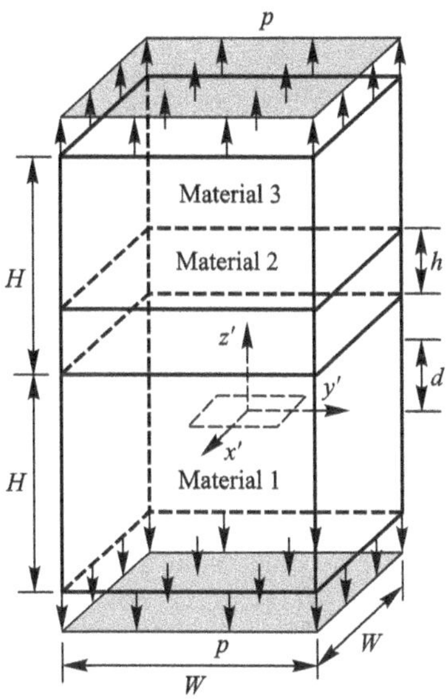

Fig. 9.25 Geometry of a finite cube with a central crack under a uniform normal stress p

that the SIFs are distributed in a similar way as for the square crack in both finite and infinite domains and these values are given in Table 9.5; here we see that the SIF values of the square crack in a finite domain are larger than those in an infinite domain.

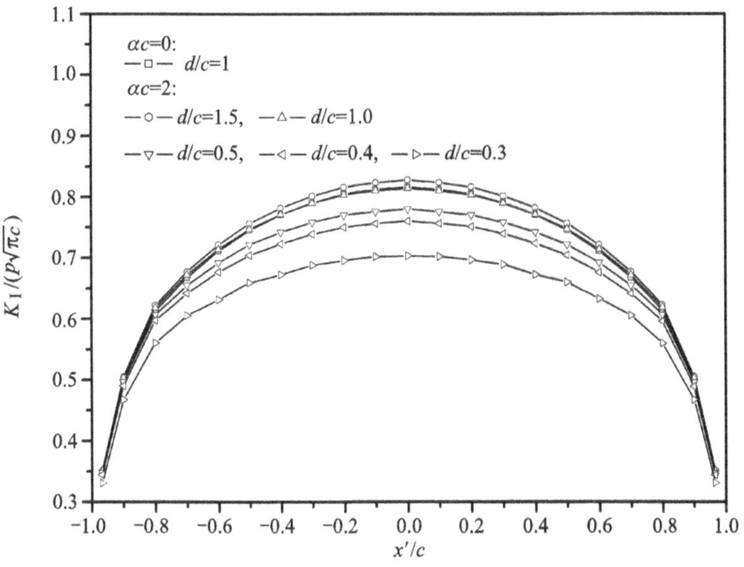

Fig. 9.26 Variations of $K_{\mathrm{I}}/(p\sqrt{\pi c})$ with the crack distance d ($\alpha c = 2$)

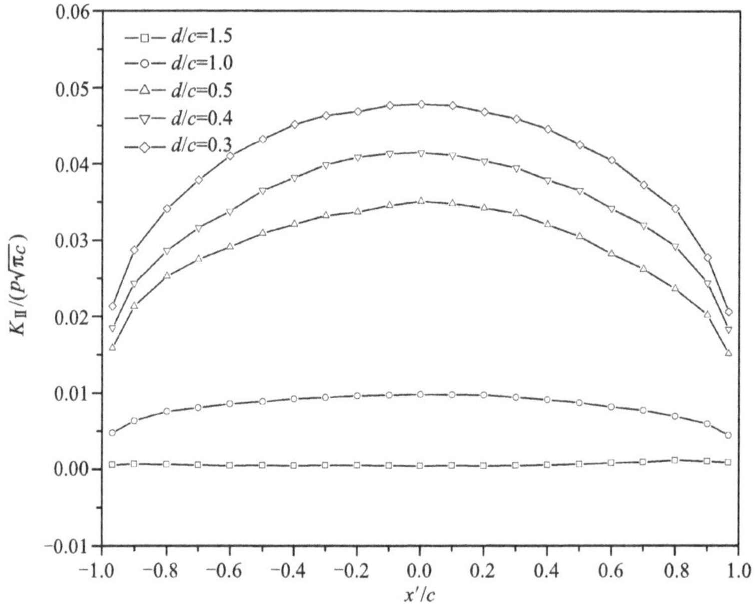

Fig. 9.27 Variations of $K_{\text{II}}/(p\sqrt{\pi c})$ with the crack distance d ($\alpha c = 2$)

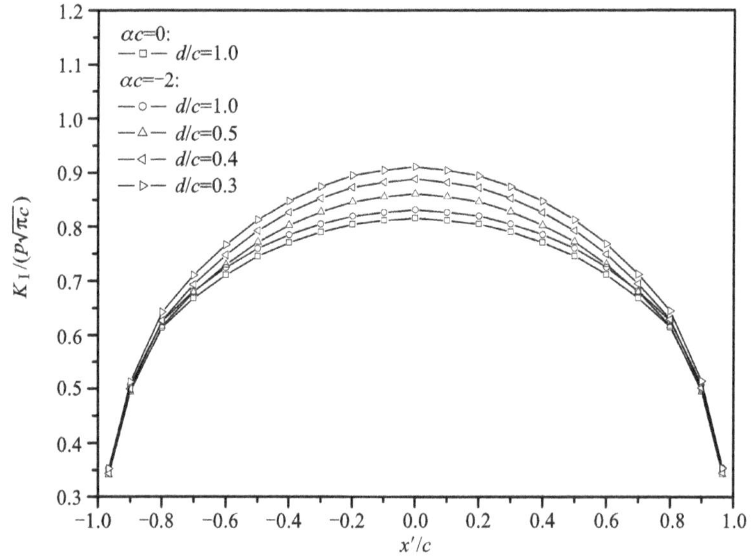

Fig. 9.28 Variations of $K_{\text{I}}/(p\sqrt{\pi c})$ with the crack distance d ($\alpha c = -2$)

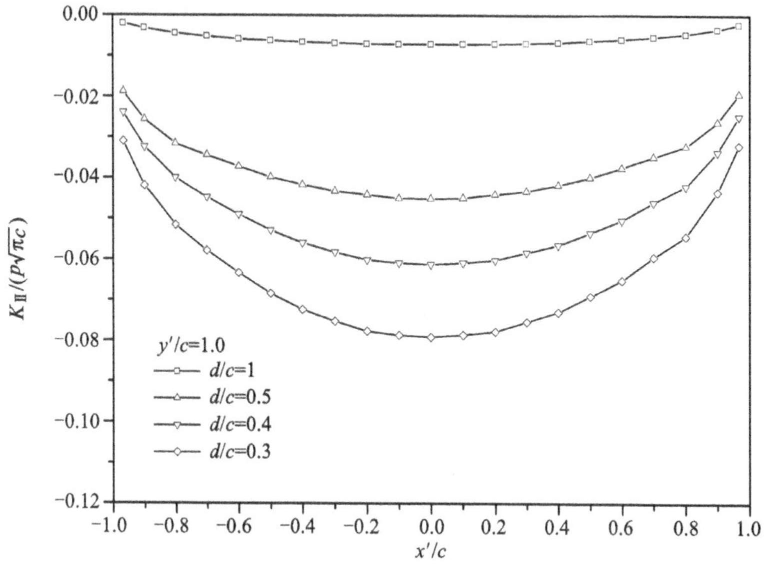

Fig. 9.29 Variations of $K_{II}/(p\sqrt{\pi c})$ with the crack distance d ($\alpha c = -2$)

Table 9.4 The SIF values along the crack front $y' = -c$ for a crack in a finite cube with a FGM interlayer

x'/c	$K_I/(p\sqrt{\pi c})$				$K_{II}/(p\sqrt{\pi c})$			
	$\alpha c = 2$		$\alpha c = -2$		$\alpha c = 2$		$\alpha c = -2$	
	d/c=0.3	0.5	0.3	0.5	d/c=0.3	0.5	0.3	0.5
−0.96667	0.330913	0.34787	0.352598	0.346046	0.021343	0.015882	−0.03098	−0.01886
−0.5	0.659524	0.72154	0.812298	0.785909	0.043209	0.03089	−0.06836	−0.03975
−0.3	0.688403	0.758184	0.874426	0.843534	0.046308	0.033183	−0.07528	−0.0433
−0.1	0.701845	0.775555	0.904686	0.871885	0.047658	0.034536	−0.07866	−0.04499
0.0	0.703359	0.77954	0.910546	0.877141	0.047811	0.035077	−0.07912	−0.04511

Table 9.5 The SIF values along the crack front $y' = -c$ for a crack in finite and infinite domains ($d/c = 0.5$)

x'/c	$K_{\rm I}/(p\sqrt{\pi c})$				$K_{\rm II}/(p\sqrt{\pi c})$			
	$\alpha c = 2$		$\alpha c = -2$		$\alpha c = 2$		$\alpha c = -2$	
	Infinite domain	Finite domain	Infinite domain	Finite domain	Infinite domain	Finite domain	Infinite domain	Finite domain
–0.96667	0.309459	0.34787	0.341389	0.346046	0.010377	0.015882	–0.01481	–0.01886
–0.5	0.640905	0.72154	0.771234	0.785909	0.022462	0.03089	–0.03346	–0.03975
–0.3	0.673369	0.758184	0.827598	0.843534	0.024838	0.033183	–0.03725	–0.0433
–0.1	0.688848	0.775555	0.855472	0.871885	0.026021	0.034536	–0.03933	–0.04499
0.0	0.692352	0.77954	0.860679	0.877141	0.026068	0.035077	–0.03955	–0.04511

9.6 Square cracks in layered rocks

9.6.1 General

Fracture mechanics of rocks has received wide attention for many years because of its direct applications to rock mechanics and engineering. Layered rocks are abundant worldwide. The interface and matrix materials in layered rocks often contain imperfections, such as faults and joints. The term crack will be used from now on to denote any such imperfection. Generally, cracks play an important role in the mechanical properties of rocks and it is therefore necessary to study the fracture mechanics of cracks in rocks.

The BEM is used in many geotechnical problems to assess the stresses and displacements around underground excavations. The displacement discontinuity method (DDM) is particularly well suited to model fractures that have relative displacements across their surface. In some cases, DBEM can be degenerated into DDM. Most of these analyses assume that the rocks are homogeneous, isotropic and linearly elastic solids, although inhomogeneity and anisotropy can be analyzed by both BEM and DDM. It should be mentioned at this point that recently there has been a renewed interest in analyzing crack problems in rocks by applying BEM and DDM. For instance, Crouch and Starfield (1983) developed a two-dimensional DDM to analyze crack problems in homogeneous rocks and Tan et al. (1998) used a DDM to simulate the development of cracks and chips by indentation tools. The BEM codes, such as Poly3D, have been used to model three-dimensional fault interaction (Crider and Pollard, 1998; Maerten and Pollard, 2002). Shou et al. (1999) developed a higher order DDM for the analysis of a circular hole in an infinite strip under tension and a pressurized crack within a three-layered system. To the best of our knowledge, a literature review completed for the present study indicates that there are few publications in the open literature on the three-dimensional DDM-based analysis of crack problems in layered rocks. In this section, the proposed DBEM is used to analyze square crack problems in layered rocks consisting of fine-grained sandstone and mudstone.

9.6.2 The crack dimensions and the parameters of layered rocks

As shown in Fig. 9.30, an elastic layer is bonded to two semi-infinite elastic domains and contains a square crack parallel to the two surfaces. It is assumed that the two elastic half-spaces have the same elastic properties. Two types of rock, i.e., fine-grained sandstone and mudstone, are selected. The elastic parameters of fine-grained sandstone are $E_1 = 56$ GPa and $\nu_1 = 0.23$, and the ones for the mudstone are $E_2 = 20$ GPa and $\nu_2 = 0.25$. Thus, there are four types of layered rocks shown in Table 9.6. The thickness of the mid-layer $h = 2$ m, the side length of the square crack is 2 m and the crack surfaces are parallel to the interfaces of the layered rocks. The crack surfaces are smooth and are subjected to uniform or non-uniform loads. The mesh of the crack surfaces shown in Fig. 8.7b are further applied.

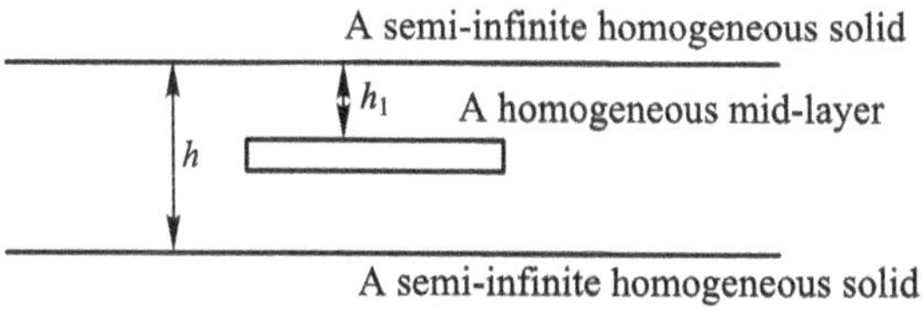

Fig. 9.30 A square crack or a rectangular crack in the interlayer of infinite extent

Table 9.6 Cases of layered rocks containing a square crack or a rectangular crack

Square crack		Rectangular crack		Layered rocks of infinite extent
Uniform load	Non-uniform load	Linear load	Non-linear load	
Case 1-1	Case 2-1	Case 3-1	Case 4-1	Homogeneous solid of infinite extent consisting of fine-grained sandstone
Case 1-2	Case 2-2	Case 3-2	Case 4-2	Sandwich solid: mudstone outerlayers of semi-infinite extent; middle layer of fine-grained sandstone
Case 1-3	Case 2-3	Case 3-3	Case 4-3	Homogeneous solid of infinite extent consisting of mudstone
Case 1-4	Case 2-4	Case 3-4	Case 4-4	Sandwich solid: fine-grained sandstone outerlayers of infinite extent; middle layer of mudstone

9.6.3 A square crack subjected to a uniform compressive load

As shown in Fig. 9.31, the crack surfaces are subjected to the uniform compressive stress, $p_0 = 8$ MPa. In the following, the discontinuous displacements of the crack surface and SIFs for four cases are analyzed.

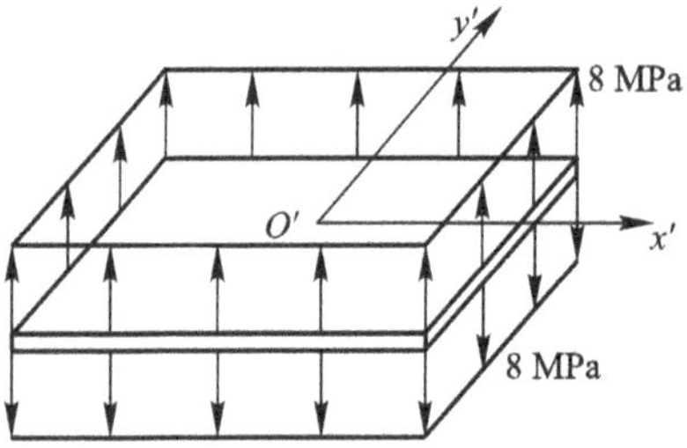

Fig. 9.31 A square crack subjected to uniform load

1. Case 1-1 and case 1-2

Cases 1-1 and 1-2 refer to a homogeneous medium and a sandwich solid, respectively. The material of case 1-1 and the mid-layer of case 1-2 is a fine-grained sandstone. For case 1-2, the materials of two semi-infinite domains are mudstone and the corresponding elastic modulus is less than that for the mid-layer.

The crack opening displacements (CODs) are presented in Fig. 9.32. It can be seen that there is a maximum COD value at the center of the crack surface, and the COD values become smaller towards the crack front and are equal to zero along the crack front. For

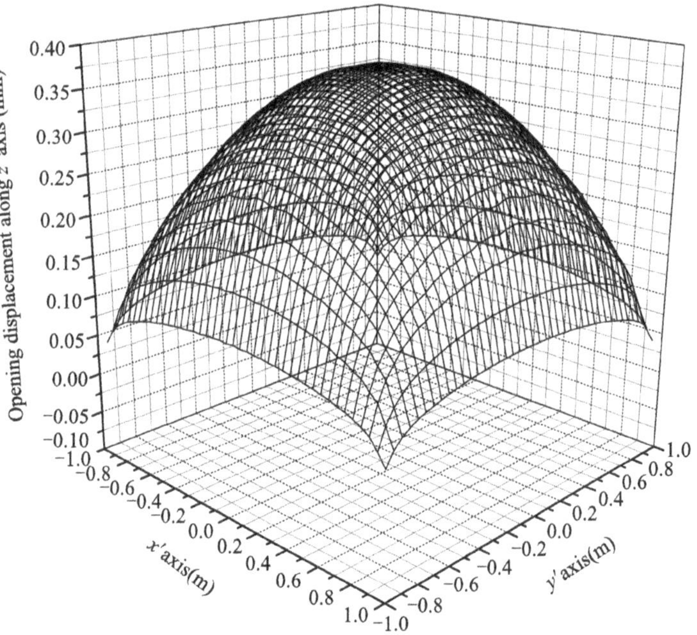

Fig. 9.32 The COD values of the crack surfaces (case 1-1)

case 1-1, the exact solution of the sliding discontinuous displacements along the crack surfaces equals zero. In the present calculation, the value is about 10^{-8}m and is nearly equal to zero. For case 1-2, if the crack is located at $h_1 = 1$ m, the same situation exists because of symmetry.

The SIF values for the crack front at the four sides are the same because of the special characteristics of material, geometry and load. Herein, only one side of the square crack is analyzed. Figures 9.33 and 9.34 present the SIF values. When compared with the values for case 1-1, the K_{I} values for case 1-2 ($h_1 = 1$ m) increase. This is because the elastic modulus of the mid-layer is larger than that of the two semi-infinite domains and the crack opening displacements are easier to obtain than those for case 1-1.

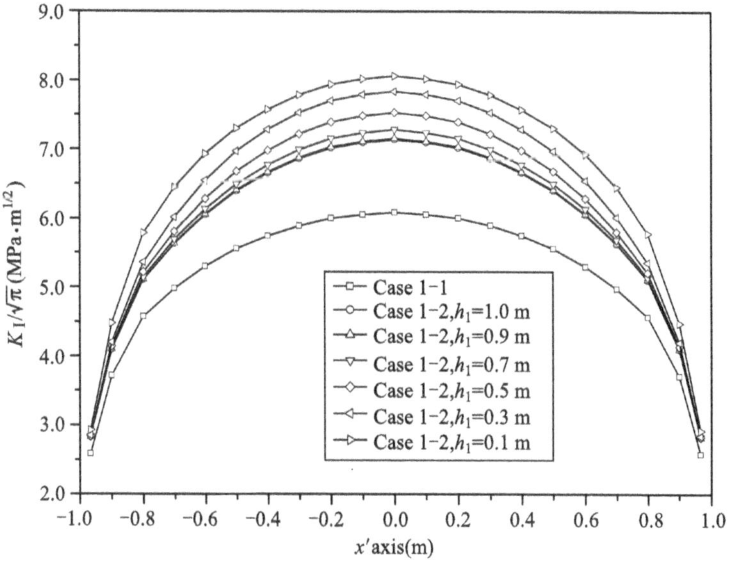

Fig. 9.33 Variations of the K_{I} values (cases 1-1 and 1-2)

For case 1-2, as the h_1 value decreases, i.e., the crack surface approaches the upper interface, the material of the upper semi-infinite domain exerts a more obvious influence on the opening and sliding of the crack surfaces. This is to say, the COD values increase and the absolute values of the sliding displacement also increase. The K_{I} values are positive and the K_{II} values are negative. As the h_1 value decreases, the K_{I} value and the absolute K_{II} value increase. Notice that there are oscillatory stress fields around the crack front when the crack surface is located at the material interface, and therefore this type of crack problem is not considered.

2. Case 1-3 and case 1-4

Cases 1-3 and 1-4 refer to a homogeneous medium and a sandwich solid, respectively. The material of case 1-3 and the mid-layer of case 1-4 is mudstone. For case 1-4, the material

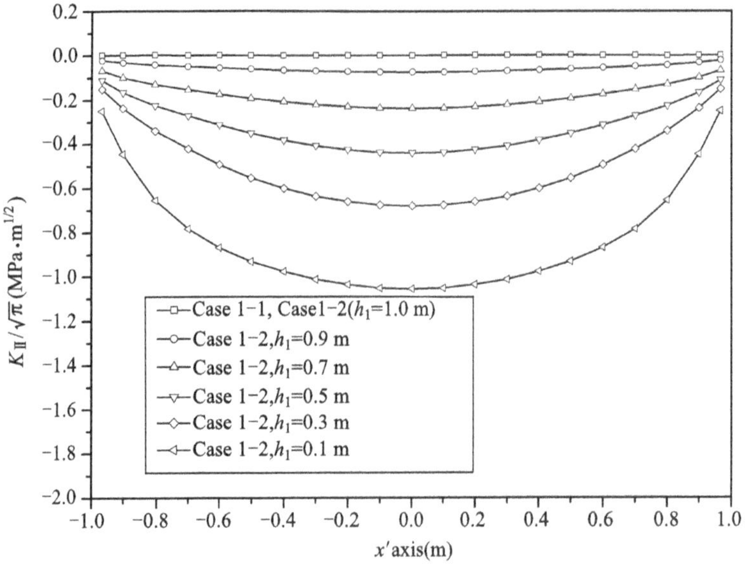

Fig. 9.34 Variations of the K_{II} values (cases 1-1 and 1-2)

of the two semi-infinite domains is a fine-grained sandstone and the corresponding elastic modulus is larger than that for the mid-layer.

Figures 9.35 and 9.36 present the variations in the SIF values for the square

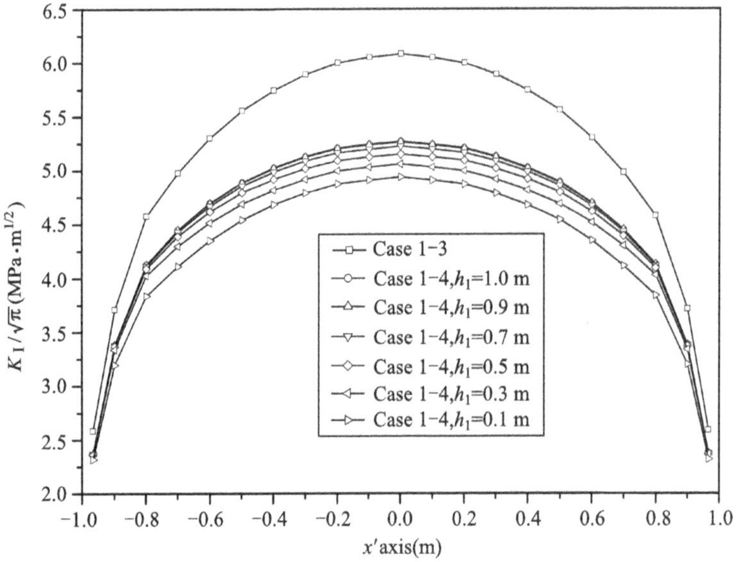

Fig. 9.35 Variations of the K_I values (cases 1-3 and 1-4)

crack; for case 1-4 ($h_1 \neq 1$ m), the mode I and II deformations are coupled together. As the crack approaches the bi-material interface, the COD values decrease and sliding discontinuous displacements increase. This is because the upper material of semi-infinite extent constrains the deformation of the crack surface. The K_I and K_{II} values are positive. As the h_1 value decreases, the K_I value decreases and the K_{II} value increases.

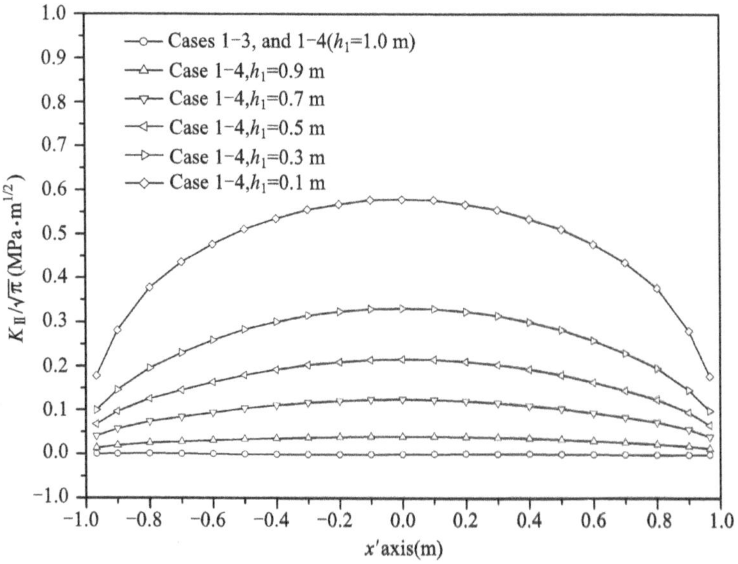

Fig. 9.36 Variations of the K_{II} values (cases 1-3 and 1-4)

9.6.4 A square crack subjected to a non-uniform compressive load

As shown in Fig. 9.37, the crack surfaces are subjected to a triangle-shaped compressive stress, where $p_0 = 0$ MPa at the side $x' = -1$ m and $p_0 = 8$ MPa at the side $x' = 1$ m. The mesh of the crack surface is the same as that given in Section 9.6.3. In the following, the COD values of crack surfaces and SIF values for the four cases are analyzed. Table 9.6 lists the cases to be analyzed.

In order to plot the SIF values along the crack front, a line coordinate L is used to measure the crack front from AB, BC to CD, as shown in Fig. 9.37. The line coordinate L starts at the corner point A of the square crack (i.e., L=0). It increases along the line AB, BC to CD. Correspondingly, L increases from $0 \sim 2$, from $2 \sim 4$, and from $4 \sim 6$, respectively.

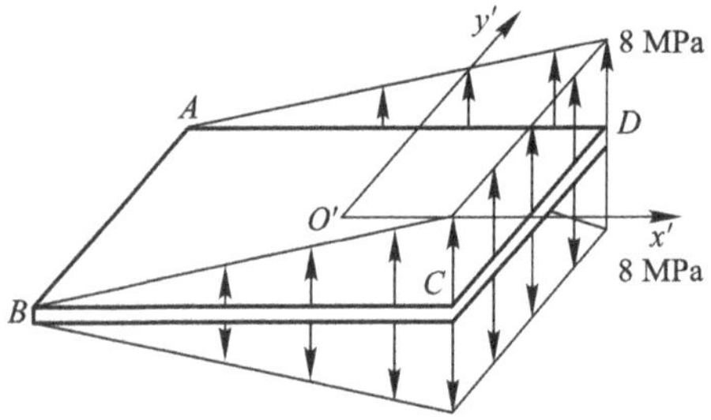

Fig. 9.37 The square crack subject to triangle compressive stresses

1. Case 2-1 and case 2-2

Figure 9.38 presents the COD values of the crack surfaces for case 2-1. It is observed that the COD values are larger near the side $x' = 1.0$ m and are smaller near the side $x' = -1$ m. For case 2-1, there is no relative sliding displacement along the crack surfaces because of the symmetry of the geometry and loads. For the same reasons, there are no relative sliding displacements for case 2-2 ($h_1 = 1$ m); in this case, the COD values increase and the K_I values also increase in comparison with case 2-1. This is because

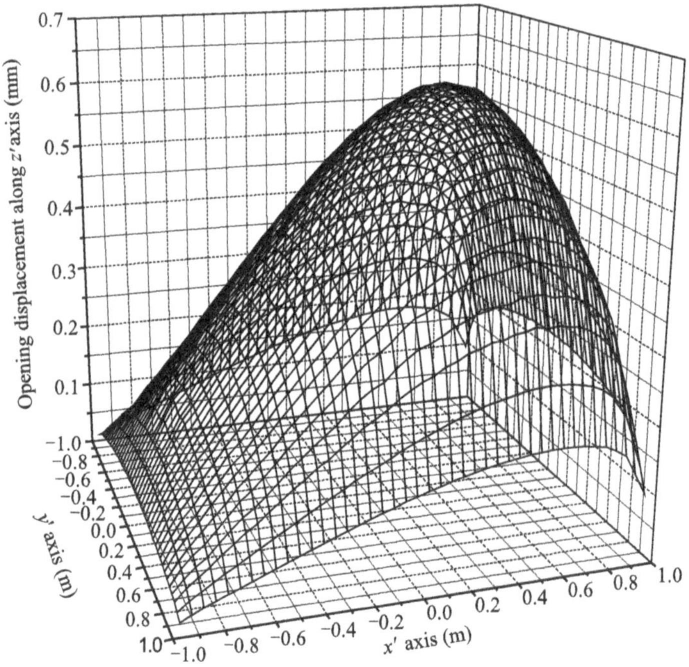

Fig. 9.38 The COD values of the crack surfaces (case 2-1)

the elastic modulus of the mudstone in the semi-infinite domain is less than that for the fine-grained sandstone.

For case 2-2 ($h_1 < 1$ m), the absolute values of displacements along the upper surface increase and the ones along the lower surface decrease. This leads to the appearance that the discontinuous displacements slide along the crack surface. Due to the non-uniform load on the crack surfaces, there are different SIF values at the crack fronts $x' = \pm 1$ m whilst there are the same SIF values at the crack fronts $y' = \pm 1$ m. In the following, the SIF values at the crack fronts $x' = \pm 1$ m and $y' = -1$ m are discussed.

Figure 9.39 shows the variation of the SIF values at the crack front $x' = \pm 1$ m and $y' = -1$ m. It can be observed that the K_I values are positive and the K_{II} values are negative. Obviously, these phenomena are related to the loads on the crack surfaces. Along the crack front $y' = -1$ m, the larger the load, the larger the absolute values of K_I and K_{II}. As the crack surface approaches the interface, the absolute values of K_I and K_{II} increase.

2. Case 2-3 and case 2-4

Figure 9.40 shows the variation of the SIF values at the crack fronts $x' = \pm 1$ m and $y' = -1$ m. Here, the K_I and K_{II} values are positive. Along the crack front $y' = -1$ m, the K_I and K_{II} values become larger near the side $x' = 1$ m. As the crack surface approaches the interface, the K_I values decrease and the K_{II} values increase.

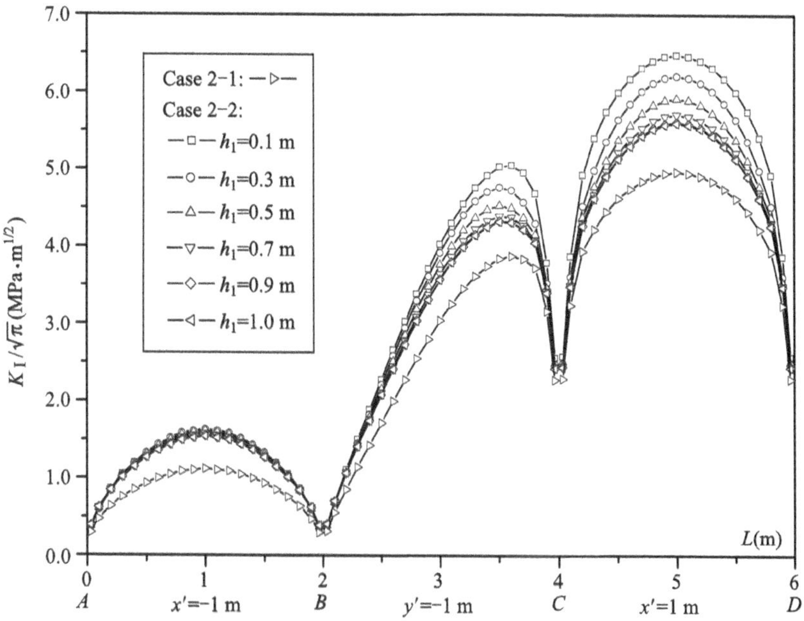

(a) K_I vs. L

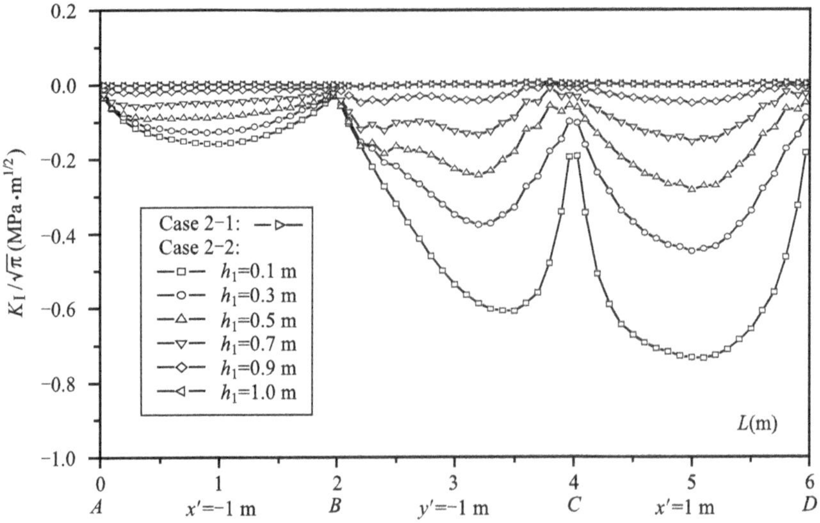

(b) K_{II} vs. L

Fig. 9.39 SIF values of the square crack along the crack front lines from AB, BC to CD in a homogeneous rock (case 2-1) and a layered rock (case 2-2)

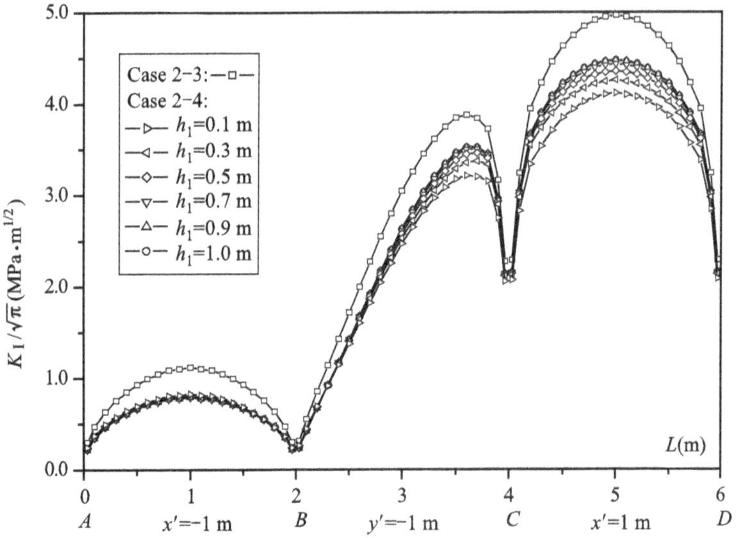

(a) K_I vs. L

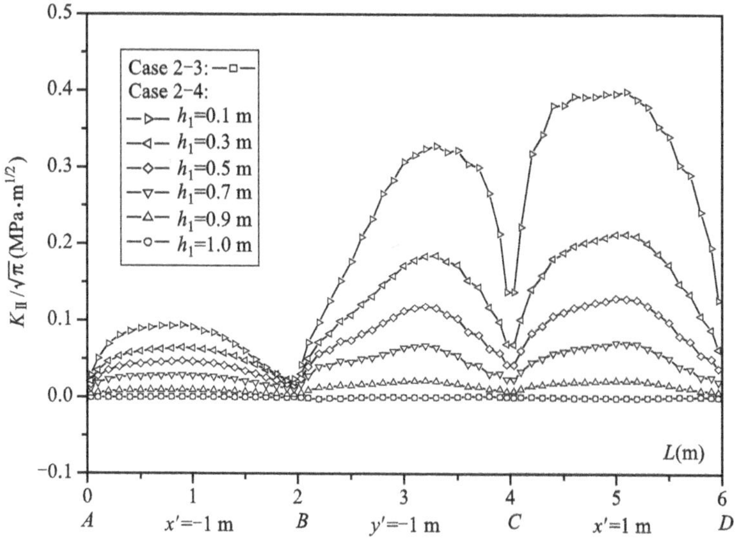

(b) K_{II} vs. L

Fig. 9.40 SIF values of the square crack along the crack front lines from *AB*, *BC* to *CD* in a homogeneous rock (case 2-3) or a layered rock (case 2-4)

9.7 Rectangular cracks in layered rocks

9.7.1 General

In this section, the proposed DBEM is used to analyze the rectangular crack problems in layered rocks. We will use the layered rock types given in Section 9.6. As shown in Fig. 9.30, a rectangular crack is located in the mid-layer and is parallel to the interfaces of the layered rocks. The short and long sides of the rectangular crack are 2 m and 4 m, respectively. The crack surfaces are smooth and are subjected to linear or nonlinear loads. There are four types of layered rocks as detailed in Table 9.6. Herein, the mesh of the crack surface given in Section 9.4 is applied.

9.7.2 A rectangular crack subjected to a linear compressive load

As shown in Fig. 9.41, the crack surfaces are subjected to a triangle-shaped compressive stress as follows

$$p = 2(x' + 2.0), \quad (-2\text{ m} \leqslant x' \leqslant 2\text{ m}, \quad -1\text{ m} \leqslant y' \leqslant 1\text{ m}) \tag{9.1}$$

where p is in units of MPa, the x' and y' are in units of m. Equation (9.1) denotes that $p = 0$ MPa at the side $x' = -2$ m and $p = 8$ MPa at the side $x' = 2$ m.

In the following we will compare the results for this rectangular crack with the ones for the square crack shown in Section 9.6.4. In this case, the CODs are distributed in a similar way to that shown in Fig. 9.38. Figures 9.42 and 9.43 present the variations in the SIF values with the crack position h_1 for cases 3-1 and 3-2, and cases 3-3 and 3-4, respectively. It can be found that there are similar variation tendencies of the SIF values between the square and rectangular cracks subjected to linear compressive loads.

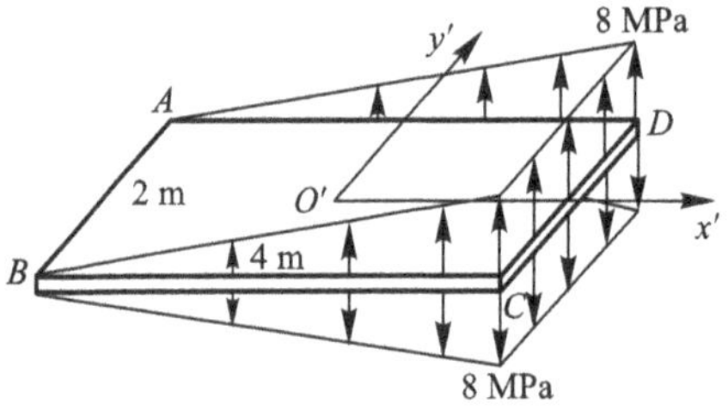

Fig. 9.41 A rectangular crack subject to linear loads

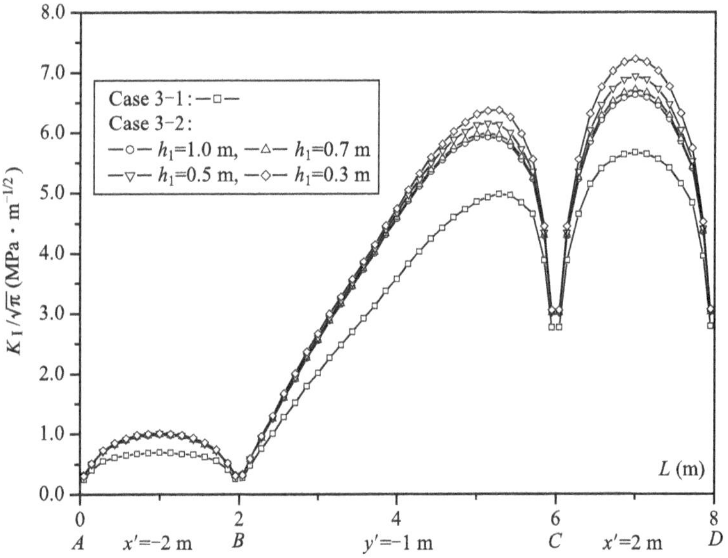

(a) K_{I} vs. L

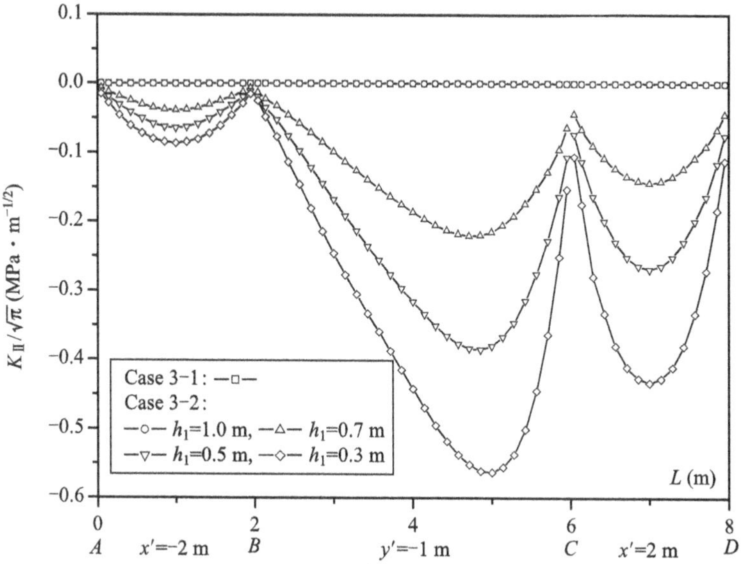

(b) K_{II} vs. L

Fig. 9.42 SIF values of rectangular crack subjected to linear loads (cases 3-1 and 3-2)

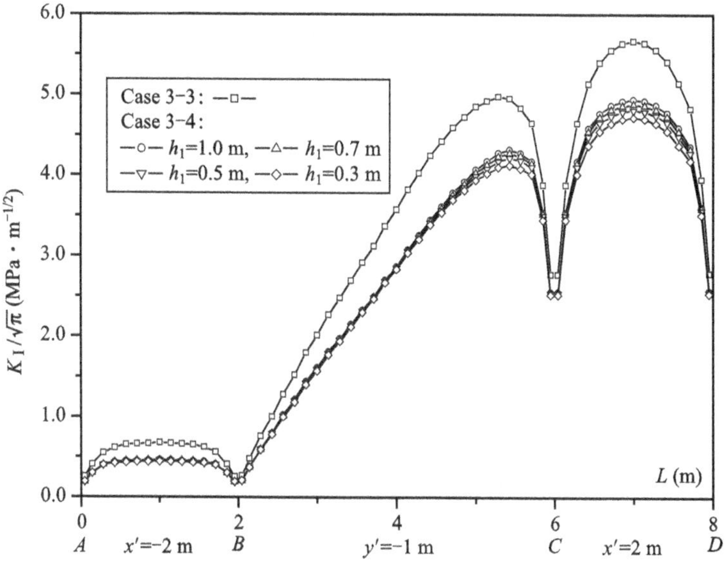

(a) K_I vs. L

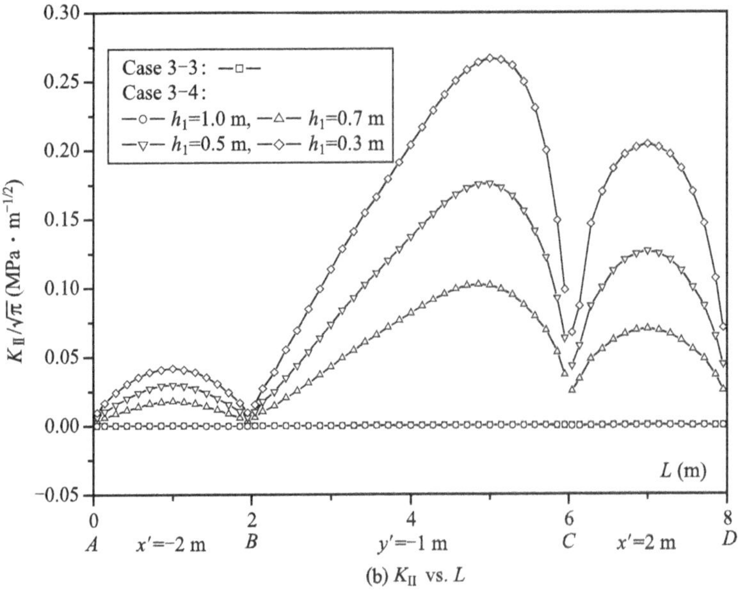

(b) K_{II} vs. L

Fig. 9.43 SIF values of rectangular crack subjected to linear loads (cases 3-3 and 3-4)

9.7.3 A rectangular crack subjected to a nonlinear compressive load

As shown in Fig. 9.44, the crack surfaces are subjected to a compressive stress that is distributed in the following form

$$p = \frac{1}{2}(x' + 2.0)^2, \quad (-2\text{ m} \leqslant x' \leqslant 2\text{ m}, \quad -1\text{ m} \leqslant y' \leqslant 1\text{ m}) \tag{9.2}$$

where the p unit is MPa, the x' and y' units are m. Equation (9.2) means that $p = 0$ MPa at the side$x' = -2$ m and $p = 8$ MPa at the side $x' = 2$ m.

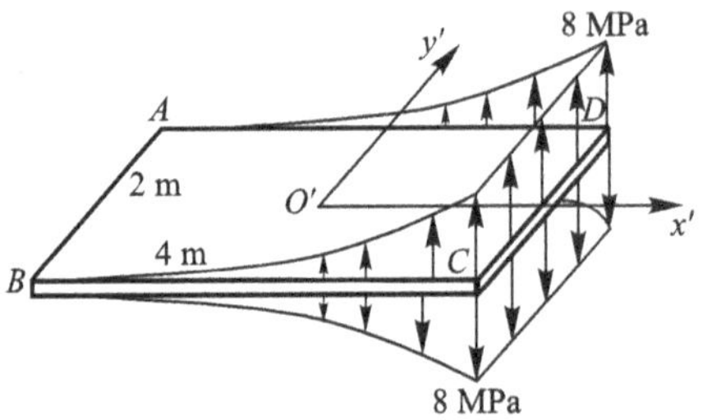

Fig. 9.44 Geometry of a rectangular crack subjected to nonlinear loads

Figures 9.45 and 9.46 illustrate the variations of the SIF values with the crack position h_1 for cases 4-1 and 4-2, and cases 4-3 and 4-4, respectively. Similar observations to those discussed for the linear load (9.1) can be seen in Figs. 9.45 and 9.46 for the non-linear load

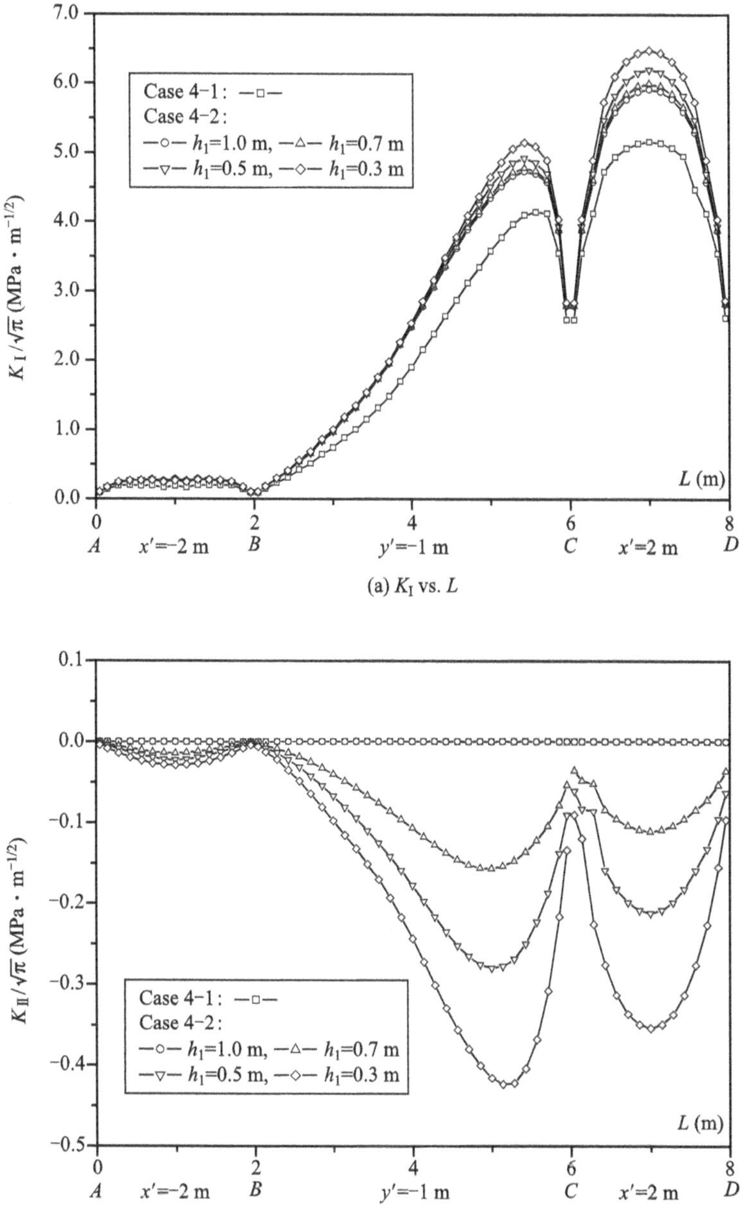

(a) K_I vs. L

(b) K_{II} vs. L

Fig. 9.45 SIF values of rectangular crack subjected to nonlinear loads (cases 4-1 and 4-2)

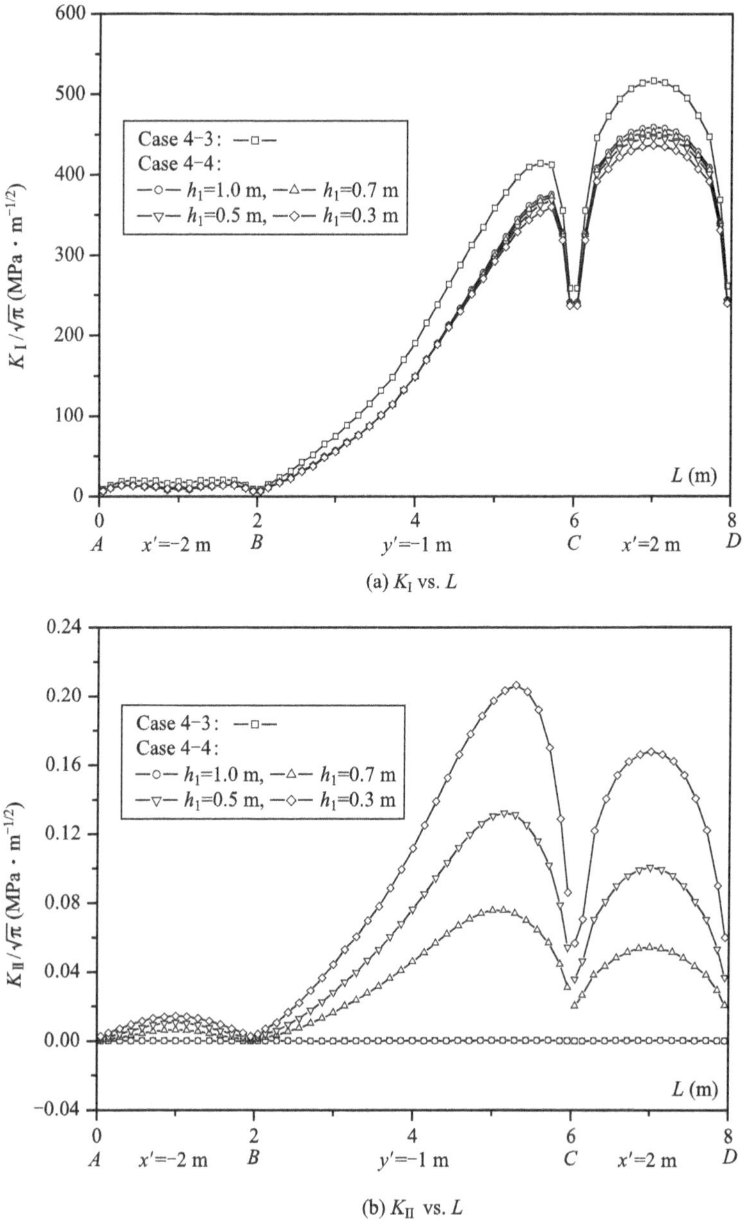

Fig. 9.46 SIF values of rectangular crack subjected to non-linear loads (cases 4-3 and 4-4)

(9.2). Furthermore, the effect of the different loads (9.1) and (9.2) on the SIF values can be found by comparing the results shown in Figs. 9.42 and 9.43 with those in Figs. 9.45 and 9.46. Basically, the SIF values for the non-linear load (9.2) become smaller than those for the linear load (9.1).

9.8 Summary

In this chapter, the SIF values for rectangular cracks in FGMs and layered rocks obtained using DBEM analyses have been reported. The numerical verifications show that very good agreement is obtained between the numerical and exact results. A great number of the SIF values of rectangular cracks can be used to assess the fracture properties of FGMs and layered rocks. These results have been published in Xiao and Yue (2008) and Xiao et al. (2011, 2012).

References

Bains R, Aliabadi MH, Rooke DP. Fracture mechanics weight functions in three dimensions: subtraction of fundamental fields. *International Journal for Numerical Methods in Engineering*, 1992, 35: 179-202.

Crider JG, Pollard DD. Fault linkage: 3D mechanical interaction between echelon normal faults. *Journal of Geophysical Research*, 1998, 103: 24373-24391.

Crouch SL, Starfield AM. Boundary Element Methods in Solid Mechanics: With applications in Rock Mechanics and Geological Engineering. George Allen & Unwin, 1983.

Delale F, Erdogan F. On the mechanical modeling of the interfacial region in bonded half-planes. ASME *Journal of Applied Mechanics*, 1988, 55: 317-324.

Itou S. Dynamic stress intensity factors around a rectangular crack in an infinite plate under impact load. *Engineering Fracture Mechanics*, 1983, 18: 145–153.

Kassir MK. A three-dimensional rectangular crack subjected to shear loading. *International Journal of Solids and Structures*, 1982, 18: 1075–1082.

Maerten L, Pollard D. Effects of local stress perturbation on secondary fault development. *Journal of Structural Geology*, 2002, 24: 145–153

Murakami Y. Stress Intensity Factors Handbook. Oxford: Pergamon Press, 1987.

Pan E, Yuan FG. Boundary element analysis of three-dimensional crack in anisotropic solids. *International Journal for Numerical Methods in Engineering*, 2000, 48: 211-237.

Qin TY, Noda NA. Stress intensity factors of a rectangular crack meeting a bimaterial interface. *International Journal of Solids and Structures*, 2003, 40: 2473-2486.

Shou KJ, Napier JAL. A two-dimensional linear variation displacement discontinuity method for three-layered elastic media. *International Journal of Rock Mechanics and Mining Sciences*, 1999, 36: 719-729.

Tan XC, Kou SQ, Lindqvist PA. Application of the DDM and fracture mechanics model on the simulation of rock breakage by mechanical tools. *Engineering Geology*, 1998, 49: 277-284.

Weaver J. Three-dimensional crack analysis. *International Journal of Solids and Structures*, 1977, 13: 321-330.

Wen PH, Aliabadi MH. Mixed-mode weight functions in 3D fracture mechanics: static. *Engineering Fracture Mechanics*, 1998, 59: 563-575.

Xiao HT, Yue ZQ. Dual boundary element analysis of rectangular-shaped cracks in graded materials (in Chinese). *Chinese Journal of Theoretical and Applied Mechanics*, 2008, 40(6): 840-847.

Xiao HT, Yue ZQ. A three-dimensional displacement discontinuity method for crack problems in layered rocks. *International Journal of Rock Mechanics and Mining Sciences*, 2011, 48(3): 412-420.

Xiao HT, Wang BJ, Yue ZQ. Analysis of rectangular cracks subjected to non-uniform loads in layered rocks (in Chinese).*Engineering Mechanics*, 2012, 12(12): 108-113.

Yue ZQ, Xiao HT, Pan E. Stress intensity factors of square crack inclined to interface of transversely isotropic bi-material. *Engineering Analysis with Boundary Elements*, 2007, 31: 50-56.

Chapter 10

Boundary element analysis of fracture mechanics in transversely isotropic bi-materials

10.1 Introduction

To simplify the discussion, it is often assumed that the physical properties of an elastic body are the same in all directions, i.e., the body is isotropic. In fact, structural materials do not satisfy the assumption completely. When, due to certain natural or technological processes, a certain orientation of microstructures in a solid prevails, the physical properties of an elastic body are different in different directions and the condition of anisotropy must be considered. A transversely isotropic material is one with physical properties symmetric about an axis that is normal to a plane of isotropy. This transverse plane has infinite planes of symmetry and thus, within this plane, the material properties are the same in all directions.

In the general case, a solid medium with anisotropic and inhomogeneous features contains various defects. Over the past decades, a good deal of research has been devoted to the fracture mechanics of a non-homogeneous medium with anisotropy. In the following, we will review the developments in this research field; Sih et al. (1965) derived the general equations for crack-tip stress fields in anisotropic bodies, and defined and evaluated the stress intensity factors (SIFs) directly from stress functions. Hoenig (1978) computed certain influence coefficients for determining the material displacements and stresses in the vicinity of the edge of an elliptic crack within an arbitrarily anisotropic elastic body. Zhang and Mai (1989) derived the SIFs at boundary points of a flat elliptical crack in a transversely isotropic solid subjected to uniform tension using a simple technique and the understanding that normal tension produces an ellipsoidal crack opening geometry. Lin and Keer (1989) developed a general mathematical formulation to analyze cracks in layered transversely isotropic media and numerically obtained the crack opening displacement for a vertical planar crack in a layered transversely isotropic medium using the boundary integral equation method. Noda et al. (2003) analyzed an elliptical crack parallel to a bi-material interface and derived the SIFs by varying the shape of the crack, the crack distance from the interface, and the elastic constants.

Pan and Chou (1976) developed the Green's function for a transversely isotropic solid with an arbitrarily oriented isotropic plane. By using this function, BEMs have been developed for the analysis of fracture mechanics in transversely isotropic materials. Saez et al. (1997) presented a boundary element formulation for three-dimensional crack problems in transversely isotropic bodies and obtained numerical solutions to several three-dimensional crack problems. Pan and Yuan (2000) presented a boundary element analysis of linear elastic fracture mechanics in three-dimensional cracks in anisotropic solids and deduced the SIFs directly from the boundary element solutions. Ariza and Dominguez (2004) presented a boundary element analysis of three-dimensional fracture mechanics problems in transversely isotropic solids based on a mixed formulation and studied cracks in boundless and finite transversely isotropic domains.

In fact, two joined transversely isotropic solids are another type of graded material. Yue (1995) presented closed-form fundamental solutions for the elastic fields in two joined dissimilar transversely semi-infinite solids induced by concentrated forces. The expressions for the fundamental solutions are presented in Appendix 5. In the past few years, the authors incorporated these fundamental solutions into multi-region and dual BEMs and analyzed penny-shaped, elliptical and rectangular cracks in the bi-materials (Xiao et al., 2005; Yue et al., 2005, 2007). In this Chapter, we will first apply the multi-region BEM to analyze penny-shaped and elliptical cracks in bi-materials and will then apply the dual BEM to analyze a square crack in the bi-materials.

10.2 Multi-region BEM analysis of cracks in transversely isotropic bi-materials

10.2.1 General

Xiao et al. (2005) and Yue et al. (2007) incorporated the fundamental solutions for two joined transversely isotropic bi-materials into the conventional displacement boundary integral equation. The corresponding numerical methods are the same as the ones used for Yue's solution-based BEM outlined in Chapters 4 and 5. Because the fundamental solutions satisfy the continuity conditions of the interface between two transversely isotropic half-spaces, it is not necessary to perform a discretization along the layer interface in this BEM formulation. In other words, only the crack surfaces and auxiliary surfaces need to be discretized when using the multi-region BEM for analyzing cracks in two transversely isotropic semi-infinite solids. On the crack surfaces, traction-singular elements are collocated in the neighbourhood of the crack front and eight-noded isoparametric elements are collocated away from the crack front. In this BEM, various types of singular integrals are solved successfully. In the following, we will discuss in detail the results of fracture mechanics analyses of penny-shaped and elliptical cracks in two joined transversely isotropic bi-materials.

10.2.2 Calculation of the stress intensity factors

Kassir and Sih (1966) obtained an asymptotic relation between the crack opening displacements (CODs) near the crack front and the SIFs for cracked transversely isotropic media. Using the leading terms of those relations, Ariza and Dominguez (2004) evaluated the SIFs from the three components of the CODs at a distance r from the crack front. Pan and Yuan (2000) established the relationship between the SIFs and the CODs. In using the formulae proposed by Pan and Yuan, the CODs at any point on the discontinuous elements can be obtained and the SIFs of the crack, which has any angle with the plane of isotropy, can be calculated with high accuracy. Herein, we will use this method to calculate the SIFs.

As shown in Fig. 10.1, let (x_1,x_2,x_3) be a local Cartesian coordinate system attached to the crack tip. The x_2-axis is normal to the crack surface, and the x_3-axis is tangential to the crack front. The x_1-axis is thus formed by the interaction of the plane normal to the crack front and the plane tangential to the crack plane. The CODs can be defined by

$$\Delta u_i(x_1,x_2,x_3) = u_i^+(x_1,x_2,x_3) - u_i^-(x_1,x_2,x_3), \quad (i=1,2,3) \tag{10.1}$$

where the superscripts + and – correspond to the crack surfaces, which are located at $x_2 = 0^+$ and $x_2 = 0^-$, respectively.

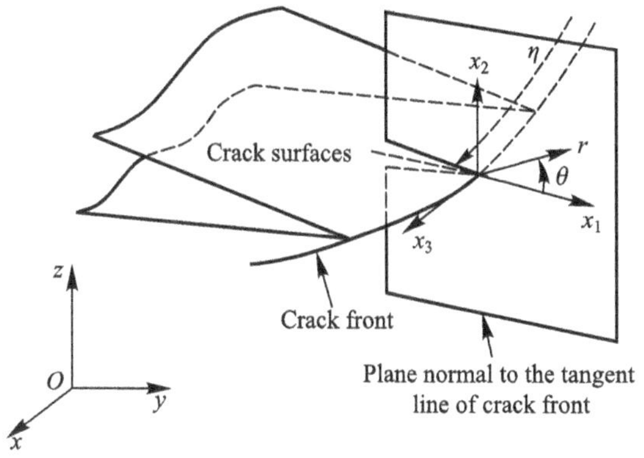

Fig. 10.1 Geometric relationship between local crack-front coordinates (x_1,x_2,x_3) and the global coordinates (x,y,z)

Suppose that the crack front is smooth and the leading singular term in the asymptotic expansion of the stress and displacement fields near the crack tip is amenable to the generalized plane strain analysis. For a crack tip in a homogeneous and transversely isotropic solid, the relation of the CODs at a distance r behind the crack tip and the SIFs can be expressed as (Ting, 1996; Pan and Yuan, 2000)

$$\Delta \boldsymbol{u} = 2\sqrt{\frac{2r}{\pi}}\boldsymbol{L}^{-1}\boldsymbol{k} \tag{10.2}$$

where $\Delta \boldsymbol{u} = (\Delta u_1, \Delta u_2, \Delta u_3)^{\mathrm{T}}$, $\boldsymbol{k} = (K_{\mathrm{I}}, K_{\mathrm{II}}, K_{\mathrm{III}})^{\mathrm{T}}$ and the mode I, II and III SIFs are defined as

$$K_{\mathrm{I}} = \lim_{r\to 0} \sqrt{2\pi r}\sigma_{22}(r,\theta,x_3)\big|_{\theta=0} \tag{10.3a}$$

$$K_{\mathrm{II}} = \lim_{r\to 0} \sqrt{2\pi r}\sigma_{12}(r,\theta,x_3)\big|_{\theta=0} \tag{10.3b}$$

$$K_{\mathrm{III}} = \lim_{r\to 0} \sqrt{2\pi r}\sigma_{23}(r,\theta,x_3)\big|_{\theta=0} \tag{10.3c}$$

Equation (10.2) is based on the Stroh formalism. $\boldsymbol{L}$ is one of the Barnett-Lothe tensors, which depends solely on the anisotropic properties of the solid in the local coordinates. Once the CODs of the crack surfaces near the crack front have been calculated, their values are substituted into Eq. (10.2) and the K_{I}, K_{II} and K_{III} can be obtained.

10.2.3 A penny-shaped crack perpendicular to the interface of transversely isotropic bi-materials

1. Material parameters and loading conditions

In the following, the penny-shaped crack perpendicular to the interface of two bonded transversely isotropic solids, shown in Fig. 10.2, will be analyzed. Assume that the (x,y) plane is the isotropic plane of a transversely isotropic solid with the z-axis being the axis of material symmetry. We introduce the following elastic constants: E_x, μ_x, ν_{xy} and E_z, μ_z, ν_{xz}. The E_x, μ_x and ν_{xy} are the elastic modulus, shear modulus and Poisson's ratio in the (x,y) plane of isotropy, respectively. The E_z, μ_z and ν_{xz} are those quantities in the transverse direction z. The five elastic constants $c_{ik}(i=1,2,3,4,5;k=1,2)$ given in Eq. (A5.1) can be related to the above-mentioned elastic constants as

$$\begin{aligned}
c_{1k} &= 2\mu_x(1-\nu_{xz}^2 E_x/E_z)/(1-\nu_{xy}-2\nu_{xz}^2 E_x/E_z),\\
c_{2k} &= E_x\nu_{xz}/(1-\nu_{xy}-2\nu_{xz}^2 E_x/E_z),\\
c_{3k} &= E_z(1-\nu_{xy})/(1-\nu_{xy}-2\nu_{xz}^2 E_x/E_z),\\
c_{4k} &= \mu_z,\\
c_{5k} &= \mu_x = E_x/2(1+\nu_{xy})
\end{aligned} \tag{10.4}$$

where the subscript k denotes the material number. For an isotropic material, the five elastic constants degenerate into two elastic constants as follows

$$c_{1k} = c_{3k} = \lambda + \mu, \quad c_{2k} = \lambda, \quad c_{4k} = c_{5k} = \mu \tag{10.5}$$

where λ and μ are Lamé's constants for an isotropic material.

As shown in Table 10.1, two sets of transversely isotropic materials are selected for numerical evaluation. Consequently, we have four cases of material combination for the two bonded solids of an infinite extent, shown in Table 10.2.

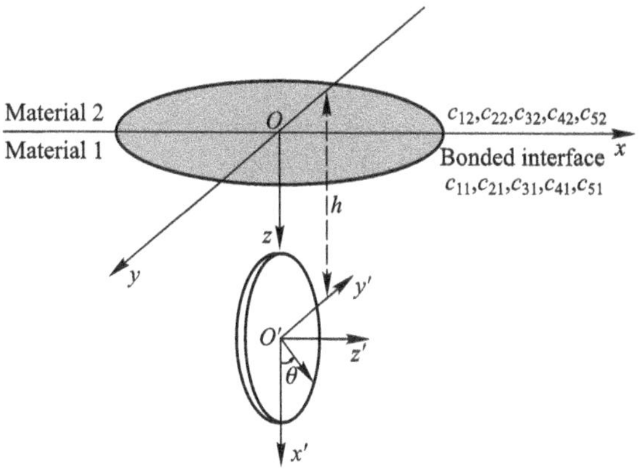

Fig. 10.2 A penny-shaped crack perpendicular to the bonded interface of two transversely isotropic materials

Table 10.1 The elastic constants of transversely isotropic materials

Material types	Elastic constants
Material 1	$E_x/E_z = 3, \nu_{xy} = 0.25, \mu_z = 0.25, \mu_z/E_z = 0.4$
Material 2	$E_x/E_z = 0.5, \nu_{xy} = 0, \mu_z = 0.4, \mu_z/E_z = 0.8$

Table 10.2 Cases of two joined transversely isotropic solids

Case	$z \geqslant 0^+$	$z \leqslant 0^-$	Note
1	Material 1	Material 1	A homogeneous solid
2	Material 2	Material 2	A homogeneous solid
3	Material 1	Material 2	Two joined solids
4	Material 2	Material 1	Two joined solids

Assume that the crack surfaces are subjected to a compressive stress as follows

$$\sigma^+_{z'z'} = \sigma^-_{z'z'} = -p, \quad (p \geqslant 0) \tag{10.6a}$$

$$\sigma^+_{x'y'} = \sigma^-_{x'y'} = 0 \tag{10.6b}$$

$$\sigma^+_{x'z'} = \sigma^-_{x'z'} = 0 \tag{10.6c}$$

where the superscripts + and – correspond to the crack surfaces, which are located at $z' = 0^+$ and $z' = 0^-$, respectively.

Let $p = p_0$ and $p_0(r/a)$, where p_0 is a constant, r is a distance from any point on the crack surface to the origin O' and $0 \leqslant r < a$ (the crack radius a). $p = p_0$ and $p_0(r/a)$ mean that the crack surfaces are subjected to uniform compressive and triangle-shaped compressive stresses, respectively.

2. Numerical verification

We first verify the accuracy and efficiency of the proposed method by comparing the existing analytical and numerical solutions for a penny-shaped crack parallel to the plane of isotropy in a homogeneous and transversely isotropic body of infinite extent. The crack surfaces are subjected to the uniform compressive stress $p = p_0$. Evidently, the cracked body is symmetric about the plane which is vertical to the plane of isotropy and passes through the crack center. Therefore, the BEM formulation of the crack problem is carried out by examining the left symmetrical part of the entire crack geometry. For convenience, we choose a hemisphere in a homogeneous material to form the two regions for the BEM analysis. Because of symmetry, it is only necessary to analyze half of the hemisphere. We only discretize this half hemisphere and do not need to discretize the plane of the symmetry into BEM meshes. The boundary element mesh, shown previously in Fig. 6.5, is employed.

For the crack problem being considered, the exact solution is $K_I = 2p_0\sqrt{a/\pi}$ (Hoenig, 1978). It should be noted that for penny-shaped or elliptical cracks parallel to the plane of isotropy of a transversely isotropic medium, the SIF values are independent of the material property. For cases 1 and 2, the numerical results for $K_I/(p_0\sqrt{\pi a})$ along the crack front vary from 0.63 to 0.64, compared with the value $2/\pi \approx 0.6366$ obtained from the analytical solutions. The largest deviation for the SIFs using the BEM is less than 1 per cent.

3. A penny-shaped crack subjected to a uniform compressive stress

Figure 10.2 shows a penny-shaped crack in material 1 that is perpendicular to the bonded interface of materials 1 and 2 and is located at a distance h from the bi-material interface. The crack surfaces are subjected to a uniform compressive stress $p = p_0$. Evidently, the cracked body is symmetric about the $O'x'z'$-plane. The boundary mesh shown in Fig. 6.5 is used. Figure 10.3 shows the variation of $K_I/(p_0\sqrt{\pi a})$ for cases 1 and 2 and Table 10.3 presents the corresponding SIF data.

Table 10.3 The SIF values ($K_I/(p_0\sqrt{\pi a})$) for the crack under the action of a uniform load

$\theta(°)$	Case 1	Case 3		Case 2	Case 4	
		$h = 1.01a$	$h = 1.2a$		$h = 1.01a$	$h = 1.2a$
90.00	0.814364	0.815765	0.8136	0.49832	0.489147	0.493119
112.50	0.761588	0.766172	0.761397	0.563141	0.547047	0.554649
135.00	0.637065	0.647633	0.637575	0.65922	0.624607	0.64375
157.50	0.515891	0.53993	0.516903	0.708049	0.624772	0.682393
180.00	0.466553	0.510983	0.467763	0.715752	0.581832	0.683984

The mode I SIF values along the crack front are symmetric with the $O'x'z'$ plane. It can be seen that the anisotropy of the materials exerts an obvious influence on the SIFs. Note that the exact solution gives $K_I/(p_0\sqrt{\pi a}) = 2/\pi$ for the penny-shaped crack in a homoge-

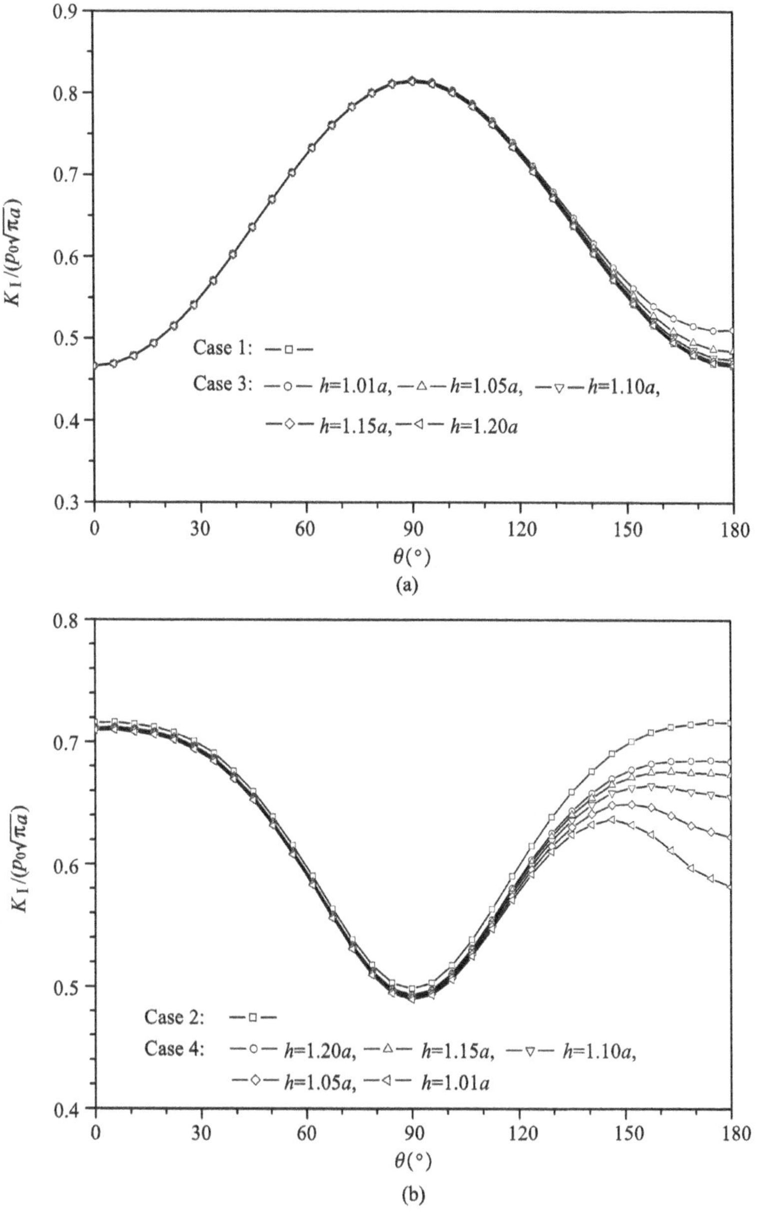

Fig. 10.3 The SIF values $(K_I/(p_0\sqrt{\pi a}))$ for the crack under the action of a uniform load

neous and isotropic medium. It is interesting that, for the case of the crack surfaces perpendicular to the plane of the isotropy, the SIF values vary along the crack front. Similar phenomena were also found by Hoenig (1978), Pan and Yuan (2000).

Figure 10.3a shows the variations of $K_I/(p_0\sqrt{\pi a})$ along the crack front with the crack distances from the material interface for cases 1 and 3. It can be found that material 2 exerts a weak influence on the SIFs along the crack front $\theta \in [90°, 180°]$ and the effect of material 2 on the SIFs along the crack front $\theta \in [0°, 90°]$ is not obvious. As the crack distances from the material interface decrease, the SIFs along the crack front around $\theta = 180°$ increase slowly. Figure 10.3b illustrates the variations of $K_I/(p_0\sqrt{\pi a})$ along the crack front with the crack distances from the material interface for cases 2 and 4; here, the material 2 exerts an obvious influence on the SIFs along the crack front $\theta \in [90°, 180°]$ and has a relatively weak influence on the SIFs along the crack front $\theta \in [0°, 90°]$. As the crack distances from the material interface decrease, the SIFs along the crack front around $\theta = 180°$ decrease.

Evidently, the above variations of the SIFs along the crack front are due to the existence of material 2. If material 2 is stiffer than material 1, material 2 tends to constrain the crack opening and the SIFs decrease. However, if material 2 is more complaint than material 1, material 2 tends to make the crack opening easier and the SIFs increase.

4. A penny-shaped crack subjected to a triangle-shaped compressive stress

As shown in Fig. 10.2, a penny-shaped crack in material 1 is perpendicular to the bonded interface of materials 1 and 2 and is located at a distance h from the bi-material interface. The crack surfaces are subjected to a triangle-shaped compressive stress $p = p_0(r/a)$. Evidently, the cracked body is symmetrical to the $O'x'z'$-plane. The boundary mesh shown in Fig. 6.5 is applied.

Figure 10.4 shows the variation of $K_I/(p_0\sqrt{\pi a})$ and Table 10.4 presents the corresponding SIF data. Table 10.5 presents a comparison of the results for $p = p_0$ and $p_0(r/a)$ at the crack front $\theta = 180°$, which were obtained by the following formula for the same loading:

[$K_I/(p_0\sqrt{\pi a})$ for a transversely isotropic solid $-K_I/(p_0\sqrt{\pi a})$ for two joined transversely isotropic solids] %.

Table 10.4 The SIF values ($K_I/(p_0\sqrt{\pi a})$) for the crack under the action of a non-uniform load

θ(°)	Case 1	Case 3		Case 2	Case 4	
		$h = 1.01a$	$h = 1.2a$		$h = 1.01a$	$h = 1.2a$
90.00	0.5967	0.59785	0.59625	0.404425	0.398865	0.40128
112.50	0.57175	0.57515	0.5717	0.45136	0.44148	0.44626
135.00	0.5031	0.51085	0.50355	0.51515	0.493125	0.50555
157.50	0.42624	0.444915	0.427105	0.5348	0.478255	0.51855
180.00	0.3925	0.429455	0.39352	0.52945	0.434415	0.50915

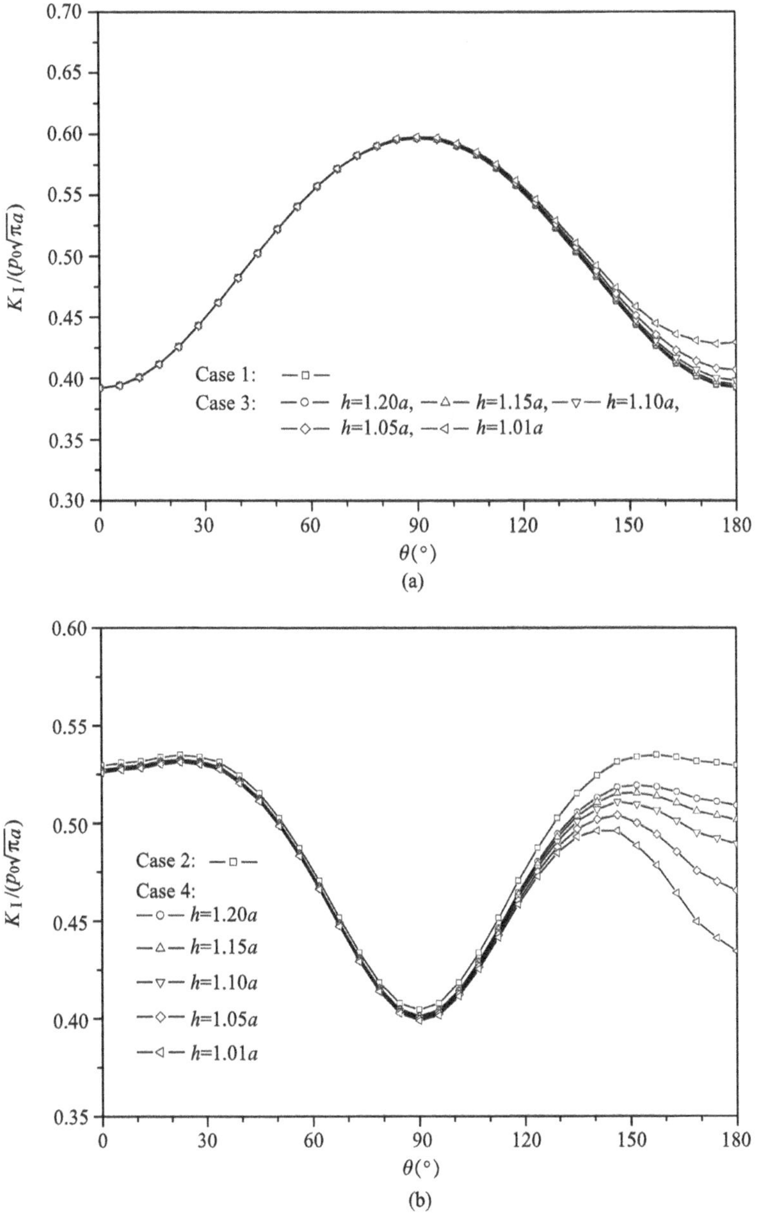

Fig. 10.4 The SIF values $(K_I/(p_0\sqrt{\pi a}))$ for the crack under the action of a non-uniform load

Table 10.5 Comparison of the SIF values between uniform and non-uniform loads

h	Case 3		Case 4	
	p_0	$p_0(r/a)$	p_0	$p_0(r/a)$
1.01a	4.44	3.70	−13.39	−9.50
1.05a	1.86	1.46	−9.32	−6.40
1.10a	0.78	0.60	−6.07	−4.03
1.15a	0.37	0.26	−4.25	−2.77
1.20a	0.12	0.10	−3.18	−2.03

In these tables and figures, the loading $p = p_0(r/a)$ causes similar variations in the SIFs to $p = p_0$ along the crack front with changes in the crack distances from the material interface. It can also be found that the absolute increments in Table 10.5 decrease from $p = p_0$ to $p = p_0(r/a)$ for the same crack distance to the material interface. This indicates that the effects of material 2 on the SIFs become weak as the load on the crack surfaces decreases.

10.2.4 An elliptical crack perpendicular to the interface of transversely isotropic bi-materials

1. Material parameters and loading conditions

Figure 10.5 shows an elliptical crack perpendicular to the interface of two bonded transversely isotropic solids and the minor axis of the elliptical crack is perpendicular to the plane of isotropy. The isotropy planes of the two joined transversely isotropic solids are both parallel to the bonded interface of the two solids. Assume that the (x, y) plane is the isotropic plane for a transversely isotropic solid with the z-axis being the axis of material symmetry. In the present study, two transversely isotropic materials are selected for the numerical evaluation and listed in Table 10.6. As such, there are four cases of material combination for the two bonded solids of infinite extent, which are given in Table 10.7.

Table 10.6 The elastic constants of two transversely isotropic materials

Materials	c_1	c_2	c_3	c_4	c_5	E_x/E_z
Cadmium (GPa)	11.6	4.14	5.1	1.95	3.685	2.7764
Rhenium (GPa)	61.2	20.6	68.3	16.2	17.1	0.8032

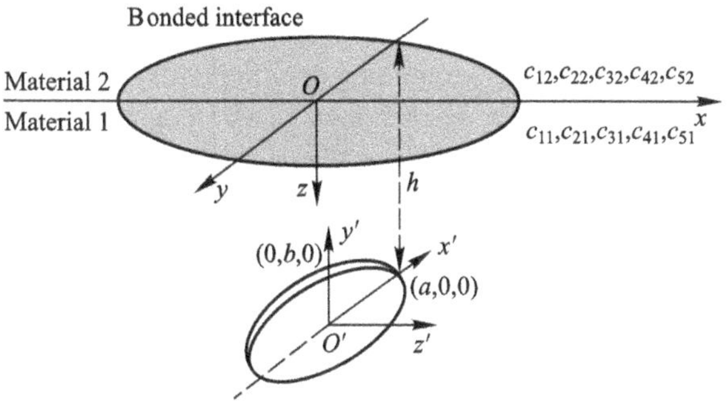

Fig. 10.5 An elliptical crack perpendicular to the bonded interface of two transversely isotropic solids $(a = 2b)$

Table 10.7 Cases of two joined transversely isotropic solids

Case	$z \geqslant 0^+$	$z \leqslant 0^-$	Note
1	Rhenium	Rhenium	A homogeneous solid
2	Cadmium	Cadmium	A homogeneous solid
3	Rhenium	Cadmium	Two joined solids
4	Cadmium	Rhenium	Two joined solids

In the crack-based coordinate system $O'x'y'z'$, the points (x', y') on the edge of the elliptical crack are described by means of the parameter θ:

$$(x', y') = (a\cos\theta, b\sin\theta) \tag{10.7}$$

where a and b are the lengths of the major and minor axes of an elliptical crack, respectively. In the present study, let $a = 2b$.

The crack surfaces are subjected to the uniform compressive stress p_0, i.e.,

$$\sigma^+_{z'z'} = \sigma^-_{z'z'} = -p_0, \quad (p_0 \geqslant 0) \tag{10.8a}$$

$$\sigma^+_{x'y'} = \sigma^-_{x'y'} = 0 \tag{10.8b}$$

$$\sigma^+_{x'z'} = \sigma^-_{x'z'} = 0 \tag{10.8c}$$

where the superscripts + and – correspond to the crack surfaces, which are located at $z' = 0^+$ and $z' = 0^-$, respectively.

2. Numerical verification

In the ensuing, the accuracy of the proposed method is verified by comparing the existing exact and numerical solutions for the elliptical crack parallel to the plane of isotropy in a transversely isotropic solid. The crack surfaces are subjected to the loads given in Eqs. (10.8). Evidently, the cracked body and the loadings are symmetric about the plane which is vertical to the plane of isotropy and passes through the crack center. So, the BEM formulation of the crack problem is carried out by examining only the left symmetrical part of the entire crack geometry. For convenience, a hemisphere in a homogeneous material is chosen to form the two regions for the BEM analysis. Because of symmetry, it is only necessary to analyze half of the hemisphere. This half hemisphere is discretized, but the plane of the symmetry is not discretized into the BEM meshes. The boundary mesh shown in Fig. 7.31 is utilized.

For an elliptical crack subjected to a uniform compressive stress p_0 on the crack surfaces and parallel to the plane of isotropy of a homogeneous and transversely isotropic medium, the mode I SIF is described in the following (Hoenig, 1978)

$$\frac{K_{\mathrm{I}}}{p_0\sqrt{\pi a}} = \frac{\sqrt{\gamma}}{E(k)}(\sin^2\theta + \gamma^2\cos^2\theta)^{1/4} \tag{10.9}$$

where

$$E(k) = \int_0^{\pi/2}\sqrt{1-k^2\sin^2\varphi}\,\mathrm{d}\varphi, \quad k=\sqrt{1-\gamma^2}, \quad \gamma = b/a$$

It should be noted that for the above elliptical crack, the SIF values are independent of the material properties. Figure 10.6 shows the numerical and exact solutions of $K_{\mathrm{I}}/(p_0\sqrt{\pi a})$ for cases 1 and 2. The maximum deviations are less than 0.8 % along the crack front for the two materials.

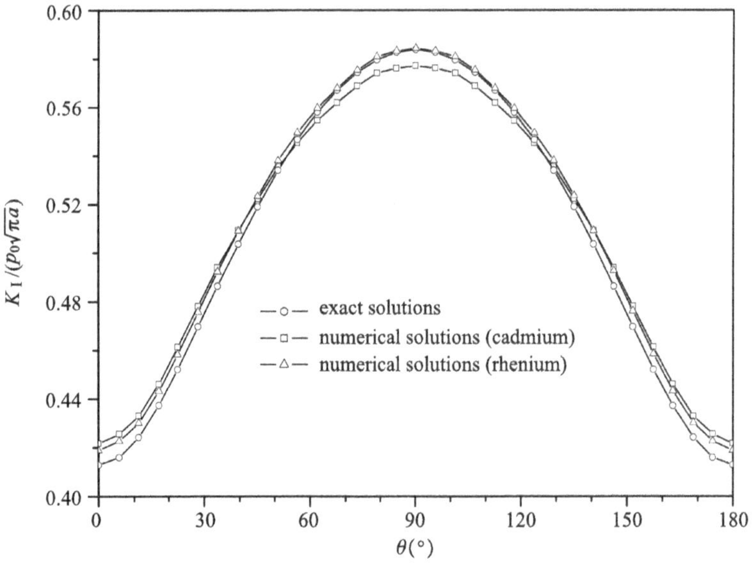

Fig. 10.6 The numerical and exact solutions for an elliptical crack in a transversely isotropic solid

3. Numerical results and discussions

As shown in Fig 10.5, the example considers an elliptical crack located in material 1 at a distance h from the bi-material interface. The crack surfaces are subject to a uniform compressive stress p_0. For this case, the cracked body and the load are symmetric with respect to the $O'y'z'$-plane. Figures 10.7 and 10.8 illustrate the variations in the K_{I} values and Table 10.8 presents the corresponding SIF data. For a convenient comparison, the SIF values of the crack in a homogeneous solid (Cases 1 and 2) are also presented. Evidently, the SIF values are symmetric with respect to the coordinate plane $O'y'z'$.

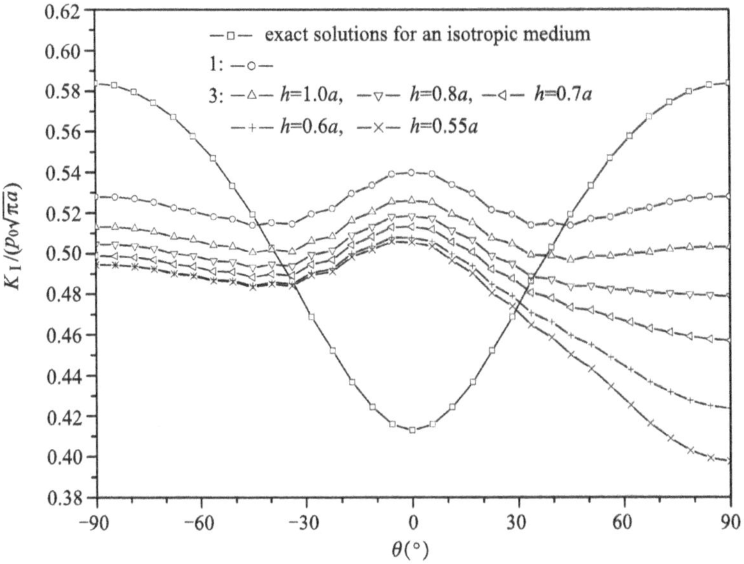

Fig. 10.7 The SIF values $(K_{\mathrm{I}}/(p_0\sqrt{\pi a}))$ for an elliptical crack, cases 1 and 3

Table 10.8 The SIF values $(K_{\mathrm{I}}/(p_0\sqrt{\pi a}))$ for cases $1 \sim 4$

$\theta(°)$	Case 1	Case 3		Case 2	Case 4	
		$h = 1.0a$	$h = 0.55a$		$h = 1.0a$	$h = 0.55a$
0.00	0.53971	0.52604	0.50549	0.40789	0.41181	0.4138
11.25	0.53382	0.51952	0.49647	0.41715	0.4216	0.42484
22.50	0.52149	0.50765	0.4805	0.4433	0.44871	0.45372
33.75	0.51385	0.49906	0.46475	0.47898	0.4857	0.49531
45.00	0.51372	0.49653	0.45015	0.51443	0.52259	0.54042
56.25	0.5179	0.49851	0.43432	0.54521	0.55498	0.58596
67.50	0.52241	0.50037	0.41616	0.56839	0.57965	0.62965
78.75	0.5269	0.50277	0.40288	0.5840	0.59633	0.66654
90.00	0.52789	0.50302	0.39747	0.58837	0.6011	0.68052

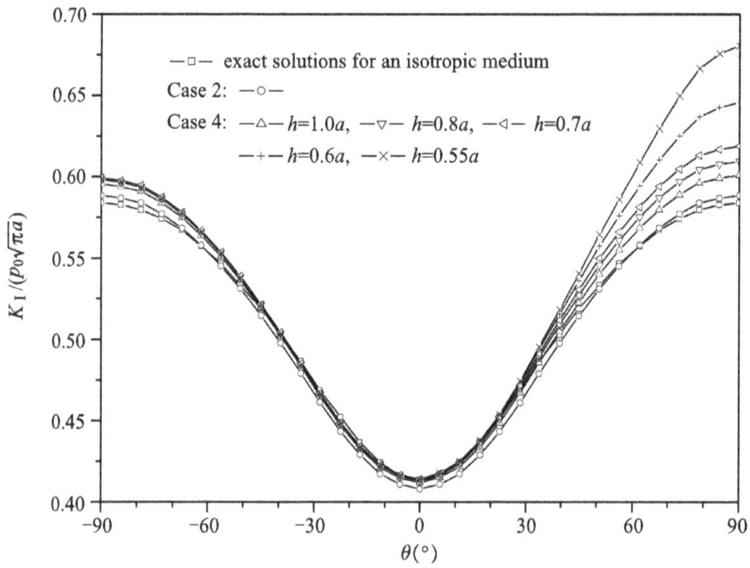

Fig. 10.8 The SIF values $(K_\mathrm{I}/(p_0\sqrt{\pi a}))$ for an elliptical crack, cases 2 and 4

For case 1, $E_x/E_z \approx 2.7764$. In Fig. 10.7, it can be seen that the SIF values for case 1 have obvious variations with reference to the ones in an isotropic solid. The SIFs around the crack front at the minor axis of the elliptical crack decrease and the SIFs around the crack front at the major axis of the elliptical crack increase. For case 2, $E_x/E_z \approx 0.8032$. Fig. 10.8 shows the variations of the SIF values for case 2. Here, the effect of the material anisotropy on the SIFs is relatively weak; the SIFs around the crack front at the minor axis of the elliptical crack increase and the SIFs around the crack front at the major axis of the elliptical crack decrease.

The variations in the K_I values along the crack front with the crack distances from the material interface for cases 1 and 3 are shown in Fig. 10.7. Material 2 exerts an obvious influence on the SIFs along the crack front $\theta \in [0°, 90°]$ while the effect of material 2 on the SIFs along the crack front $\theta \in [-90°, 0°]$ is relatively weak. At the crack front $\theta = 90°$, material 2 exerts the maximum effect on the SIFs. As the distance of the crack from the material interface decreases, the SIFs along the crack front also decrease.

Figure 10.8 also illustrates the variations in the K_I values along the crack front with the crack distances from the material interface for cases 2 and 4. It can be found that material 2 exerts an obvious influence on the SIFs along the crack front $\theta \in [0°, 90°]$ but has a relatively weak influence on the SIFs along the crack front $\theta \in [-90°, 0°]$. At the crack front $\theta = 90°$, the maximum effect imposed by material 2 on the SIFs occurs. When the crack distances from the material interface decrease, the SIF values along the crack front increase.

10.3 Dual boundary element analysis of a square crack in transversely isotropic bi-materials

10.3.1 General

By applying the fundamental solution of two joined transversely isotropic solids, Yue et al. (2007) developed the dual boundary element method (DBEM) and studied a rectangular crack in bi-materials. The numerical methods in this DBEM are the same as the ones outlined in Chapter 8 only except that different fundamental solutions are utilized.

The new DBEM is particularly useful for analyzing crack problems; only the crack surface and the non-crack boundaries need to be discretized. In order to obtain numerical solutions with a high degree of accuracy, the discontinuous elements in the vicinity of the crack front are collocated. The weakly singular, strongly singular and hypersingular integrals are solved successfully. By solving the linear equations using the DBEM, the CODs of the crack surfaces are obtained and applied to calculate the SIF values. Herein, we will apply the DBEM to analyze two examples: square cracks in a two-bonded transversely isotropic cube, shown in Fig. 10.9, and in a two-bonded transversely isotropic medium of infinite extent.

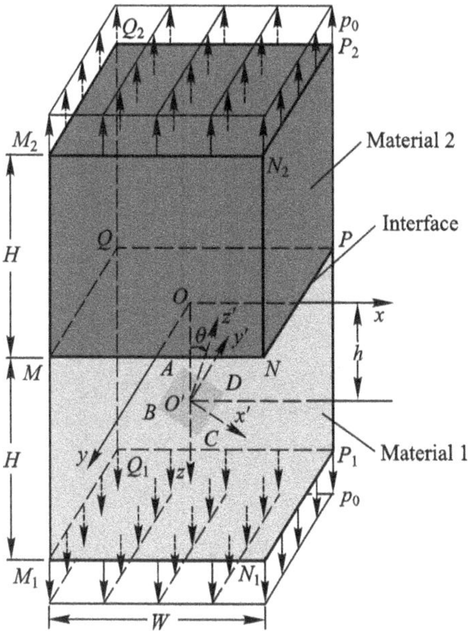

Fig. 10.9 A square crack ($ABCD: 2c \times 2c$) in a transversely isotropic bi-material occupying a finite cube ($2H \times W \times W$)

10.3.2 Numerical verification

In order to verify the proposed numerical method, we analyze a square crack in a homogeneous medium of infinite extent. Assume that the side length of the crack is $2c$ and the crack surfaces are subjected to a uniform compressive stress p_0. The crack surface is discretized into 100 nine-noded elements among which there are the 64 type I elements, the 32 type Ⅱ discontinuous elements and the 4 type Ⅲ elements. Figure 10.10 shows the boundary mesh for the crack surface *ABCD*.

When the crack surfaces are parallel to the plane of isotropy, the SIF values are not related to the material properties and are the same as the ones in a homogeneous medium. Weaver (1977) obtained numerical results for a rectangular crack in a homogeneous medium of infinite extent. The maximum SIF values appear at the middle of the square sides and the SIF values decrease to zero at the corners. The numerical solution of $K_{\mathrm{I}}/(p_0\sqrt{\pi c})$ at the middle of the square side is 0.7605 and the ones given by Weaver (1977) and Pan and Yuan (2000) are 0.74 and 0.7626, respectively. Evidently, the proposed DBEM is capable of producing numerical results with a high degree of accuracy.

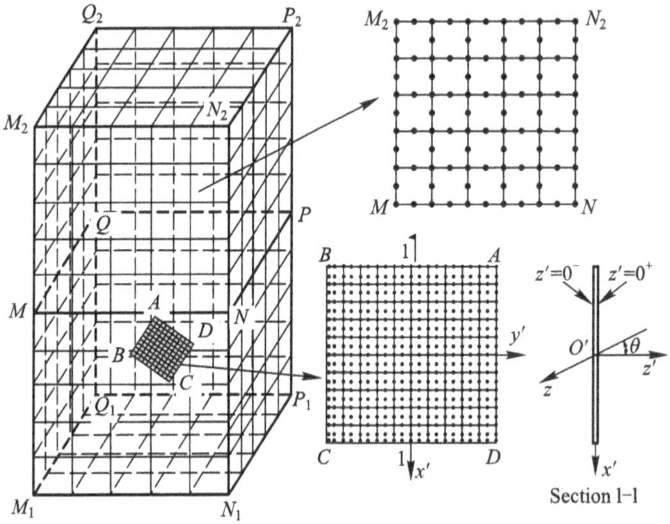

Fig. 10.10 Boundary element mesh of the cracked finite cube

10.3.3 Numerical results and discussions

In the ensuing, we will analyze a square crack in two-bonded transversely isotropic solids, shown in Fig. 10.9. Two transversely isotropic materials, rhenium and cadmium, were selected for the numerical evaluation. The elastic constants of these two materials are listed in Table 10.6. Accordingly, there are four different material combinations for the bi-material, as shown in Table 10.7.

1. A square crack in a transversely isotropic bi-material of finite cubic space

(1) General conditions

Figure 10.9 shows a square crack in transversely isotropic bi-materials occupying a finite cubic space. The finite cube with a square crack is subjected to a uniform tension p_0 at the top and bottom faces along the direction of the z-axis.

The cube has a height $2H$ and a width W. In particular, $W = 4c$ and $W = H$. The side length of the inclined square crack $ABCD$ is $2c$. The origin of the global coordinate system $Oxyz$ is located at the center of the bi-material interface of the cube. The origin O' and the crack center are located at the same point $(0,0,h)$ in the global coordinate system. Without loss of generality, the square crack is always located in material 1 ($z \geqslant 0^+$) for the cases under consideration.

Figure 10.10 shows the boundary element mesh of the cracked finite cube. The non-crack boundaries $M_1N_1P_1Q_1$, $M_2N_2P_2Q_2$, $M_1N_1M_2N_2$, $N_1P_1P_2N_2$, $P_1Q_1Q_2P_2$ and $Q_1M_1M_2Q_2$ are discretized into a mesh system with 250 eight-noded isoparametric elements and 752 nodes. The square crack is discretized in the same way as mesh 4 in Fig. 8.7b (Chapter 8) for the square crack in an infinite space. In order to plot the SIF values along the crack front lines, a line coordinate L is used to measure the crack front lines from AB, BC, CD to DA (Fig. 10.10). The line coordinate L starts at the corner point A of the square crack (i.e., $L/c = 0$) and increases along the lines AB, BC, CD to DA. Correspondingly, L/c increases from $0 \sim 2$, from $2 \sim 4$, from $4 \sim 6$, and from $6 \sim 8$, respectively.

(2) General results

Theoretically, it can be derived that the inclined square crack in the homogeneous material has $K_{\mathrm{I}}(x',y',z')$ for the mode I SIF, $K_{\mathrm{II}}(x',y',z')$ for the mode II SIF and $K_{\mathrm{III}}(x',y',z')$ for the mode III SIF. Due to the symmetry of loading, the material properties and the geometry, the following results are valid

$$K_{\mathrm{I}}(x',-c,0) = K_{\mathrm{I}}(x',c,0) \tag{10.10a}$$

$$K_{\mathrm{II}}(x',-c,0) = K_{\mathrm{II}}(x',c,0) = 0 \tag{10.10b}$$

$$K_{\mathrm{III}}(x',-c,0) = K_{\mathrm{III}}(x',c,0) \tag{10.10c}$$

$$K_{\mathrm{III}}(-c,y',0) = K_{\mathrm{III}}(c,y',0) = 0 \tag{10.10d}$$

Therefore, the results along the crack front line DA (i.e., $6 < L/c < 8$) are not plotted in the following figures. If $\theta = 0°$, the following results are valid

$$K_{\mathrm{I}}(x',y',0) \neq 0 \tag{10.11a}$$

$$K_{\mathrm{II}}(x',y',0) = 0 \tag{10.11b}$$

$$K_{\mathrm{III}}(x',y',0) = 0 \tag{10.11c}$$

Moreover, if $\theta = 90^\circ$, the following results are valid

$$K_{\mathrm{I}}(x',y',0) = 0 \tag{10.12a}$$

$$K_{\mathrm{II}}(x',y',0) = 0 \tag{10.12b}$$

$$K_{\mathrm{III}}(x',y',0) = 0 \tag{10.12c}$$

In addition, the numerical calculations give the values of $K_{\mathrm{II}}(x',-c,0)$ and $K_{\mathrm{II}}(x',c,0)$ at the 42 points along the crack front lines *BC* and *DA*. For all the calculated results presented in Fig. 10.11b, $K_{\mathrm{II}}(x',\pm c,0)/(p_0\sqrt{\pi c})$ have values between –0.00928 and 0.00929 with an average value of 8.95×10^{-5}. This average value is almost equal to zero, i.e., the theoretical value.

Similarly, the numerical calculations give the values of $K_{\mathrm{III}}(-c,y',0)$ and $K_{\mathrm{III}}(-c,y',0)$ at the 42 points along the crack front lines *AB* and *CD*. For all the calculation results presented in Fig. 10.11c, $K_{\mathrm{III}}(\pm c,y',0)/(p_0\sqrt{\pi c})$ have values between –0.00561 and 0.00561 and have an average value of -3.52×10^{-7}. This average value is almost equal to zero, i.e., the theoretical value.

(3) A square crack with different angles θ in a homogeneous medium

In this section, both the materials 1 and 2 shown in Fig. 10.9 are assumed to be rhenium (case 1). The crack problem degenerates to an inclined square crack in a homogeneous and transversely isotropic solid of cubic extent. This degenerated crack problem can also be used to further verify the accuracy of the present DBEM.

If $\theta = 0^\circ$ and $h = 0$, the crack surface becomes parallel to the top and bottom surfaces of the cube. The maximum SIF values $(K_{\mathrm{I}}/(p_0\sqrt{\pi c}))$ along the square side of the finite cube is calculated to be 0.8180. This value compares well with the value 0.8183 given in Pan and Yuan (2000).

Figure 10.11 shows the numerical results for $K_{\mathrm{I}}/(p_0\sqrt{\pi c})$, $K_{\mathrm{II}}/(p_0\sqrt{\pi c})$, and $K_{\mathrm{III}}/(p_0\sqrt{\pi c})$ associated with the crack along the line coordinate L/c at $\theta = 0^\circ, 30^\circ, 45^\circ, 60^\circ$. From Fig. 10.11, the following can be observed:

- At any given crack front point $(x',y',0)$, $K_{\mathrm{I}}/(p_0\sqrt{\pi c})$ decreases as θ increases from 0°, 30°, 45° to 60° (Fig. 10.11a). At any given crack front point $(\pm c,y',0)$, $K_{\mathrm{II}}/(p_0\sqrt{\pi c})$ increase as θ increases from 30° to 45° while it decrease as θ increase from 45° to 60°. At any given crack front $(x',-c,0)$, $K_{\mathrm{III}}/(p_0\sqrt{\pi c})$ increases as θ increases from 30° to 45° while it decreases as θ increases from 45° to 60°.
- For any given angle θ, $K_{\mathrm{I}}/(p_0\sqrt{\pi c})$ monotonically increases as L/c increases from $0 \sim 1$, reaches its peak at the center point $L/c = 1$, and monotonically decreases as L/c increases from $1 \sim 2$ along the crack front line *AB*. Similar results are found along the crack front lines *BC* to *CD*.

- For any given angle θ, $K_{\mathrm{II}}/(p_0\sqrt{\pi c})$ monotonically increases as L/c increases from $0 \sim 1$, reaches its peak at the center point $L/c = 1$, and monotonically decreases as L/c increases from $1 \sim 2$ along the crack front line *AB*. Similar results are also found along the crack front line *CD*.
- For any given angle θ, $K_{\mathrm{III}}/(p_0\sqrt{\pi c})$ monotonically increases as L/c increases from $2 \sim 3$, reaches its peak at the center point $L/c = 3$, and monotonically decreases as L/c increases from $3 \sim 4$ along the crack front line *BC*.
- For $\theta = 30°$, the peak values of $K_{\mathrm{I}}/(p_0\sqrt{\pi c})$ are 0.6252, 0.6093 and 0.6202 at $L/c = 1,3,5$, respectively. The differences in $K_{\mathrm{I}}/(p_0\sqrt{\pi c})$ along the crack front lines *AB*, *BC* and *CD* could be due to the rhenium anisotropy $E_x/E_z = 0.8032$.

(4) A square crack with $\theta = 30°$ in the bi-materials

In this section, the four cases of the bi-material combination given in Table 10.7 are examined for the square crack inclined at $\theta = 30°$ in Fig. 10.9. The calculated values of $K_{\mathrm{I}}/(p_0\sqrt{\pi c})$, $K_{\mathrm{II}}/(p_0\sqrt{\pi c})$ and $K_{\mathrm{III}}/(p_0\sqrt{\pi c})$ along the crack front lines *AB*, *BC*, and *CD* are plotted against the coordinate L/c in Figs. 10.12 and 10.13. The depth h of the inclined square crack is assumed to be c for the homogeneous materials (cases 1 and 2) and to be $h = 0.8c, 0.9c, c, 1.1c, 1.2c$, respectively, for the bi-materials (cases 3 and 4). Some calculated values are selected and are given in Table 10.9 for future comparison.

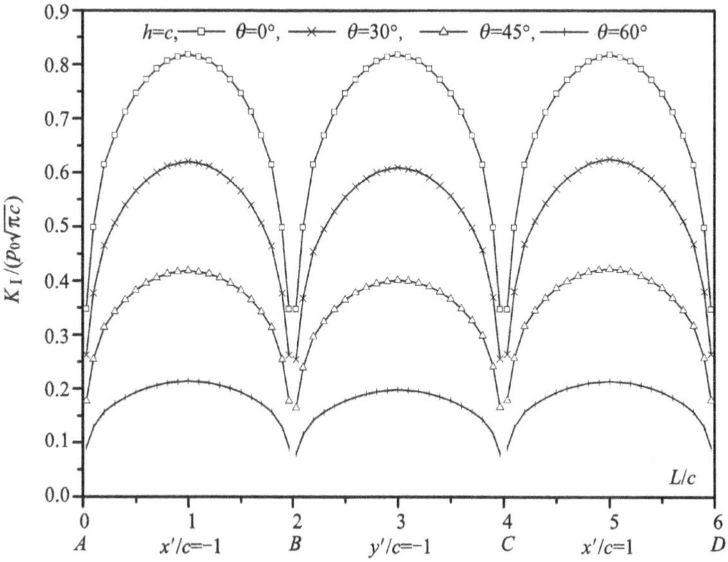

(a) K_{I} vs. L

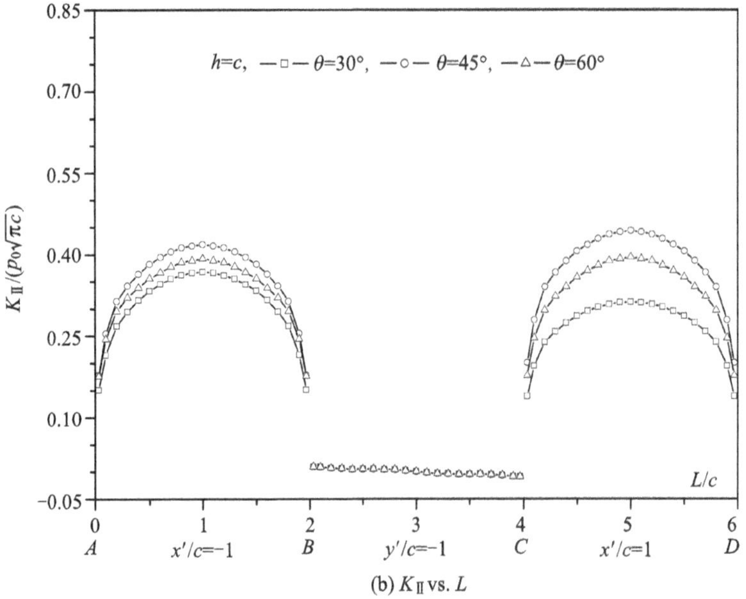

(b) K_{II} vs. L

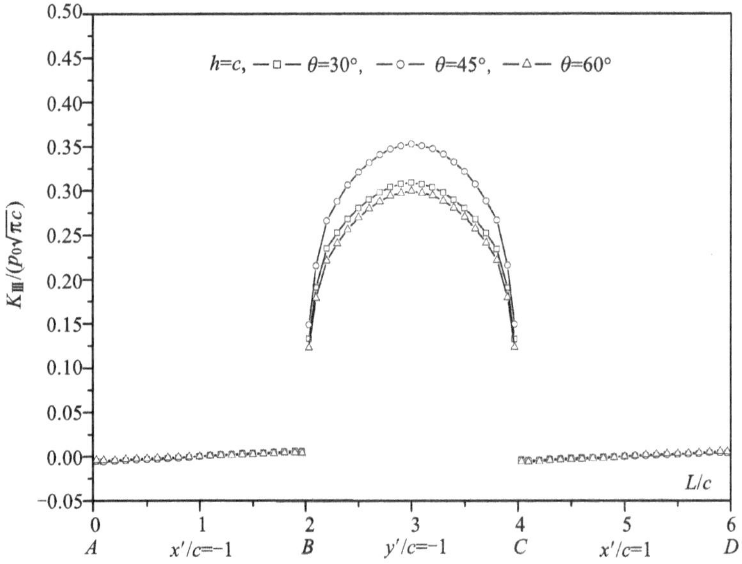

(c) K_{III} vs. L

Fig. 10.11 The SIF values for the inclined square crack along the crack front line from *AB*, *BC* to *CD* in a cubic rhenium (case 1)

Table 10.9 The SIF values for the square crack in a finite cube at the crack front $x'/c = -1$ for $\theta = 30°$ and $h = c$

y'/c	$K_I/(p_0\sqrt{\pi c})$				$K_{II}/(p_0\sqrt{\pi c})$			
	Case 1	Case 2	Case3	Case 4	Case 1	Case 2	Case 3	Case 4
–0.9	0.3770	0.3600	0.2960	0.3507	0.2387	0.2413	0.2149	0.2119
–0.7	0.5067	0.4846	0.4218	0.4604	0.3124	0.3163	0.2942	0.2716
–0.5	0.5661	0.5437	0.4934	0.5063	0.3443	0.3655	0.3331	0.3134
–0.3	0.6001	0.5785	0.5388	0.5316	0.3628	0.4031	0.3560	0.3470
–0.1	0.6165	0.5954	0.5615	0.5436	0.3720	0.4342	0.3672	0.3765
0.0	0.6201	0.6004	0.5657	0.5480	0.3728	0.4405	0.3683	0.3827

As shown in Table 10.7, for case 1, material 1 where the crack is located is rhenium and in this case is the same as material 2. For case 3, material 1 where the crack is located is rhenium and is stiffer than material 2 (cadmium). Figure 10.12 shows the calculated results of $K_I/(p_0\sqrt{\pi c})$, $K_{II}/(p_0\sqrt{\pi c})$ and $K_{III}/(p_0\sqrt{\pi c})$ along the line coordinate L/c, respectively. From this figure, the following can be observed:

- For $h = c$, the differences in the SIF values between cases 1 and 3 are mainly due to the different material properties of material 2. For each given L/c, the values of $K_I/(p_0\sqrt{\pi c})$, $K_{II}/(p_0\sqrt{\pi c})$ and $K_{III}/(p_0\sqrt{\pi c})$ are greater than those for case 3, respectively. This result shows that a softer material 2 can cause an increase in the SIF values in the bi-material.

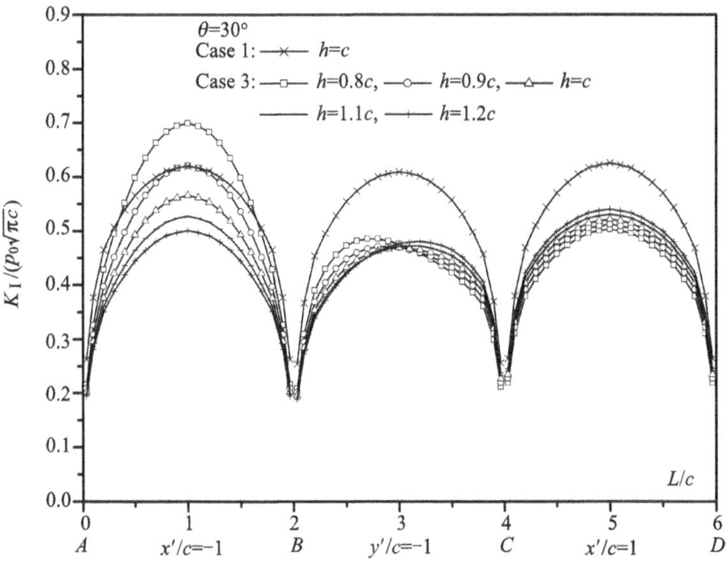

(a) K_I vs. L

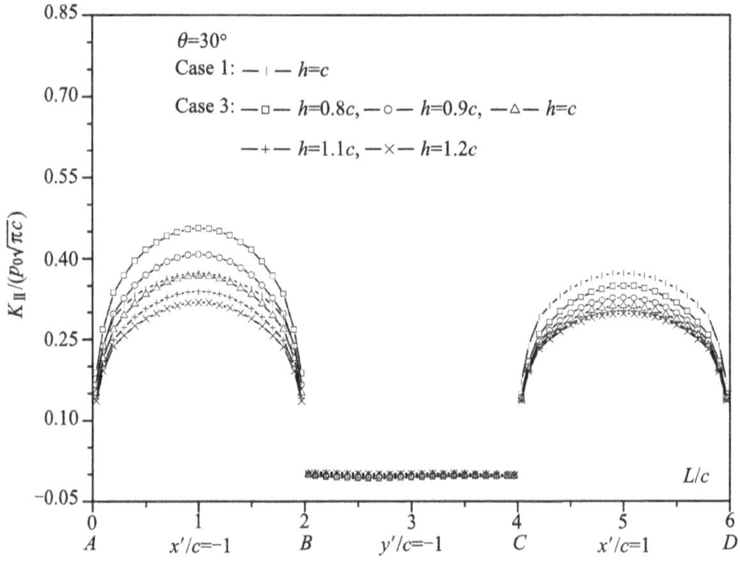

(b) K_{II} vs. L

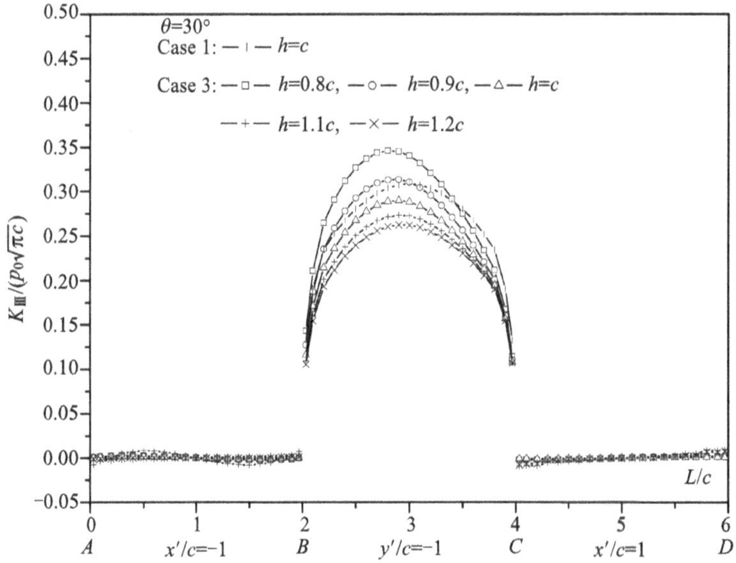

(c) K_{III} vs. L

Fig. 10.12 The SIF values for the inclined square crack with $\theta = 30^\circ$ along the crack front lines from AB, BC to CD in the cubic rhenium (Case 1) or the cubic bi-material (Case 3)

- In particular, when $0.4 < L/c < 1.6$, $K_I/(p_0\sqrt{\pi c})$ for case 3 at $h = 0.8c$ can be greater than the corresponding value for case 1 at $h = c$. Similarly, when $0 < L/c < 2$, $K_{II}/(p_0\sqrt{\pi c})$ for case 3 at $h = 0.8c$ or $h = 0.9c$ can be greater than the corresponding value for case 1 at $h = c$. When $2 < L/c < 3.5$, $K_{III}/(p_0\sqrt{\pi c})$ for case 3 at $h = 0.8c$ or $h = 0.9c$ can be greater than the corresponding value for case 1 at $h = c$. These results are due to the presence of the softer material 2.

Similarly, as shown in Table 10.7, for case 2, material 1 where the crack is located is cadmium and this is the same as material 2. For case 4, material 1 where the crack is located is cadmium, which is softer than material 2 (rhenium). Figure 10.13 shows the calculated results for the values of $K_I/(p_0\sqrt{\pi c})$, $K_{II}/(p_0\sqrt{\pi c})$ and $K_{III}/(p_0\sqrt{\pi c})$ along the line coordinate L/c, respectively. From this figure, the following can be observed:

- For $h = c$, the differences in the SIF values between cases 2 and 4 are mainly due to the different material properties of material 2. As L/c increases from $0 \sim 6$, the $K_I/(p_0\sqrt{\pi c})$ associated with case 4 becomes less and less changeable as the depth h decreases. For $0 < L/c < 2$, the $K_I/(p_0\sqrt{\pi c})$ associated with case 4 decreases significantly as the depth h decreases from $1.2c$ to $0.8c$. Also, when $0 < L/c < 2$, the $K_I/(p_0\sqrt{\pi c})$ associated with case 4 for $0.8c < h < 1.2c$ is less than that associated with case 2. However, when $2 < L/c < 6$, the $K_I/(p_0\sqrt{\pi c})$ associated with case 4 for $0.8c < h < 1.2c$ is greater than that associated with case 2 for $h = c$.
- $K_{II}/(p_0\sqrt{\pi c})$ associated with case 4 decrease as the depth h decreases from $1.2c$ to $0.8c$ and is less than that associated with case 2 for $h = c$, where $0 < L/c < 2$ or $4 < L/c < 6$.

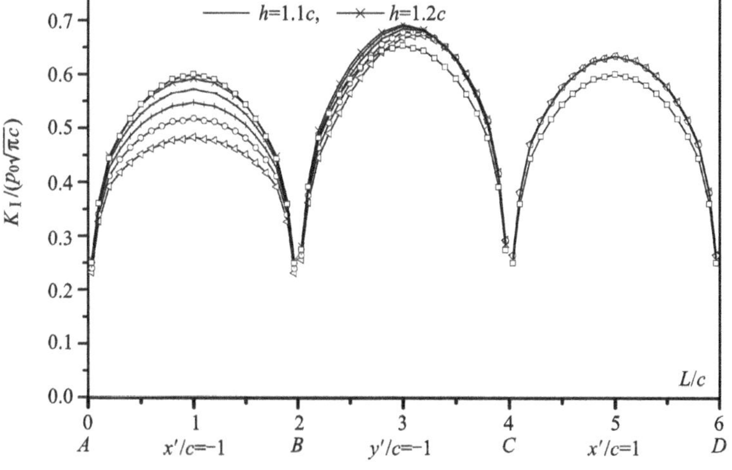

(a) K_I vs. L

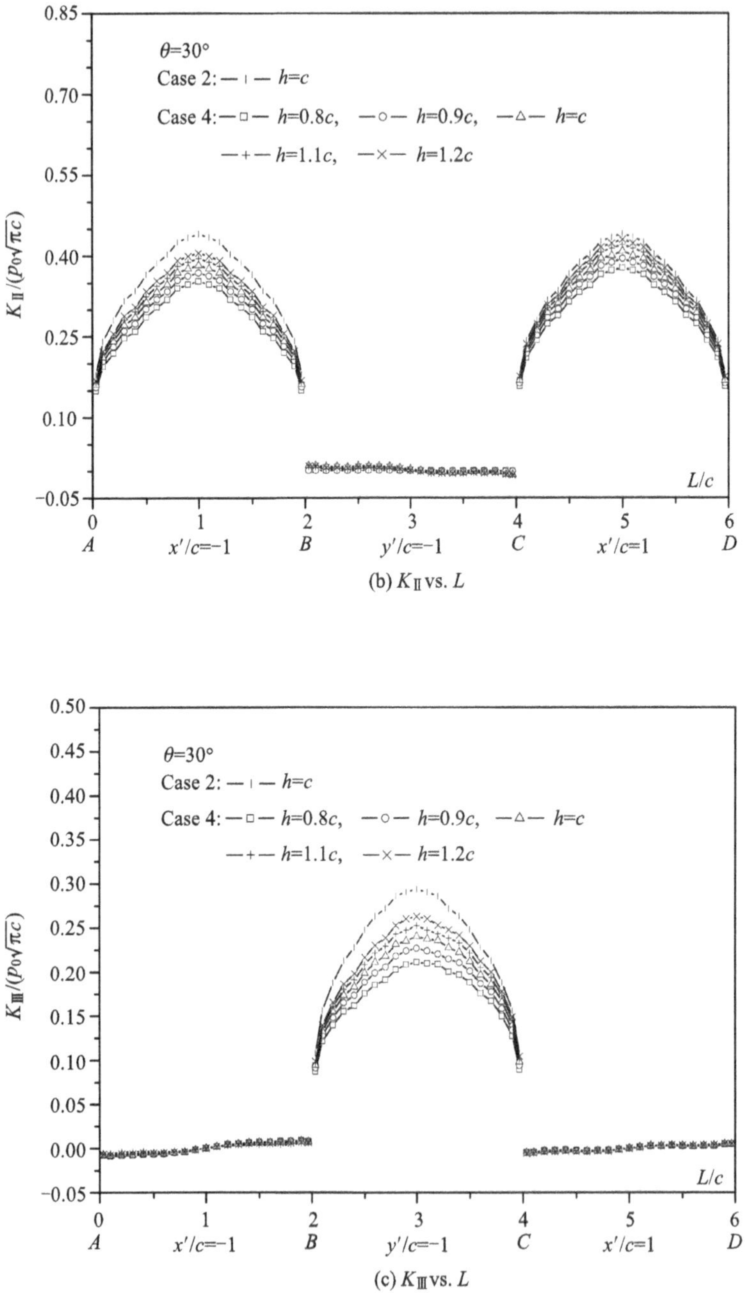

Fig. 10.13 The SIF values for the inclined square crack with $\theta = 30^\circ$ along the crack front lines from AB, BC to CD in the cubic rhenium (Case 2) or the cubic bi-material (Case 4)

- $K_{\mathrm{III}}/(p_0\sqrt{\pi c})$ associated with case 4 decrease as the depth h decreases from $1.2c$ to $0.8c$ and is less than that associated with case 2 for $h = c$, where $2 < L/c < 4$.

Based on the above analysis, it can be concluded that when the crack is located in a stiffer material, the softer material 2 tends to increase the magnitude of the SIF along the crack front. If the crack is situated in the softer material 2, the magnitude of the SIF along the crack front is reduced. However, the K_{I} SIFs along the crack front lines BC and CD can have different variation patterns with respect to the relative stiffness of materials 1 and 2.

(5) A square crack with $\theta = 45°$ in the bi-materials

In this section, the four cases of the bi-material combination given in Table 10.7 are examined for a square crack inclined at $\theta = 45°$, as shown in Fig. 10.9. The calculated values of $K_{\mathrm{I}}/(p_0\sqrt{\pi c})$, $K_{\mathrm{II}}/(p_0\sqrt{\pi c})$ and $K_{\mathrm{III}}/(p_0\sqrt{\pi c})$ along the crack front lines AB, BC, and CD are plotted against the coordinate L/c in Figs. 10.14 and 10.15.

The depth h of the inclined square crack is assumed to be c for the homogeneous materials (cases 1 and 2) and to be $h = 0.8c, 0.9c, c, 1.1c, 1.2c$ for the bi-materials (cases 3 and 4). Some calculated values are selected in Table 10.10 for future comparison.

Similar observations as those discussed for $\theta = 30°$ can be found in Figs. 10.14 and 10.15 for $\theta = 45°$. Furthermore, the effect of the inclination angle θ of the square crack on its values in the bi-material cube can be found by comparing the results in Figs. 10.12 and 10.13 with the corresponding results in Figs. 10.14 and 10.15. Basically, $K_{\mathrm{I}}/(p_0\sqrt{\pi c})$ for $\theta = 45°$ is smaller than that for $\theta = 30°$ and the $K_{\mathrm{II}}/(p_0\sqrt{\pi c})$ for $\theta = 45°$ become much larger than those for $\theta = 30°$.

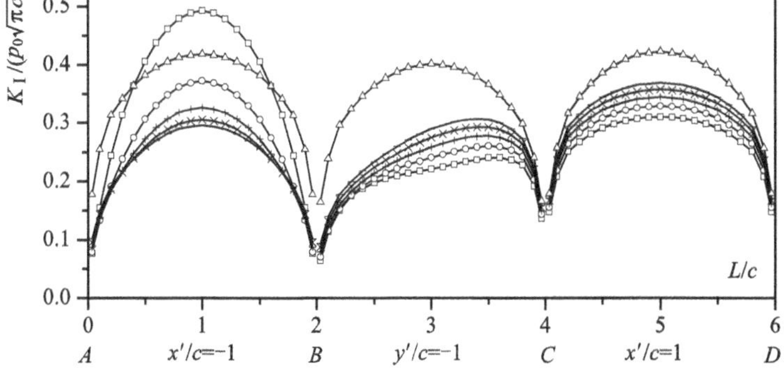

(a) K_{I} vs. L

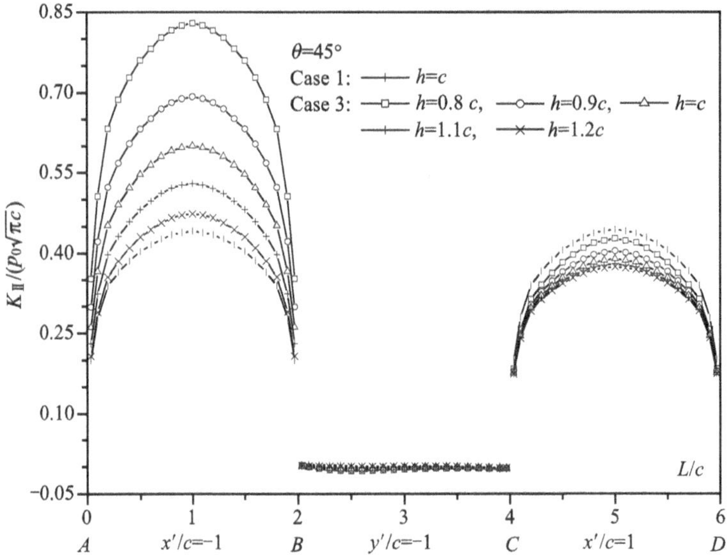

(b) K_{II} vs. L

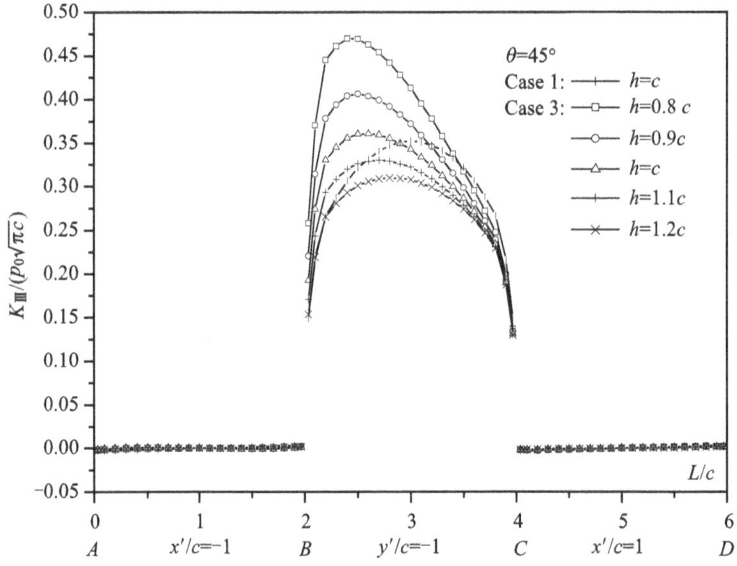

(c) K_{III} vs. L

Fig. 10.14 The SIF values for the inclined square crack with $\theta = 45^\circ$ along the crack front lines from AB, BC to CD in the cubic rhenium (case 1) or the cubic bi-material (case 3)

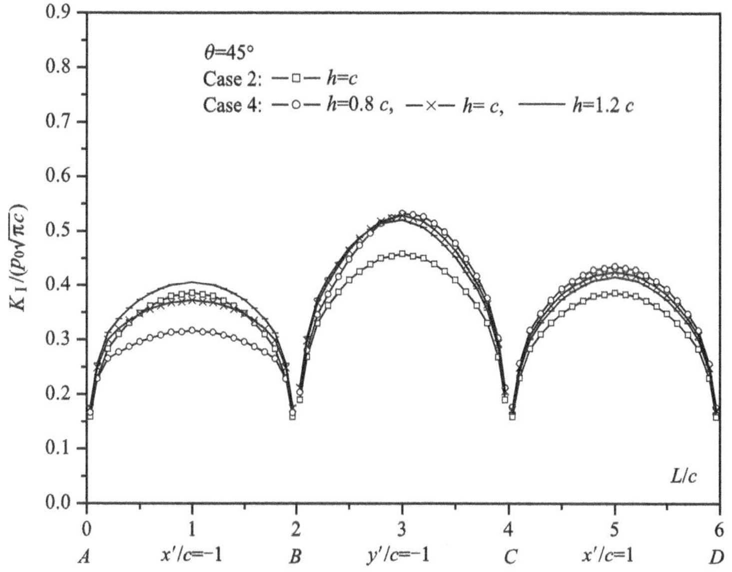

(a) K_I vs. L

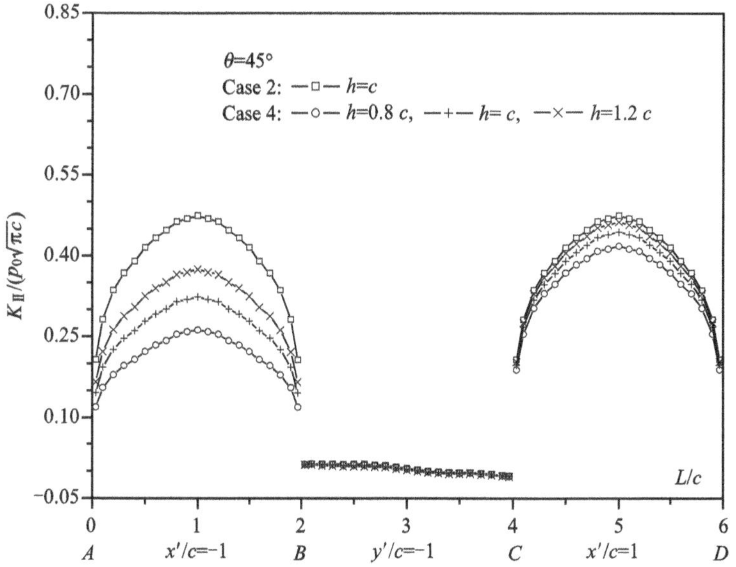

(b) K_{II} vs. L

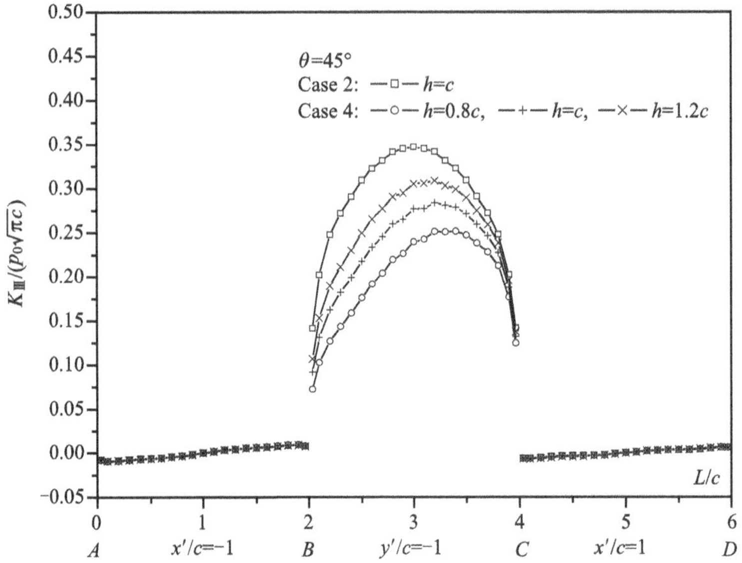

(c) K_{III} vs. L

Fig. 10.15 The SIF values for the inclined square crack with $\theta = 45°$ along the crack front lines from AB, BC to CD in the cubic rhenium (case 2) or the cubic bi-material (case 4)

Table 10.10 The SIF values for the square crack in a finite cube at the crack front $x'/c = -1$ for $\theta = 45°$ and $h = c$

y'/c	$K_{\mathrm{I}}/(p_0\sqrt{\pi c})$				$K_{\mathrm{II}}/(p_0\sqrt{\pi c})$			
	Case 1	Case 2	Case 3	Case 4	Case 1	Case 2	Case 3	Case 4
–0.9	0.2547	0.2304	0.1366	0.2495	0.2777	0.2812	0.3661	0.1928
–0.7	0.3423	0.3100	0.2191	0.3206	0.3649	0.3679	0.4902	0.2470
–0.5	0.3825	0.3484	0.2728	0.3478	0.4036	0.4149	0.5470	0.2783
–0.3	0.4055	0.3715	0.3072	0.3628	0.4266	0.4472	0.5804	0.3014
–0.1	0.4164	0.3828	0.3239	0.3700	0.4385	0.4687	0.5976	0.3185
0.0	0.4187	0.3862	0.3268	0.3729	0.4412	0.4744	0.6007	0.3233

2. A square crack in a transversely isotropic bi-material of infinite space

In this section, a square crack in a transversely isotropic bi-material of infinite space is examined. The objective of this examination is to study the effect of the external non-crack boundary on the SIF values. The effect is found by comparing the SIF values for the square crack in the finite cubic bi-material with those in the bi-material occupying an infinite space.

The square crack is shown in Fig. 10.9, where the external boundary points M_1, N_1, P_1, Q_1, M, N, P, Q, M_2, N_2, P_2 and Q_2 are extended to infinity. Let $\theta = 45^\circ$ and $h = c$. The square crack surfaces Γ^+ ($z' = 0^+$) and Γ^- ($z' = 0^-$) are subjected to a uniform compressive stress p_0. The bi-materials correspond to cases 3 and 4 in Table 10.7. The crack surface mesh *ABCD* shown in Fig. 10.10 is used.

Fig. 10.16 shows the calculated results of $K_\mathrm{I}/(p_0\sqrt{\pi c})$, $K_\mathrm{II}/(p_0\sqrt{\pi c})$ and $K_\mathrm{III}/(p_0\sqrt{\pi c})$ with respect to the line coordinate L/c along the crack front lines *AB*, *BC*, and *CD* in either the cubic bi-materials or the infinite bi-materials. From this figure, the following can be observed

- The pattern of variations in $K_\mathrm{I}/(p_0\sqrt{\pi c})$, $K_\mathrm{II}/(p_0\sqrt{\pi c})$ and $K_\mathrm{III}/(p_0\sqrt{\pi c})$ with respect to the line coordinate L/c are similar for both the cubic and the infinite bi-materials, which indicates that the external non-crack boundary does not have any effect on the variation patterns.
- However, for each L/c, the calculated values of $K_\mathrm{I}/(p_0\sqrt{\pi c})$, $K_\mathrm{II}/(p_0\sqrt{\pi c})$ and $K_\mathrm{III}/(p_0\sqrt{\pi c})$ for the crack in the infinite bi-material are slightly less than those in the finite cubic bi-materials.
- In particular, for $0 < L/c < 2$ or on the crack front line *AB* ($x'/c = -1$), $K_\mathrm{I}/(p_0\sqrt{\pi c})$ has its maximum value as follows: (a) 0.3268 for the cubic bi-material case 3; (b) 0.2988 for the infinite bi-material case 3; (c) 0.3729 for the cubic bi-material case 4; and (d) 0.3379 for the infinite bi-material case 4.

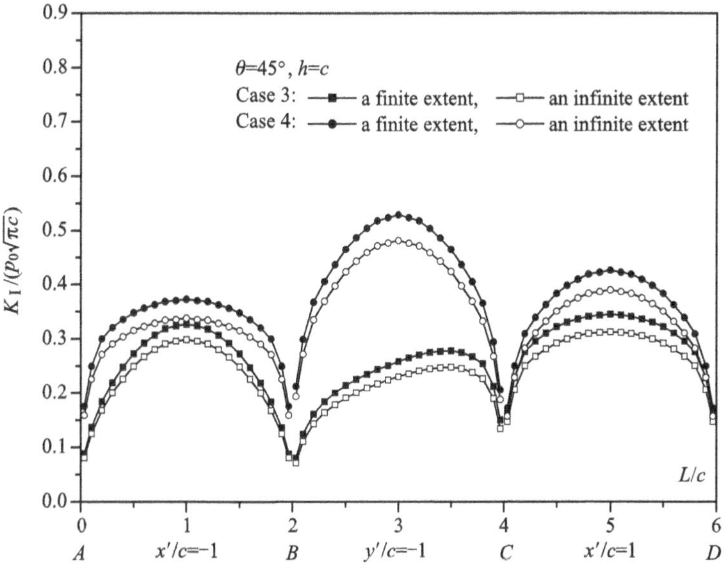

(a) K_I vs. L

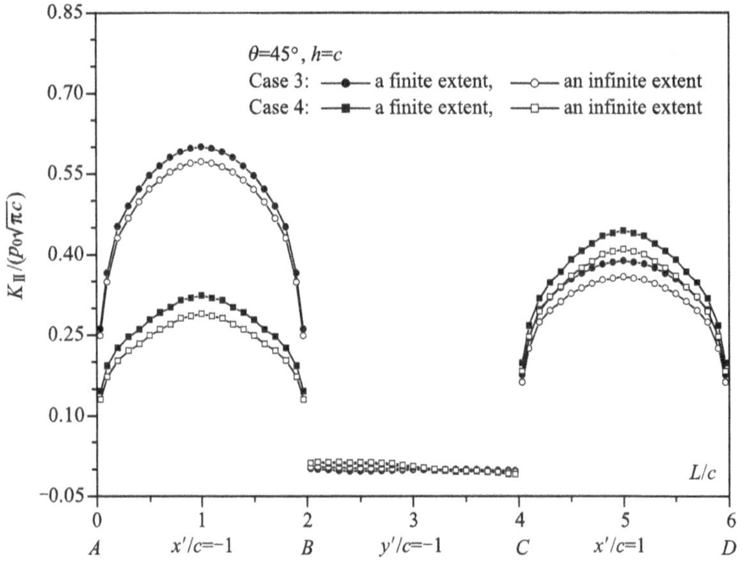

(b) K_{II} vs. L

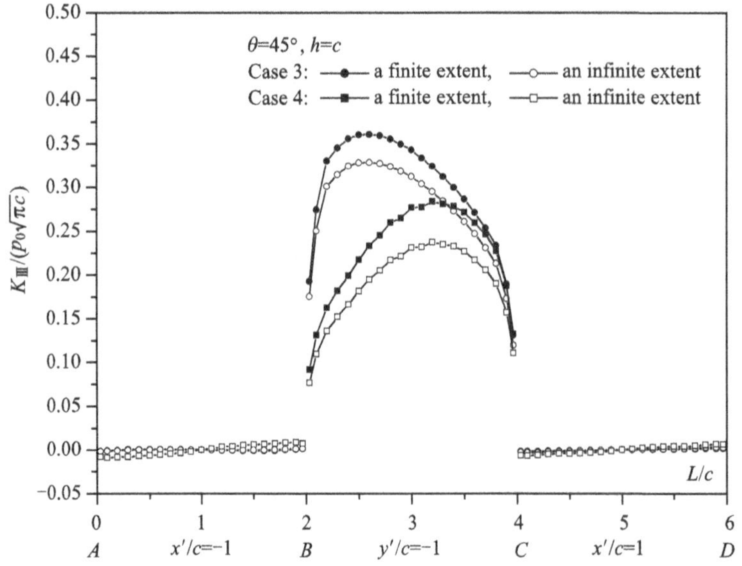

(c) K_{III} vs. L

Fig. 10.16 The SIF values for the inclined square crack with $\theta = 45^{\circ}$ along the crack front lines from AB, BC to CD in the cubic rhenium (Case 1) or the cubic bi-material (Case 3)

- In particular, for $0 < L/c < 2$ or on the crack front line AB $(x'/c = -1)$, $K_{\mathrm{II}}/(p_0\sqrt{\pi c})$ has its maximum value as follows: (a) 0.6007 for the cubic bi-material case 3; (b) 0.5730 for the infinite bi-material case 3; (c) 0.3233 for the cubic bi-material case 4; and (d) 0.2893 for the infinite bi-material case 4.

The above findings indicate that the larger the external boundary of the cracked bi-material, the higher the constraint of the external boundary of the cracked bi-material on the opening and sliding of the square crack.

10.4 Summary

In this chapter, we have introduced a simple multi-region BEM based on the fundamental solution of two joined transversely isotropic bi-materials, analyzed penny-shaped and elliptical cracks in this type of bi-materials, and presented the variations in the SIF values. Then we have introduced a simple dual BEM, based on this fundamental solution, analyzed the square crack in this type of bi-materials, and presented the variations in the SIF values for the cracks in infinite or finite cubic bi-materials.

The above results have been published in Xiao et al. (2005) and Yue et al. (2007). Many researchers, such as Benedetti et al. (2009), Chen (2008), Chen et al. (2008, 2009), Dong et al. (2011) and Omer and Yosibash (2008), have cited our work and developed their own BEMs for the analysis of crack problems.

Appendix 5 The fundamental solution of transversely isotropic bi-materials

For a transversely isotropic bi-material solid with the z-axis being the axis of material symmetry, the (x,y) plane is the plane of isotropy, as shown in Fig. A5.1. It is assumed that the planar interface between the two solids that occupy infinite spaces is perfectly bonded. The isotropic planes of the two anisotropic solids are parallel to the interface. The elastic constants of materials 1 and 2 are $c_{11}, c_{21}, c_{31}, c_{41}, c_{51}$ and $c_{12}, c_{22}, c_{32}, c_{42}, c_{52}$, respectively. Among the subscripts of the elastic constants, the first letter denotes the number of five material constants and the second letter denotes which of two materials is being considered.

The constitutive relationship between the stress σ_{ij} and the strain ε_{ij} can be expressed in terms of the contracted stiffness matrix as

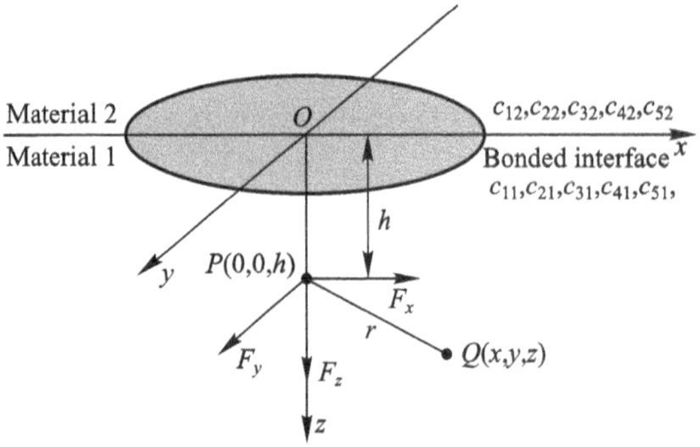

Fig. A5.1 Two perfectly bonded dissimilar solids with transversely isotropic properties subjected to concentrated body forces in the directions of the x, y and z axes. P is the source point and Q the field point

$$\begin{bmatrix} \sigma_{xx} \\ \sigma_{yy} \\ \sigma_{zz} \\ \sigma_{yz} \\ \sigma_{xz} \\ \sigma_{xy} \end{bmatrix} = \begin{bmatrix} c_{1k} & c_{1k}-2c_{5k} & c_{2k} & 0 & 0 & 0 \\ c_{1k}-2c_{5k} & c_{1k} & c_{2k} & 0 & 0 & 0 \\ c_{2k} & c_{2k} & c_{3k} & 0 & 0 & 0 \\ 0 & 0 & 0 & 2c_{4k} & 0 & 0 \\ 0 & 0 & 0 & 0 & 2c_{4k} & 0 \\ 0 & 0 & 0 & 0 & 0 & 2c_{5k} \end{bmatrix} \begin{bmatrix} \varepsilon_{xx} \\ \varepsilon_{yy} \\ \varepsilon_{zz} \\ \varepsilon_{yz} \\ \varepsilon_{xz} \\ \varepsilon_{xy} \end{bmatrix} \tag{A5.1}$$

where c_{1k}, c_{2k}, c_{3k}, c_{4k} and c_{5k} are the five elastic constants for the k-th transversely isotropic solid of the bi-material system, where $k = 1$ for $0^+ \leqslant z < \infty$ or 2 for $-\infty < z \leqslant 0^-$.

The vectors of displacements, vertical stresses and plane strains are, respectively, defined by

$$\boldsymbol{u} = (u_x \quad u_y \quad u_z)^{\mathrm{T}}, \quad \boldsymbol{T}_z = (\sigma_{xz} \quad \sigma_{yz} \quad \sigma_{zz})^{\mathrm{T}}, \quad \boldsymbol{\Gamma}_p = (\varepsilon_{xx} \quad \varepsilon_{xy} \quad \varepsilon_{yy})^{\mathrm{T}} \tag{A5.2}$$

where the superscript T denotes the transpose of a matrix.

Without loss of generality, we assume that the body forces are in the half region $k = 1$, i.e., $h \geqslant 0^+$. The vector of the concentrated body forces is located at $P(0,0,h)$ and defined by

$$\boldsymbol{F} = (F_x \quad F_y \quad F_z)^{\mathrm{T}} \tag{A5.3}$$

and the field point is located at $Q(x,y,z)$. The solutions for four different cases are presented in the ensuing. Let $\Delta_k = \sqrt{c_{1k}c_{3k}} - c_{2k} - 2c_{4k}$, and Δ_{1k} and Δ_{2k} correspond to two different transversely isotropic solids, respectively. By using the classical Fourier transform techniques, Yue (1995) developed the fundamental solution for the elastic fields in two joined transversely isotropic solids subject to concentrated point forces. The fundamental solution for the displacements and stresses is presented in the form of elementary harmonic functions and complete elliptical integrals. For easy reference, the formulations of the fundamental solution are briefly outlined in the following.

Case A: $\Delta_1 \neq 0$ and $\Delta_2 \neq 0$
(i) In the solid $k = 1 (z \geqslant 0)$, we have

$$\boldsymbol{u} = \Big\{ \boldsymbol{G}_v[0, z_{01}, \boldsymbol{\Phi}_{01}] + \sum_{n=1}^{4} \boldsymbol{G}_v[0, z_{n1}, \boldsymbol{\Phi}_{an1}] + \boldsymbol{G}_v[0, \gamma_{01}|z|, \boldsymbol{\Phi}_v] + \boldsymbol{G}_v[0, \gamma_{11}|z|, \boldsymbol{\Phi}_u(\gamma_{11})] - \boldsymbol{G}_v[0, \gamma_{21}|z|, \boldsymbol{\Phi}_u(\gamma_{21})] \Big\} \boldsymbol{F},$$

$$\boldsymbol{T}_z = \Big\{ \boldsymbol{G}_v[1, z_{01}, \boldsymbol{\Psi}_{01}] + \sum_{n=1}^{4} \boldsymbol{G}_v[1, z_{n1}, \boldsymbol{\Psi}_{an1}] + \boldsymbol{G}_v[1, \gamma_{01}|z|, \boldsymbol{\Psi}_v] + \boldsymbol{G}_v[1, \gamma_{11}|z|, \boldsymbol{\Psi}_u(\gamma_{11})] - \boldsymbol{G}_v[1, \gamma_{21}|z|, \boldsymbol{\Psi}_u(\gamma_{21})] \Big\} \boldsymbol{F}, \tag{A5.4}$$

$$\boldsymbol{\Gamma}_p = \Big\{ \boldsymbol{G}_p[1, z_{01}, \boldsymbol{\Phi}_{01}] + \sum_{n=1}^{4} \boldsymbol{G}_p[1, z_{n1}, \boldsymbol{\Phi}_{an1}] + \boldsymbol{G}_p[1, \gamma_{01}|z|, \boldsymbol{\Phi}_v] + \boldsymbol{G}_p[1, \gamma_{11}|z|, \boldsymbol{\Phi}_u(\gamma_{11})] - \boldsymbol{G}_p[1, \gamma_{21}|z|, \boldsymbol{\Phi}_u(\gamma_{21})] \Big\} \boldsymbol{F}$$

(ii) In the solid $k = 2 (z \leqslant 0)$, we have

$$\boldsymbol{u} = \left\{ \boldsymbol{G}_v[0, z_{02}, \boldsymbol{\Phi}_{02}] + \sum_{n=1}^{4} \boldsymbol{G}_v[0, z_{n2}, \boldsymbol{\Phi}_{an2}] \right\} \boldsymbol{F},$$

$$\boldsymbol{T}_z = \left\{ \boldsymbol{G}_v[1, z_{02}, \boldsymbol{\Psi}_{02}] + \sum_{n=1}^{4} \boldsymbol{G}_v[1, z_{n2}, \boldsymbol{\Psi}_{an2}] \right\} \boldsymbol{F},$$

$$\boldsymbol{\Gamma}_p = \left\{ \boldsymbol{G}_p[1, z_{02}, \boldsymbol{\Phi}_{02}] + \sum_{n=1}^{4} \boldsymbol{G}_p[1, z_{n2}, \boldsymbol{\Phi}_{an2}] \right\} \boldsymbol{F} \tag{A5.5}$$

where $z_{01} = \gamma_{01}(z+h)$, $z_{11} = \gamma_{11}(z+h)$, $z_{21} = \gamma_{21}z + \gamma_{11}h$, $z_{31} = \gamma_{11}z + \gamma_{21}h$, $z_{41} = \gamma_{21}(z+h)$, $z_{02} = \gamma_{01}h - \gamma_{02}z$, $z_{12} = \gamma_{11}h - \gamma_{12}z$, $z_{22} = \gamma_{11}h - \gamma_{22}z$, $z_{32} = \gamma_{21}h - \gamma_{12}z$, $z_{42} = \gamma_{21}h - \gamma_{22}z$, γ_{0k}, γ_{1k} and $\gamma_{2k} (k = 1, 2)$ are given as follows,

$$\gamma_{0k} = \sqrt{c_{5k}/c_{4k}},$$

$$\gamma_{1k} = \frac{1}{2\sqrt{c_{3k}c_{4k}}} \left[\sqrt{(\sqrt{c_{1k}c_{3k}} - c_{2k})(\sqrt{c_{1k}c_{3k}} + c_{2k} + 2c_{4k})} + \sqrt{(\sqrt{c_{1k}c_{3k}} + c_{2k})\Delta_k} \right],$$

$$\gamma_{2k} = \frac{1}{2\sqrt{c_{3k}c_{4k}}} \left[\sqrt{(\sqrt{c_{1k}c_{3k}} - c_{2k})(\sqrt{c_{1k}c_{3k}} + c_{2k} + 2c_{4k})} - \sqrt{(\sqrt{c_{1k}c_{3k}} + c_{2k})\Delta_k} \right]$$

Case B: $\Delta_1 = 0$ and $\Delta_2 \neq 0$
(i) In the solid $k = 1 (z \geqslant 0)$, we have

$$\begin{aligned}
\boldsymbol{u} &= \{\boldsymbol{G}_v[0,z_{01},\boldsymbol{\Phi}_{01}]+\boldsymbol{G}_v[0,z_{a0},\boldsymbol{\Phi}_{b11}]+z\boldsymbol{G}_v[1,z_{a0},\boldsymbol{\Phi}_{b21}]+h\boldsymbol{G}_v[1,z_{a0},\boldsymbol{\Phi}_{b31}]+\\
&\quad zh\boldsymbol{G}_v[2,z_{a0},\boldsymbol{\Phi}_{b41}]+\boldsymbol{G}_v[0,\gamma_{01}|\underline{z}|,\boldsymbol{\Phi}_v]+\boldsymbol{G}_v[0,\gamma_a|\underline{z}|,\boldsymbol{\Phi}_x]+\underline{z}\boldsymbol{G}_v[1,\gamma_a|\underline{z}|,\boldsymbol{\Phi}_y]\}\boldsymbol{F},\\
\boldsymbol{T}_z &= \{\boldsymbol{G}_v[1,z_{01},\boldsymbol{\Psi}_{01}]+\boldsymbol{G}_v[1,z_{a0},\boldsymbol{\Psi}_{b11}]+z\boldsymbol{G}_v[2,z_{a0},\boldsymbol{\Psi}_{b21}]+h\boldsymbol{G}_v[2,z_{a0},\boldsymbol{\Psi}_{b31}]+\\
&\quad zh\boldsymbol{G}_v[3,z_{a0},\boldsymbol{\Psi}_{b41}]+\boldsymbol{G}_v[1,\gamma_{01}|\underline{z}|,\boldsymbol{\Psi}_v]+\boldsymbol{G}_v[1,\gamma_a|\underline{z}|,\boldsymbol{\Psi}_x]+\underline{z}\boldsymbol{G}_v[2,\gamma_a|\underline{z}|,\boldsymbol{\Psi}_y]\}\boldsymbol{F},\\
\boldsymbol{\Gamma}_p &= \{\boldsymbol{G}_p[1,z_{01},\boldsymbol{\Phi}_{01}]+\boldsymbol{G}_p[1,z_{a0},\boldsymbol{\Phi}_{b11}]+z\boldsymbol{G}_p[2,z_{a0},\boldsymbol{\Phi}_{b21}]+h\boldsymbol{G}_p[2,z_{a0},\boldsymbol{\Phi}_{b31}]+\\
&\quad zh\boldsymbol{G}_p[3,z_{a0},\boldsymbol{\Phi}_{b41}]+\boldsymbol{G}_p[1,\gamma_{01}|\underline{z}|,\boldsymbol{\Phi}_v]+\boldsymbol{G}_p[1,\gamma_a|\underline{z}|,\boldsymbol{\Phi}_x]+\\
&\quad \underline{z}\boldsymbol{G}_p[2,\gamma_a|\underline{z}|,\boldsymbol{\Phi}_y]\}\boldsymbol{F} \qquad \text{(A5.6)}
\end{aligned}$$

(ii) In the solid $k = 2(z \leqslant 0)$, we have

$$\begin{aligned}
\boldsymbol{u} &= \{\boldsymbol{G}_v[0,z_{02},\boldsymbol{\Phi}_{02}]+\boldsymbol{G}_v[0,z_{a1},\boldsymbol{\Phi}_{b12}]+h\boldsymbol{G}_v[1,z_{a1},\boldsymbol{\Phi}_{b22}]+\boldsymbol{G}_v[0,z_{a2},\boldsymbol{\Phi}_{b32}]+\\
&\quad h\boldsymbol{G}_v[1,z_{a2},\boldsymbol{\Phi}_{b42}]\}\boldsymbol{F},\\
\boldsymbol{T}_z &= \{\boldsymbol{G}_v[1,z_{02},\boldsymbol{\Psi}_{02}]+\boldsymbol{G}_v[1,z_{a1},\boldsymbol{\Psi}_{b12}]+\\
&\quad h\boldsymbol{G}_v[2,z_{a1},\boldsymbol{\Psi}_{b22}]+\boldsymbol{G}_v[1,z_{a2},\boldsymbol{\Psi}_{b32}]+h\boldsymbol{G}_v[2,z_{a2},\boldsymbol{\Psi}_{b42}]\}\boldsymbol{F},\\
\boldsymbol{\Gamma}_p &= \{\boldsymbol{G}_p[1,z_{02},\boldsymbol{\Phi}_{02}]+\boldsymbol{G}_p[1,z_{a1},\boldsymbol{\Phi}_{b12}]+h\boldsymbol{G}_p[2,z_{a1},\boldsymbol{\Phi}_{b22}]+\boldsymbol{G}_p[1,z_{a2},\boldsymbol{\Phi}_{b32}]+\\
&\quad h\boldsymbol{G}_p[2,z_{a2},\boldsymbol{\Phi}_{b42}]\}\boldsymbol{F} \qquad \text{(A5.7)}
\end{aligned}$$

where $z_{a0} = \gamma_a(z+h)$, $z_{a1} = \gamma_a h - \gamma_{12}z$, $z_{a2} = \gamma_a h - \gamma_{22}z$, $\gamma_{11} = \gamma_{21} = \gamma_a$ and $\gamma_{12} = \gamma_{22} = \gamma_b$.

Case C: $\Delta_1 \neq 0$ and $\Delta_2 = 0$

In this case, the solutions for $\boldsymbol{u}, \boldsymbol{T}_z$ and $\boldsymbol{\Gamma}_p$ in the solid $k = 1(z \geqslant 0)$ can be obtained by substituting γ_b for γ_{22} in Eqs. (A5.4) for case A. In the solid $k = 2(z \leqslant 0)$, we have

$$\begin{aligned}
\boldsymbol{u} &= \{\boldsymbol{G}_v[0,z_{02},\boldsymbol{\Phi}_{02}]+\boldsymbol{G}_v[0,z_{b1},\boldsymbol{\Phi}_{c12}]+z\boldsymbol{G}_v[1,z_{b1},\boldsymbol{\Phi}_{c22}]+\boldsymbol{G}_v[0,z_{b2},\boldsymbol{\Phi}_{c32}]+\\
&\quad z\boldsymbol{G}_v[1,z_{b2},\boldsymbol{\Phi}_{c42}]\}\boldsymbol{F},\\
\boldsymbol{T}_z &= \{\boldsymbol{G}_v[1,z_{02},\boldsymbol{\Psi}_{02}]+\boldsymbol{G}_v[1,z_{b1},\boldsymbol{\Psi}_{c12}]+z\boldsymbol{G}_v[2,z_{b1},\boldsymbol{\Psi}_{c22}]+\boldsymbol{G}_v[1,z_{b2},\boldsymbol{\Psi}_{c32}]+\\
&\quad z\boldsymbol{G}_v[2,z_{b2},\boldsymbol{\Psi}_{c42}]\}\boldsymbol{F},\\
\boldsymbol{\Gamma}_p &= \{\boldsymbol{G}_p[1,z_{02},\boldsymbol{\Phi}_{02}]+\boldsymbol{G}_p[1,z_{b1},\boldsymbol{\Phi}_{c12}]+z\boldsymbol{G}_p[2,z_{b1},\boldsymbol{\Phi}_{c22}]+\boldsymbol{G}_p[1,z_{b2},\boldsymbol{\Phi}_{c32}]+\\
&\quad h\boldsymbol{G}_p[2,z_{b2},\boldsymbol{\Phi}_{c42}]\}\boldsymbol{F} \qquad \text{(A5.8)}
\end{aligned}$$

where $z_{b1} = \gamma_{11}h - \gamma_b z$ and $z_{b2} = \gamma_{21}h - \gamma_b z$.

Case D: $\Delta_1 = 0$ and $\Delta_2 = 0$

In this case, the solutions for $\boldsymbol{u}, \boldsymbol{T}_z$ and $\boldsymbol{\Gamma}_p$ in the solid $k = 1(z \geqslant 0)$ can be obtained by substituting γ_b for γ_{12} and γ_{22} in Eqs. (A5.6) for case B. In the solid $k = 2(z \leqslant 0)$, we have

$$\begin{aligned}
\boldsymbol{u} &= \{\boldsymbol{G}_v[0,z_{02},\boldsymbol{\Phi}_{02}]+\boldsymbol{G}_v[0,z_{ab},\boldsymbol{\Phi}_{d12}]+z\boldsymbol{G}_v[1,z_{ab},\boldsymbol{\Phi}_{d22}]+\\
&\quad h\boldsymbol{G}_v[1,z_{ab},\boldsymbol{\Phi}_{d32}]+zh\boldsymbol{G}_v[2,z_{ab},\boldsymbol{\Phi}_{d42}]\}\boldsymbol{F},\\
\boldsymbol{T}_z &= \{\boldsymbol{G}_v[1,z_{02},\boldsymbol{\Psi}_{02}]+\boldsymbol{G}_v[1,z_{ab},\boldsymbol{\Psi}_{d12}]+z\boldsymbol{G}_v[2,z_{ab},\boldsymbol{\Psi}_{d22}]+\\
&\quad h\boldsymbol{G}_v[2,z_{ab},\boldsymbol{\Psi}_{d32}]+zh\boldsymbol{G}_v[3,z_{ab},\boldsymbol{\Psi}_{d42}]\}\boldsymbol{F},\\
\boldsymbol{\Gamma}_p &= \{\boldsymbol{G}_p[1,z_{02},\boldsymbol{\Phi}_{02}]+\boldsymbol{G}_p[1,z_{ab},\boldsymbol{\Phi}_{d12}]+z\boldsymbol{G}_p[2,z_{ab},\boldsymbol{\Phi}_{d22}]+\\
&\quad h\boldsymbol{G}_p[2,z_{ab},\boldsymbol{\Phi}_{d32}]+zh\boldsymbol{G}_p[3,z_{ab},\boldsymbol{\Phi}_{d42}]\}\boldsymbol{F}
\end{aligned}\tag{A5.9}$$

where $z_{ab}=\gamma_a h-\gamma_b z$.

In the above equations, the fundamental solution matrices $\boldsymbol{G}_v[n,z,\boldsymbol{\Phi}]$ and $\boldsymbol{G}_p[n,z,\boldsymbol{\Phi}]$ $(n=0,1,2,3;z\geqslant 0)$ are defined in the following

$$\begin{aligned}
4\pi\boldsymbol{G}_v[n,z,\boldsymbol{\Phi}] &= \phi_{22}\begin{pmatrix} g_{n02}(z) & -g_{n11}(z) & 0\\ -g_{n11}(z) & g_{n20}(z) & 0\\ 0 & 0 & 0\end{pmatrix}+\\
&\begin{pmatrix} \phi_{11}g_{n20}(z) & \phi_{11}g_{n11}(z) & \phi_{13}g_{n10}(z)\\ \phi_{11}g_{n11}(z) & \phi_{11}g_{n02}(z) & \phi_{13}g_{n01}(z)\\ -\phi_{31}g_{n10}(z) & -\phi_{31}g_{n01}(z) & \phi_{33}g_{n00}(z)\end{pmatrix},\\
4\pi\boldsymbol{G}_p[n,z,\boldsymbol{\Phi}] &= \phi_{22}\begin{pmatrix} g_{n12}(z) & -g_{n21}(z) & 0\\ \frac{1}{2}[g_{n03}(z)-g_{n21}(z)] & \frac{1}{2}[g_{n30}(z)-g_{n12}(z)] & 0\\ -g_{n12}(z) & g_{n21}(z) & 0\end{pmatrix}+\\
&\begin{pmatrix} \phi_{11}g_{n30}(z) & \phi_{11}g_{n21}(z) & -\phi_{13}g_{n20}(z)\\ \phi_{11}g_{n21}(z) & \phi_{11}g_{n12}(z) & -\phi_{13}g_{n11}(z)\\ \phi_{11}g_{n12}(z) & \phi_{11}g_{n03}(z) & -\phi_{13}g_{n02}(z)\end{pmatrix}
\end{aligned}\tag{A5.10}$$

where $z>0$, $n=0,1,2,3$, and the harmonic functions $g_{0lm}(z)$ are given by

$$\begin{aligned}
&g_{000}(z)=\frac{1}{R},\quad g_{002}(z)=\frac{1}{R_z}\left[1-\frac{y^2}{RR_z}\right],\quad g_{010}(z)=-\frac{x}{RR_z},\\
&g_{030}(z)=\frac{x}{2R_z^2}\left[\frac{2x^2}{RR_z}-3\right],\quad g_{001}(z)=-\frac{y}{RR_z},\quad g_{011}(z)=-\frac{xy}{RR_z^2},\\
&g_{021}(z)=\frac{y}{2R_z^2}\left[\frac{2x^2}{RR_z}-1\right],\quad g_{020}(z)=\frac{1}{R_z}\left[1-\frac{x^2}{RR_z}\right],\quad g_{012}(z)=\frac{x}{2R_z^2}\left[\frac{2y^2}{RR_z}-1\right],\\
&g_{003}(z)=\frac{y}{2R_z^2}\left[\frac{2y^2}{RR_z}-3\right]
\end{aligned}\tag{A5.11}$$

where $R=\sqrt{x^2+y^2+z^2}$, $R_z=R+z$.

For $n\geqslant 1$, the harmonic function $g_{nlm}(z)$ $(0\leqslant l+m\leqslant 3)$ can be obtained by using the following transfer formula

$$g_{nlm}(z)=-\frac{\partial g_{(n-1)lm}(z)}{\partial z}\tag{A5.12}$$

In Eq. (A5.10), the constant matrix $\boldsymbol{\Phi}$ is defined by

$$\boldsymbol{\Phi} = \begin{pmatrix} \phi_{11} & 0 & \phi_{13} \\ 0 & \phi_{22} & 0 \\ \phi_{31} & 0 & \phi_{33} \end{pmatrix} \tag{A5.13}$$

Each element of the constant matrix $\boldsymbol{\Phi}$ depends only on the 10 elastic constants $c_{jk}(j = 1,2,3,4,5; k = 1,2)$. Their specific forms can be found in Yue (1995).

References

Ariza MP, Dominguez J. Boundary element formulation for 3D transversely isotropic cracked bodies. *International Journal for Numerical Methods in Engineering*, 2004, 60: 719-753.

Benedetti I, Milazzo A, Aliabadi MH. A fast dual boundary element method for 3D anisotropic crack problems. *International Journal for Numerical Methods in Engineering*, 2009, 80: 1356-1378.

Chen CH. Application of Boundary Element Method on Fracture Mechanics Analysis of Anisotropic Rocks. Taiwan: National Cheng Kung University, 2008.

Chen CS, Chen CH, Pan E. Three-dimensional stress intensity factors of central square crack in a transversely isotropic cuboid with arbitrary material orientations. *Engineering Analysis with Boundary Elements*, 2009, 33: 128-136.

Chen CH, Chen CS, Pan E, Tseng HC. Boundary element analysis of mixed-mode stress intensity factors in an anisotropic cuboid with an inclined surface crack. *Engineering Computations*, 2009, 26: 1056-1073.

Dong CY, Yang X, Pan E. Analysis of cracked transversely isotropic and inhomogeneous solids by special BIE formulation. *Engineering Analysis with Boundary Elements*, 2011, 35: 200-206.

Hoenig A. The behavior of a flat elliptical crack in an anisotropic elastic body. *International Journal of Solids and Structures*, 1978, 14: 925-934.

Kassir MK, Sih GC. Three-dimensional stress distribution around an elliptical crack under arbitrary loadings. ASME *Journal of Applied Mechanics*, 1966, 33: 601-611.

Lin W, Keer LM. Three-dimensional analysis of cracks in layered transversely isotropic media.*Proceedings of Royal Society* (London), 1989, A424: 307-322.

Noda NA, Ohzono RM, Chen C. Analysis of an elliptical crack parallel to a bimaterial interface under tension. *Mechanics of Materials*, 2003, 35: 1059-1076.

Omer N, Yosibash Z. Edge singularities in 3-D elastic anisotropic and multi-material domains. *Computer Methods in Applied Mechanics and Engineering*, 2008, 197: 959-978.

Pan E, Yuan FG. Boundary element analysis of three-dimensional cracks in anisotropic solids. *International Journal for Numerical Methods in Engineering*, 2000, 48: 211-237.

Pan YC, Chou TW. Point force solution for an infinite transversely isotropic solid. ASME *Journal of Applied Mechanics*, 1976, 43: 608-612.

Saez A, Ariza MP, Dominguez J. Three-dimensional fracture analysis in transversely isotropic solids. *Engineering Analysis with Boundary Elements*, 1997, 20: 287-298.

Sih GC, Paris PC, Irwin GR. On crack in rectilinearly anisotropic bodies. *International Journal of Fracture*, 1965,1: 189-203.

Ting TCT. Anisotropic Elasticity: Theory and Applications. New York: Oxford University Press, 1996.

Weaver J. Three-dimensional crack analysis. *International Journal of Solids and Structures*, 1977, 13: 321-330.

Xiao HT, Yue ZQ, Tham LG. Analysis of elliptical cracks perpendicular to the interface of two joined transversely isotropic solids. *International Journal of Fracture*, 2005, 133: 329-354.

Yue ZQ. Elastic fields in two joined transversely isotropic solids due to concentrated forces. *International Journal of Engineering Sciences*, 1995, 33: 351-369.

Yue ZQ, Xiao HT, Tham LG, Lee CF, Pan E. Boundary element analysis of 3D crack problems in two joined transversely isotropic solids. *Computational Mechanics*, 2005,36: 459-474.

Yue ZQ, Xiao HT, Pan E. Stress intensity factors of square crack inclined to interface of transversely isotropic bi-material. *Engineering Analysis with Boundary Elements*, 2007, 31: 50-65.

Zhang ZG, Mai YW. A simple solution for the stress intensity factor of a flat elliptical crack in a transversely isotropic solid. *Engineering Fracture Mechanics*, 1989, 34: 645-628.